ENVIRONMENTAL SYSTEMS

ENVIRONMENTAL SYSTEMS

Henry J. Cowan • Peter R. Smith

Department of Architectural Science, University of Sydney

VNR VAN NOSTRAND REINHOLD COMPANY
New York Cincinnati Toronto London Melbourne

Library of Congress Catalog Card Number 82-23753
ISBN 0-442-21490-1 (cloth)
ISBN 0-442-21489-8 (paper)

Printed in the United States of America
Designed by Loudan Enterprises

Published by Van Nostrand Reinhold Company Inc.
135 West 50th Street
New York, New York 10020

Van Nostrand Reinhold
480 Latrobe Street
Melbourne, Victoria 3000, Australia

Van Nostrand Reinhold Company Limited
Molly Millars Lane
Wokingham, Berkshire RG11 2PY, England

16 15 14 13 12 11 10 9 8 7 6 5 4 3 2 1

Library of Congress Cataloging in Publication Data

Cowan, Henry J.
Environmental systems.

Includes bibliographies and index.
1. Buildings—Environmental engineering. I. Smith, Peter R. II. Title.
TH6021.C68 1983 696 82-23753
ISBN 0-442-21490-1
ISBN 0-442-21489-8 (pbk.)

To

John Ward

Vice-Chancellor of the University of Sydney

Contents

ABOUT THE AUTHORS

HENRY J. COWAN graduated from the University of Manchester, England, with first class honours in civil and mechanical engineering, and he has doctorates in engineering and in philosophy from the University of Sheffield, England. He was appointed professor and head of the Department of Architectural Science of the University of Sydney, Australia, in 1953 and was Dean of Architecture in 1966 and 1967. He has lectured at more than forty universities throughout the world and has been a visiting professor in the United States, Turkey, and Ghana.

Dr. Cowan is a Fellow of the American Society of Civil Engineers, a Fellow and Chapman Medallist of the Institution of Engineers, Australia, a Fellow and Corresponding Member of Council of the Royal Society of Arts, and an Honorary Fellow of the Royal Australian Institute of Architects. He is a former president of the Building Science Forum of Australia, and a former President of Architecture and Engineering Sections of the ANZ Association for the Advancement of Science. In 1983 he was appointed an Officer of the Order of Australia. Dr. Cowan has written more than 200 articles and 18 books.

PETER R. SMITH graduated from the University of Sydney with bachelor and master degrees in architecture, and doctor of philosophy. After a period in architectural practice he joined the staff of the Department of Architectural Science at the University of Sydney, where he is Associate Professor. He is a member of committees of the Royal Australian Institute of Architects, Building Science Forum of Australia, and the Energy Authority of New South Wales, dealing with energy in buildings and related subjects.

Preface

This book deals with the application of physical science to the design of the interior environment in buildings. The subject has a respectable prehistory. The Ancient Greeks and Romans developed a number of rules for thermal and acoustic design that are still valid today. Socrates said that the house in which the owner can find a pleasant retreat in all seasons is the most beautiful, whether it is decorated or not.

However, environmental design has always run a poor third to structural design and the study of materials. The collapse of a building cannot be ignored, and the deterioration of a building's surfaces requires attention; but people may endure thermal discomfort, glary illumination, poor acoustics, and excessive noise without taking any action.

Since the late eighteenth century it has to an increasing extent become possible to overcome many failures of environmental design by providing more building services and expending more energy. Heating is clearly essential for thermal comfort in cold weather, and artificial light is needed at night. However, the authors have assumed in writing this book that the fabric of the building should always be designed to provide the most favorable interior environment that can be achieved at a reasonable cost, and that mechanical and electrical equipment should be regarded as a necessary supplement when the building fabric is unable to provide the required environment without the aid. Thus the emphasis throughout this book is on energy conservation and on the utilization of solar energy.

Environmental Systems is primarily intended for architects, who normally dèsign sunshades, insulation, natural ventilation, daylight, and all the other environmental aspects of a building without the aid of consultants; but architects retain a consultant for the design of the building services, which are, therefore, discussed only briefly. We would like to refer readers who require more information to one of the excellent books on that subject, for example, *Mechanical and Electrical Equipment of Buildings,* by W.J. McGuiness, B. Stein and J.S. Reynolds (Wiley, New York, 1980).

The book deals with a number of essentially separate design topics that contribute to the environmental system of a building, and there is no undisputed hierarchy for their order of presentation. The largest single topic is the heating, cooling, and ventilation of buildings, and this is closely related to the supply of water, since water is used for heating and cooling, and the hot water supply is itself a large consumer of heat. Following the historical introduction in Chapter 1, Chapter 2 therefore considers the supply and reticulation of water, and the related topic of waste disposal.

Chapter 3 described the movement of the sun and the devices for sun control that are important both for the thermal and the luminous environment. Chapters 4, 5, and 6 deal with the design of the thermal environment and the related building services. Chapter 7 discusses a heat-related problem, the control of fire.

Chapters 8, 9, and 10 are devoted to the design of the luminous environment, which interacts with the design of the thermal environment. Chapters 11 and 12 deal with noise control and acoustics. Chapter 13 considers the organization of space and movement within the building and the services required for communication and safety.

We have assumed that some readers will have only a slight knowledge of physics and mathematics. The examples, which occur in most chapters, contain only elementary arithmetic. The physical concepts are explained when they are introduced, and in addition the Glossary contains definitions of all scientific and architectural terms used.

SI units are used throughout this book, but the British/American and the old metric units of measurement are explained in Appendix B, and conversion figures are given between the various units. Fortunately, SI units are already employed throughout the world, including the United States, for lighting and acoustics, and they are gradually being adopted also for thermal design; however, the Btu and the calorie are likely to survive alongside the joule for some time.

Chapters 1, 3, 4, 5, 7, 8, 9, 10, 11, 12, 14, and part of Chapter 13 were written mainly by H.J. Cowan, and Chapters 2, 6, and part of Chapter 13 by P.R. Smith. Unless otherwise stated, drawings are by P.R. Smith and photographs by H.J. Cowan.

We are indebted to Miss Aphrodite Poulos, Miss Anne Woods, and Miss Jenny Whyte for typing the manuscript; to Miss Estelle Lazer for some of the research undertaken for this book; and to the following for reading the whole or part of the manuscript, and commenting critically on it: Mr. L. Challis, Mr. B. Forwood, Dr. F. Fricke, Mr. E.L. Harkness, Dr. V. Havyatt, Mr. S. Hayman, Mr. W. Julian, Mr. J. Keough, Miss E. Lazer, Prof. R.K. Macpherson, Dr. A. Radford, and Mr. J. Whittemore.

HJC
PRS

Chapter 1 An Historical Introduction

The Ancient Greeks had some environmental prescriptions that would not look out of place in a contemporary book. The Romans developed effective heating, water-supply, and sewage-disposal systems, but these fell into disuse in Europe during the Middle Ages. Although the Renaissance revived Roman ideas of environmental design, the Roman building services were not surpassed until the nineteenth century. Progress has been rapid during the last hundred years, but at the cost of an enormous increase in energy usage. To conserve energy, a much greater emphasis is now placed on the contribution of the building fabric to the creation of a satisfactory interior environment.

1.1 The Ancient World

Architectural problems were mentioned by several Greek and Roman authors. For example, Xenophon (Ref. 1.1, Book 3, Chapter 8, Section 8-10, pp. 221–23) in the fourth century B.C. credited Socrates, in a discussion with Aristippus, with this pronouncement on sun control and noise pollution:

> Again his dictum about houses, that the same house is both beautiful and useful, was a lesson in the art of building houses as they ought to be.
>
> He approached the problem thus:
>
> "When one means to have the right sort of house, must he contrive to make it as pleasant to live in and as useful as can be?"
>
> And this being admitted, "Is it pleasant," he asked, "to have it cool in summer and warm in winter?"
>
> And when they agreed with this also, "Now in houses with a south aspect, the sun's rays penetrate into the porticoes in winter, but in summer the path of the sun is right over our heads and above the roof, so that there is shade.
>
> "If, then, this is the best arrangement, we should build the south side loftier to get the winter sun and the north side lower to keep out the cold winds. To put it shortly, the house in which the owner can find a pleasant retreat at all seasons and can store his belongings safely is presumably at once the pleasantest and the most beautiful. As for paintings and decorations, they rob one of more delights than they give."
>
> For temples and altars the most suitable position, he said, was a conspicuous site remote from traffic; for it is pleasant to approach them filled with holy thoughts.

However, most of our information about Ancient Greek and Roman buildings comes from Vitruvius' *Ten Books of Architecture,* written about two thousand years ago, the only book specifically on architecture that survives from Ancient times.

Fig. 1.1.1. Roman theater dug into the slope of the Acropolis. The seating has been restored, and it is used at the present time in summer, occasionally for grand opera. It can seat 6000 people. The heavy wall, of original Roman construction, provides a satisfactory barrier to the traffic noise of the highway.

Vitruvius gives an explanation of the nature of sound that agrees with modern concepts:

> Voice is a flowing breath of air, perceptible to the hearing by contact. It moves in an endless number of circular rounds, like the innumerably increasing circular waves which appear when a stone is thrown into smooth water, and which keep on spreading indefinitely from the centre unless interrupted by narrow limits, or by some obstruction which prevents such waves from reaching their end in due formation.
>
> When they are interrupted by obstructions, the first waves, flowing back, break up the formation of those which follow.
>
> In the same manner the voice executes its movements in concentric circles; but while in the case of water the circles move horizontally on a plane surface, the voice not only proceeds horizontally, but also ascends vertically by regular stages. Therefore, as in the case of the waves formed in the water, so it is the case of the voice: the first wave, when there is no obstruction to

interrupt it, does not break up the second or the following waves, but they all reach the ears of the lowest and the highest spectators without an echo.

Hence the ancient architects, following in the footsteps of nature, perfected the ascending rows of seats in theatres from their investigations of the ascending voice, and, by means of the canonical theory of the mathematicians and that of the musicians, endeavoured to make every voice uttered on the stage come with greater clearness and sweetness to the ears of the audience. For just as musical instruments are brought to perfection of clearness in the sound of their strings by means of bronze plates or horn echeia, so the ancients devised methods of increasing the power of the voice in theatres through the application of harmonics. (Ref. 1.2, Book 5, Chapter 4, pp. 138–39)

Greek and Roman theaters survive in countries that once formed part of the Roman Empire, and several are used today. Some have excellent acoustics by comparison with modern *open-air* theaters (Fig. 1.1.1). Roman concepts of theater design were revived in the Renaissance, and some were still used in the nineteenth century, when the modern science of acoustics was developed [Chapter 12].

Vitruvius suggested how the various rooms in the house should be oriented:

Winter dining rooms and bathrooms should have a southwestern exposure, for the reason that they need the evening light, and also because the setting sun, facing them with all its splendour but with abated heat, lends a gentler warmth to that quarter in the evening. Bedrooms and libraries ought to have an eastern exposure, because their purposes require morning light, and also because books in such libraries will not decay. . . .

Dining rooms for spring and autumn to the east; for when the windows face that quarter, the sun, as he goes on his career from over against them to the west, leaves such rooms at the proper temperature at the time when it is customary to use them.

Summer dining rooms to the north, because that quarter is not, like the others, burning with heat during the solstice, for the reason that it is unexposed to the sun's course, and hence it always keeps cool, and makes the use of the rooms both healthy and agreeable. (Ref. 1.2, Book 6, Chapter 4, pp. 180–81)

We still build bedrooms facing east, when possible, to receive the morning sun, but separate seasonal dining rooms are a luxury few people could afford today.

Most Renaissance texts on architecture were based on Virtuvius' book, which was still highly regarded in the eighteenth century. However, not all of his recommendations for environmental design accord with modern ideas. For example, he did not believe in natural ventilation:

Fig. 1.1.2. Ruins of heated floor of the Roman baths at Ephesus, now Efes in Turkey. The floor was supported on brick pillars, and hot gases passed from the furnace (hypocaust) through the passages between the pillars to heat the floor.

Cold winds are disagreeable, hot winds are enervating, moist winds unhealthy. . . . By shutting out the winds from our dwellings we shall not only make the place healthful for people who are well, but also in the case of diseases due perhaps to unfavourable situations elsewhere, the patients, who in other healthy places might be cured by a different form of treatment, will here be cured more quickly by the mildness that comes from shutting out of the winds. (Ref. 1.2, Book 1, Chapter 6, pp. 24–25)

Building services to improve the environment artificially are mainly a development of the nineteenth and twentieth centuries, because the technology did not exist in earlier times, but the Romans invented in the first century B.C. an ingenious heating system which they used for the hot rooms of their baths and also for heating houses in colder climates. It consisted of a tile floor raised on short brick pillars and heated by smoke and hot air from a slow-burning furnace called a hypocaust. The gases passed under the floor and were exhausted by a chimney at the far end. Remains of this heating system have been found in the ruins of many country houses and baths (Fig. 1.1.2). It was still a new technique when Vitruvius described it:

The hanging floors of hot bath rooms are to be constructed as follows. First the surface of the ground should be laid with tiles a foot* and a half square, sloping towards a furnace in such a way that, if a ball is thrown in it cannot stop inside but must return of itself to the furnace room; thus the heat of the fire will more readily spread under the hanging flooring. Upon them, pillars made of half-foot square bricks are built, and set at such a distance apart that two-foot square tiles may

* The Roman foot in Vitruvius' time was approximately 11½ American inches, or 295 mm.

be used to cover them. These pillars should be two feet in height, laid with clay mixed with hair, and covered on top with the two-foot square tiles which support the floor. (Ref. 1.2, Book 5, Chapter 10, p. 157)

After the fall of the Roman Empire hypocausts ceased to be used in Europe, but they continued to be employed in baths in Byzantine and Islamic buildings. They were also independently invented by the Koreans several centuries later, and the Korean heating system was occasionally used in northern China and in Japan.

Water-supply and sewage-disposal systems were built by several ancient civilizations, but generally only for the ruling family and its household. The Romans provided water-supply and sewage-disposal systems for the entire population of the city of Rome and for many other cities of the Empire, and these functioned for about four centuries [see Section 2.1]. In some Eastern cities these were maintained by Byzantine and Islamic rulers, but in Europe sanitation comparable to that of Ancient Rome was reestablished only after the Industrial Revolution in the eighteenth century. In 1842, when Britain had the most advanced technology of any country, Edwin Chadwick produced for the British government a three-volume *Report on the Sanitary Conditions of the Labouring Population of Great Britain*. He described the water supply and sewage disposal in contemporary Britain as greatly inferior to those of imperial Rome.

1.2 Vernacular Architecture

Vernacular methods of construction survive in many parts of the world, particularly where labor is cheap and imported building materials are expensive. Some of these methods produce houses that are comfortable under extreme climatic conditions, sometimes more so than air-conditioned tourist hotels.

Fig. 1.2.1. Circular hut in the New Guinea highlands under construction. Once the materials have been assembled, the hut can be erected by two men in one day. The structure consists of bamboo tied with vegetable fibers; the walls are plaited mats.

The first contact with the people occupying the New Guinea highlands, who had only stone tools at the time, was made earlier in the twentieth century. Apart from the use of steel tools for cutting the materials, the method of construction has probably not changed significantly.

In the twentieth century vernacular architecture has been confined mainly to the hotter parts of the world. These may be divided into hot-arid and hot-humid zones, of which the hot-humid is the more hospitable. It produces an ample supply of bamboo, reed, and tree saplings. These can be cut at leisure. When there is an adequate supply of structural materials of the right length, a small house can easily be built in one day.

The resulting house is not entirely weatherproof, may become infested with vermin, and is easily destroyed by fire. However, it does not cost much to build, even in comparison to the resources of a developing country, and it is easily replaced. Indeed, in many regions it is treated as a disposable building, to be abandoned or destroyed when it no longer meets the sanitary standards of its owners. It is interesting to note that in the 1960s, when consumerism was fashionable, a disposable house was regarded as a desirable modern objective.

The construction is environmentally suited to the climate. It provides ventilation [Section 5.11] in a hot-humid climate; insulation would be of

Fig. 1.2.2. Large bamboo hut in Fiji, which has two vertical poles and a ridge beam.

little value when the difference between the daily maximum and minimum temperatures is small.

The type of building illustrated in Fig. 1.2.1 is still erected in many parts of the Pacific region, even on islands, such as American Samoa and Fiji, that have been greatly influenced by American concepts. There are regional differences, notably the use of two vertical poles and a ridge pole to produce a larger hut (Fig. 1.2.2). There are modern improvements, such as the use of concrete floors in American Samoa (Fig. 1.2.3). However, the traditional method has been totally abandoned on some islands, such as Rarotonga, where timber-framed huts with corrugated iron roofs have acquired the status of an indigenous architecture.

The vernacular architecture commonly associated with hot-arid regions has thick walls of local materials, such as mud, adobe (see Glossary), brick, or stone, which provide good thermal inertia [Section 5.6]. The roof poses a structural problem. Curved roofs of the same material as the walls can be used, but the attainable span is very small if mud or mudbrick is used, and the interior space is limited accordingly. Another solution is a timber roof, where timber is available. A roof of light material largely negates the insulation and thermal storage of massive walls; however, in some parts of the world with great temperature variations, notably the Middle East, timber roofs are covered with a thick layer of soil, and sometimes also with grass (Fig. 1.2.4), to provide the roof with the necessary thermal insulation and heat-storage capacity [see also Section 5.6].

In the 1930s and again in the 1960s, it was fashionable to proclaim vernacular architecture as the true guide to the correct use of local materials in conformity with the local climate. Primitive peoples, if given many centuries to experiment, would unerringly discover solutions in conformity with the modern precepts of building science. The surviving methods of vernacular construction offer much evidence in support of this thesis. But when a primitive society comes into contact with another that has a higher technology, it naturally adopts many of its techniques, for example, houses built from sawn timber instead of bamboo. The old methods are more likely to survive if they perform satisfactorily.

Even with that proviso there is a good deal of vernacular architecture that is not well suited to its geographical region. Central Ghana may serve as an example. The northern part of the country is hot-arid, and the southern part is hot-humid and has ample resources of timber and reed. However, the people of a large part of the forest country live in mud huts that are ill-suited to the

Fig. 1.2.3. Traditional Polynesian houses in Samoa improved by the addition of a concrete floor.

climate. It may be that these people migrated from the arid north, but this would have happened several centuries ago. Attempts made by various foreign experts to persuade the people in the forests to use the readily available timber for more comfortable buildings have not so far been successful.

This is one example of the role played by cultural traditions, which often exercise more influence than climate on vernacular design (Ref. 1.3).

Vernacular architecture therefore does not invariably provide the correct solution to design in conformity with the climate. Traditional construction can also be more expensive than modern methods. Thus adobe, whose use has been advocated in recent years for passive solar construction [see Sections 5.14 and 5.15] because of its high thermal inertia, is a cheap and comparatively simple method of building if the owner is willing to erect the walls himself, without counting the value of his labor. It is expensive if used by an American building contractor because of the relatively high labor content.

Vernacular design can generally be improved by calculation. If the thermal inertia of massive walls is too low or too high, the maximum and minimum internal temperatures will not occur at the most favorable times [Section 5.6]. If roof overhangs are too great (Fig. 1.2.5), the interior becomes unnecessarily dark.

Fig. 1.2.4. House in Erzerum, in eastern Turkey, with a grass-covered roof. Many of the older houses still have roofs built from heavy timber beams and covered with soil and grass. The weather is very cold in winter and very hot in summer.

Fig. 1.2.5. Experiment Farm Cottage in Parramatta, New South Wales, latitude 34° S, built in 1798. It is one of the oldest surviving buildings in Australia, which was first settled in 1788. The veranda gives excellent protection from the summer sun, which would have struck a British immigrant as particularly fierce, and it provides a large outdoor living space. However, the interior is too dark at all times without artificial lighting, which, at the time of construction, was limited to candles and oil lamps. Furthermore, the sun cannot penetrate in winter to give warmth to any of the rooms.

Until mineral oil, gas light, and electric light were discovered in the nineteenth century, all sources of artificial light, except for the expensive beeswax, were edible fats or oils. Hence artificial light was not used to a great extent except by the rich. Natural light was thus of particular importance. This was still true when the Renaissance was exported from its birthplace in Italy to the colder and cloudier British Isles (Fig. 1.2.6).

Traffic noise was already a matter for concern in Ancient Rome, and laws were enacted at various times limiting traffic at night. The same happened in Paris and London as they became big cities in the seventeenth century. It is unlikely that the sound insulation of buildings received serious consideration until the nineteenth century, but the thick walls traditionally used both in cool and in hot-arid climates provide excellent insulation against airborne sound [Section 11.4], if the windows are not too large.

Fig. 1.2.6. Two Renaissance buildings erected in the sixteenth century.
(a) The Farnese Palace in Rome has small windows that restrict the admission of solar radiation in summer.
(b) Longleat House in Wiltshire, England, has much larger windows to admit more daylight in winter.

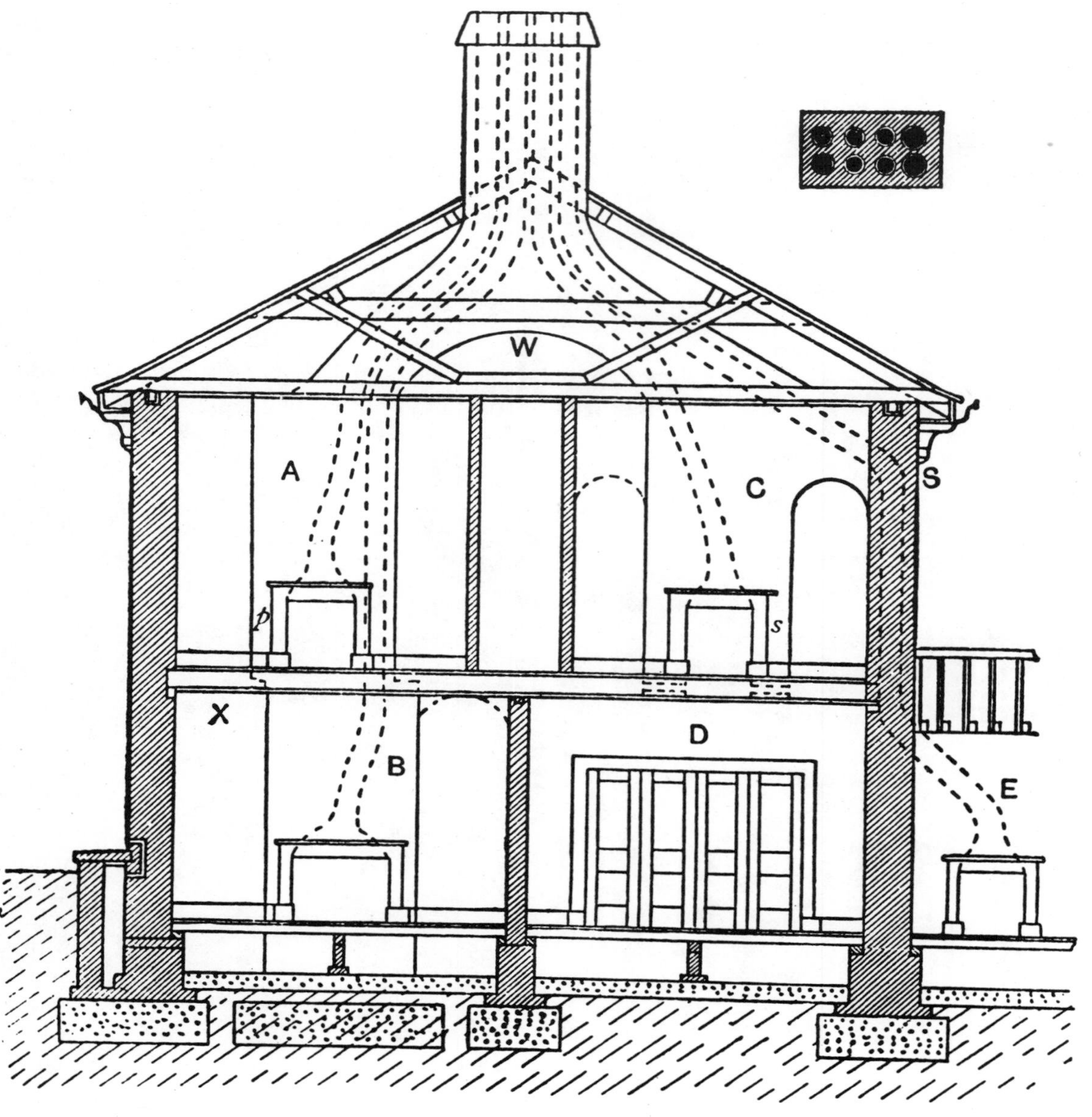

Fig. 1.3.1. Each fireplace in nineteenth-century British houses was provided with its own flue to ensure proper exhaustion of the smoke. The individual flues were usually collected into a single chimney. (*From Rivington's* Rivington's Notes on Building Construction, *published in London in 1874.*)

1.3 The Effect of Mechanical and Electrical Services on Architectural Design

Hypocaust heating [Section 1.1] was forgotten in Europe after the fall of Rome. Heat came from a large fire in the center of the Great Hall, and smoke escaped through a hole in the roof, which was protected from wind and rain by a lantern equipped with louvers. Hampton Court, the palace of Henry VIII of England, still had a central hearth in the sixteenth century. Heating was a luxury in the Middle Ages, and the Great Hall served as sleeping quarters for all but a privileged few. In rural cottages the only heat generally came from the kitchen fire.

During the Renaissance chimneys became more common, and fireplaces were increasingly built in the bedrooms of the wealthy. Tiled stoves were used in northern Europe beginning in the eighteenth century.

In the late eighteenth century there were significant advances in the art of heating. Fireplaces were improved to ensure proper exhaustion of smoke, and a fireplace became a normal part of the equipment of every room in the better houses (Fig. 1.3.1).

In the new factories waste heat from steam engines was sometimes used for heating, and in the early nineteenth century steam heating and warm-water heating were used in some British hospitals and other public buildings.

In North America the climate was more severe, servants were scarce and expensive, and the general standard of living during the nineteenth century overtook that of Europe. Domestic heating was normally supplied by a boiler that in time became more and more automatic. The actual heating was done by cast-iron hot water or steam radiators, which were frequently decorated with flower patterns or geometric curves (Fig. 1.3.2).

American architects in the late nineteenth and early twentieth centuries were more receptive to innovations in heating than their European colleagues. In the Baker House, built by Frank Lloyd Wright at Wilmette, Illinois, in 1908, hot water pipes were concealed in the wainscoting to supply radiators under the window seats that were slatted to permit the warm air to circulate. Toward the end of the nineteenth century some of the wealthier American homes were both heated and ventilated by hot air, supplied to the rooms through plenum ducts (see Glossary) terminating in grilles in the floor and skirting. Because the air was fed into the rooms by pressure, there were no drafts.

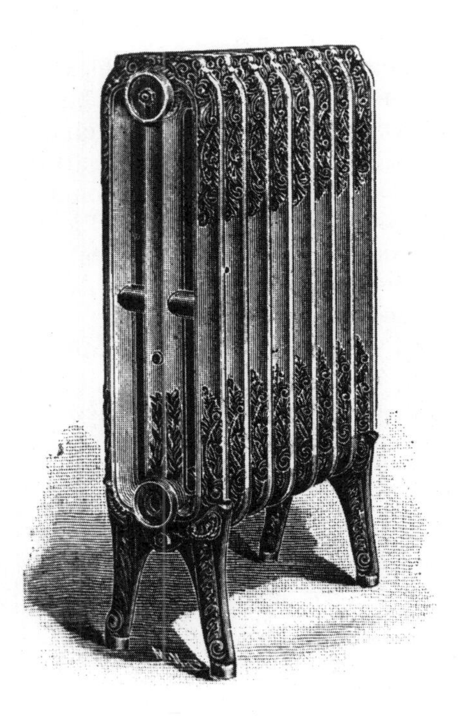

Fig. 1.3.2. Radiator for domestic heating, made in Buffalo, N.Y. (*From an advertisement published in 1895.*)

By comparison an open fire produced a large flow of air up the chimney. Because there were, in general, insufficient grilles, or none at all, the air intake was balanced by an inflow of air under the doors and through any badly fitting windows. This created the drafts for which many of England's stately homes were famous.

Cleaning of the air was first introduced in British hospitals. The Victoria Hospital in Glasgow had a wet airscreen in its ventilating system before 1895. The air, after being drawn in by propellers, was passed through a mesh of cords made of horsehair and hemp, closely wound over a top rail of wood and under a bottom rail, to form a close screen 16 ft by 12 ft (4.9 m by 3.7 m). A constant stream of water, by which dust and soot particles were removed and carried away, trickled down the screen. By an instant automatic flush tank 20 imperial gallons (90 liters) of water were instantaneously discharged over the screen every hour.

Removing the humidity is more difficult. In 1906 in Charlotte, North Carolina, Stuart W. Cramer used a spray of chilled water to clean and cool the air and to control humidity. He called this method air conditioning. In the same year William H. Carrier invented the dewpoint method of control. According to his own account, Carrier observed fog at the railroad station in Pittsburgh:

> Here air is approximately 100% saturated with moisture. The temperature is low so, even though saturated, there is not much moisture. There could not be, at so low a temperature. Now, if I can saturate the air and control the temperature at saturation, I can get air with any amount of moisture I want in it. . . .
>
> I can do it by drawing the air through a fine spray of water to create actual fog. By controlling the water temperature I can control the temperature at saturation. When very moist air is desired, I'll heat the water. When very dry air is desired, that is, air with a small amount of moisture, I'll use cold water to get low-temperature saturation. The cold-water spray will actually be the condensing surface. I certainly will get rid of the rusting difficulties that occur when using steel coils for condensing vapour in air. Water won't rust. (Ref. 1.4, p. 15)

The uncomfortable conditions on hot days, particularly in factories and theaters, had caused many complaints, and attempts were made during the nineteenth century to improve them. But refrigeration was needed to get satisfactory results, and this did not become available in quantity at a reasonable cost until the 1920s. The first air-conditioned building was Graumann's Metropolitan Theater in Los Angeles, built in 1922 with a Carrier-made installation. The air was cooled by a refrigerating plant to a temperature whose dewpoint (see Glossary) corresponded to the required humidity. It was then heated to the required temperature.

Although there was little new construction between 1930 and 1945, there were many advances in the technology of air conditioning. Gases used for refrigeration before 1930 had been more or less toxic. The freons, a group of chemicals based on fluorine, first manufactured in 1930, are stable and physiologically harmless, and they are still used for air conditioning. World War II produced innovations in air conditioning machinery because of its usefulness in naval operations in tropical waters. By 1945 the industry was ready for a major expansion. Air conditioning, which had been a luxury, became normal practice in office buildings all over the United States, and it was used to varying extents in other countries.

The 1950s were a period of great prosperity and confidence in the future. Incomes were rising, investment capital was plentiful, oil was abundant, and energy was cheap. It seemed the right time for buildings that had a perfect interior environment throughout the year, created by a large air conditioning plant.

The facades of many of these buildings were selected for visual and structural reasons. Environmental rules that had been used successfully since the Renaissance were abandoned, since the desired environment could be created by mechanical and electrical plants. The largest of the new buildings each used more energy for their environmental services than some countries in Africa use for their entire population. When the energy crisis occurred in 1973, this not only became an embarrassment in relations with the developing world but also created economic problems because of the unexpectedly rapid increase in the cost of energy.

1.4 Assessment of Thermal Comfort

Temperatures that would be regarded today as much too high (in summer) or too low (in winter) have been accepted in the past. To take an extreme example, in a northern Canadian winter the igloo can create an interior temperature 20 degrees Celsius (36 degrees Fahrenheit) or more above the outside temperature, but it can rise only slightly above the freezing point; otherwise, the walls would melt. The igloo is an ingenious invention, both structurally and environmentally, but Eskimos would today build one only in an emergency. They prefer the higher indoor temperatures obtainable in timber huts.

About 1840, when the British House of Commons was rebuilt, the consulting engineer, Dr. Boswell Reid, conducted an inquiry among the members of the House, mostly people who could afford every comfort at home, to ascertain the preferred temperature. The preferences ranged from 11°C (52°F) to 22°C (71°F). Sir Douglas Galton, who contributed the article on "Heating" to the Ninth Edition of the *Encyclopaedia Britannica*, published in 1880, expressed a similar view. He recommended a temperature for living rooms ranging from 12°C to 20°C (54°F to 68°F) and a temperature for bedrooms of "not less than 4°C (40°F)."

People wore heavier clothes in those days, and they were accustomed to temperatures which we would today find far too low.

Until the beginning of the twentieth century every technological advance in heating, cooling, ventilating, and artificial lighting made it possible to produce a more comfortable environment; but it could be argued that some design temperatures used since 1950 have been set beyond the limits of thermal comfort.

The optimum indoor temperature for a person who never leaves the controlled environmment is different from that for a person who from time to time goes outdoors where the temperature is either much higher or much lower. In 1960 it seemed a reasonable and desirable objective that in a more prosperous future everyone would be able to travel from an air-conditioned home to an air-conditioned workplace, restaurant, or place of entertainment in an air-conditioned car, transferring to and from the car in an air-conditioned garage.

This no longer seems either a realistic or a desirable objective, and design temperatures therefore need to be suitable for persons who leave the heated or cooled interior from time to time to encounter a very different outdoor temperature. The question of subjective assessment is discussed further in Chapter 4.

1.5 The Influence of Science on the Environmental Design of Buildings

As we have seen, some environmental concepts used today, particularly in the fields of thermal comfort and of acoustics, are of very ancient origin, but quantitative rules based on theory were developed only during the last hundred years. Xenophon's recommendation to design a portico so that it allowed sunlight penetration in winter but excluded the sun in summer is in accordance with modern concepts of passive solar design [Section 5.14]. It is likely that a good builder in Xenophon's time would have been able to construct this portico for a given latitude from observations on previous buildings. But it is unlikely that the dimensions were calculated from the position of the sun, although the astronomical data and the mathematical theory were probably available in the fourth century B.C.; they certainly existed in the second century A.D., when Ptolemy wrote *Mathematike Syntaxis,* also known from its Arabic translation as the *Almagest*.

The scientific design of sunshades came into use only in the 1920s [Section 3.5], but not many architects employed quantitative methods before the 1970s. *Brise soleil* (see Glossary) became fashionable after Le Corbusier recommended its use for the Ministry of Education in Rio de Janeiro in 1936, but its precise shading effect was rarely calculated. On some buildings it was decorative rather than functional.

Qualitative design for natural lighting received particular attention in England, encouraged by a judicial ruling in 1922 that defined a daylight factor [Section 9.3] of 0.2 as the border between adequacy and inadequacy of daylight in a room.

In English law the owner of a building acquires a right to prevent the owner of adjoining land from obstructing the light received through a window if he has had the use of it for twenty years. This ruling could be used to object to high-density developments. Elsewhere daylight calculations were rare before the significance of daylight design for energy conservation was appreciated [Section 10.1].

Music progressed quickly during the seventeenth and eighteenth centuries, but the science of acoustics, which had a significant early development in Ancient Greece and Rome [see Section 1.1], grew at a slower pace. A great deal was written on architectural acoustics during the nineteenth century, but it was mostly speculative. Charles Garnier started the construction of the Paris Opéra in 1861, seventy years after the death of Wolfgang Amadeus Mozart. He found the existing literature of little help in designing one of the world's most prestigious and successful opera houses. Leo Beranek (Ref. 1.5, p. 237) has said that the Paris Opéra has "very good acoustics — not the finest, but very good." Garnier wrote a book on its design in which he stated:

> The credit is not mine. I merely wear the marks of honor. . . . It is not my fault that acoustics and I can never come to an understanding. I gave myself great pains to master this bizarre science, but after 15 years of labor, I found myself hardly in advance of where I stood the first day. . . . I had read diligently in my books, and conferred industriously with philosophers—nowhere did I find a positive rule of action to guide me; on the contrary, nothing but contradictory statement. For long months, I studied, questioned everything, but after this travail, finally I made this discovery: a room to have good acoustics must either be long or broad, high or low, of wood or stone, round or square, and so forth. (Quoted in Ref. 1.5, p. 237)

Acoustics as a quantitative science dates from 1900, when Wallace Clement Sabine, a Harvard professor of mathematics and natural philosophy, designed the Boston Symphony Hall. The hall was immediately recognized as a masterpiece, and it is still ranked by most musicians as one of the world's finest concert halls. Acoustic design had consisted mainly of the determination of sight lines and first reflections. Sabine demonstrated the importance of reflecting surfaces and gave a formula, which still bears his name, for predicting reverberation time.

As we pointed out in Sections 1.1–1.3, heating and artificial lighting have been used since time immemorial. Mechanical ventilation was developed in the sixteenth century. These building services progressed quickly during the nineteenth century as byproducts of the Industrial Revolution. As a result, the calculations that have been used for their design since the early nineteenth century have gradually become more refined.

Gas lighting was invented in 1765, the passenger elevator in 1854, electric lighting in 1878, and air conditioning in 1906. As we pointed out in Section 1.3, the use of building services for the control of the thermal and luminous environments has proliferated particularly since the 1950s. In this book the emphasis is on the design of the building fabric, and the building services are regarded as a necessary supplement when the building fabric is unable to provide the required environment without that aid.

References

1.1 XENOPHON (Trans. E.C. MARCHANT): *Memorabilia and Oeconomicus*. Harvard University Press, Cambridge, 1968. 525 pp.

1.2 MARCUS VITRUVIUS POLLO (Trans. M. MORGAN): *The Ten Books of Architecture*. Dover, New York, 1960. 331 pp.

1.3 AMOS RAPAPORT: *House, Form and Culture*. Prentice-Hall, Englewood Cliffs, N.J., 1969. 150 pp.

1.4 MARGARET INGELS: *William Haviland Carrier, Father of Air Conditioning*. Country Life Press, Garden City, N.Y., 1952. 176 pp.

1.5 L.L. BERANEK: *Music, Acoustics, and Architecture*. Wiley, New York, 1962. 586 pp.

Suggestions for Further Reading

1.6 H.J. COWAN: *The Masterbuilders*. Wiley, New York, 1977. 299 pp.

1.7 H.J. COWAN: *Science and Building*. Wiley, New York, 1978. 374 pp.

1.8 REYNER BANHAM: *The Architecture of the Well-Tempered Environment*. Architectural Press, London, 1969. 295 pp.

Chapter 2 Water Supply and Waste Disposal

In this chapter we discuss the need for purity in a water supply. Waste-disposal management and the treatment of the water supply both contribute to the quality of the water. We look at the requirements for hot and cold water in buildings and at methods of producing hot water. Solar hot water systems are described, since this is one of the most practical applications of solar energy. Sanitary drainage and sewerage are discussed, and a consideration of the handling and disposal of solid wastes concludes the chapter.

2.1 How Clean Is Clean Water?

A reliable water supply is essential to any permanent human settlement. The small populations of ancient times either lived by a stream or dug wells to an underground water-bearing stratum. Fortresses and walled cities had a well within them to supply water during a siege.

None of these primitive people, except for the Romans, who had a highly organized system of aqueducts to bring water to their cities, paid much attention to the quality of the water as long as it was palatable. The Romans produced a pure water supply by taking it from a pure source and conveying it in aqueducts that could be cleaned from time to time. They had no means of purifying water, and they consequently distinguished between aqueducts carrying drinking water from a pure source, for example the Aqua Marcia in Rome, and aqueducts with water of lesser quality, for example the Aqua Anio Novus. Water from the different aqueducts was kept separate (Ref. 2.1, p. 74). It is only in the last centuries that we have known about bacteria, and gradually water-supply engineers have taken steps to ensure that harmful organisms from human wastes are not allowed to percolate back into the water supply.

Although it is now taken for granted that the water supply to any major community in a developed country is safe to drink, this does not hold true for a large part of the world, as tourists frequently discover to their discomfort. As recently as 1976 a United Nations conference on human settlements noted that safe and adequate water supply was not available to nearly two thirds of the populations in the less-developed countries (Ref. 2.2, p. 50).

Fresh water originates as rain or snow. The sun's heat evaporates it, and it later condenses and falls as rain. Rain is therefore distilled water, but as it passes through the atmosphere it becomes acidic by dissolving the oxides of carbon, sulfur, and other elements from the air. It also collects some dust particles. Surface water (collected from roofs or streams) is therefore often muddy and slightly acidic. The condition of water that percolates into the ground and is then collected from underground wells depends on the composition of the soil and rock through which it has passed. Usually most of the solid particles are filtered out. Alkalis in the ground neutralize most of the acidity or make the water slightly alkaline. Calcium and magnesium carbonates in solution make the water "hard." "Soft" water contains only small amounts of these minerals; it is neutral or slightly acidic.

Hard water does not produce a lather with soap, and the dissolved chemicals are deposited on the insides of pipes and boilers, greatly reducing their life. Acidic water avoids these problems but causes rapid corrosion of pipes, boilers, and tanks.

The methods for treating water have been developed mainly during the last hundred years. Settlement tanks allow the larger suspended solids (coarse silt, grit, and debris) to settle to the bottom. Sand filters remove the finer particles. Chlorine is added to kill the bacteria. Other chemicals are added to flocculate the fine particles, to make it easier to remove them. Excessive hardness or acidity can be removed by chemical treatment, but this adds greatly to the cost of the water.

The population of urban areas, as well as the per-capita demand for water, is increasing rapidly. We have to find more water from the land around our cities. At the same time the water in urban areas is becoming increasingly polluted by detergents, industrial wastes, and the things that society discards. Some of the industrial pollutants, in particular, are in solution and are therefore difficult to remove.

To avoid unnecessary treatment of the whole water supply, we should insist that industrial waste and sewage are effectively treated before they are discharged. Alternatively, we could recognize that only a small part of the total water supply is used for drinking and washing, while the majority is for purposes that can accept a reasonably high level of pollution. A dual water supply could be piped through the city and the "dirty water" used for flushing toilets, washing cars, watering gardens, firefighting, and any use that does not require cleanliness. Alternatively, a "dirty-water" supply could be used in conjunction with water-purification devices in each building to produce water for human consumption. The cost and possible health risk of such a system make it unacceptable at present, except for private water supplies in remote areas.

2.2 Demand for Cold and for Hot Water

A person requires only a few liters of drinking water per day to survive. However, many other uses of water are taken for granted, and we use several hundred liters per day per person as part of our normal way of life. The largest and most variable items of water usage are the watering of gardens and the process water used in industry. Fortunately, the designer of a large building need only provide the water used inside the building; this usually includes water required for firefighting.

In an apartment building, both hot and cold water are needed for:

- cooking and drinking
- dishwashing
- personal washing (shower, washing hands, brushing teeth, etc.)
- laundry (unless it is sent out)

Cold water only is used for:

flushing toilets
hosing patios
washing cars
watering house plants

For a suburban house we must add the water used in the garden, while for an office or factory building we require less for cooking and dishwashing, which is only needed in the cafeterias, and for laundry, as industrial clothing is probably laundered off-site. The usage for drinking, toilets, and personal washing is reduced because of the shorter hours of occupancy. Showers are required for factory workers whose work is hard or dirty. Hospitals use more water because of the nature of the occupancy and the extended hours of use.

Although it is not possible to predict water consumption precisely, Table 2.1 gives some guidance.

Estimates of water usage are required by city authorities to design reservoirs and reticulation systems. They are needed to size tanks for cold water to hold at least one day's supply. Hot water tanks often are designed to hold one day's water, particularly if the electricity company offers a cheap rate for heating water at night when there is little other demand for electricity. If a smaller tank is used, we need to consider the peak demand and the time required to reheat the tank once all the hot water is used up.

Individual families can have different patterns of hot water use. A parent at home with small children may do some laundry every day. A couple who both work outside the home might do a week's laundry on Saturday morning; they would face serious inconvenience if they had a hot water system designed only for their average daily needs, particularly if water could be reheated only at night.

Table 2.1

Approximate Requirements for Hot and Cold Water
(liters per person per day)

Type of Building	Cold Water	Hot Water
House (with garden)	400–800[a]	100–200
Apartment, hotel	200–400	100–200
Office	50–100	10–15
Factory	60–120[b]	15–25

[a]Average figures; higher in hot, dry weather.
[b]Plus process water for industrial use.

2.3 Supply of Cold Water

A city water supply is reticulated along the streets in pipes fed by an elevated storage tank, or sometimes by a low-level pressure pump. Late at night, when there is little or no water use, the pressure in the pipes reaches its maximum value. Whenever water is used from the system, the normal frictional resistance to flow in the pipes causes a drop in pressure between the storage tank and the point of use. One person taking a bath does not affect the pressure in the city's mains, but many people all watering their gardens at once, or the fire department fighting a large fire, can do so.

Within a building, where the size of the pipes is much smaller than the mains in the street, the pressure also drops when water is used from anywhere in the system. This is most annoying when one person is showering and another turns on a hot or cold water faucet in another room. A change in the available pressure upsets the balance between hot and cold, which had been carefully adjusted to give a comfortable shower. The problem cannot be completely avoided, except by installing thermostatically controlled showers; however, it is greatly reduced by using adequate pipe sizes for all but the final branch to each fitting; this can be the smallest size to suit that fitting.

The water supply in the street usually has a pressure of about 300 to 500 kPa. As this drops by 100 kPa for every 10 m increase in height, there is an upper limit to the height of a building that can be fed from the mains. Most outlets require a pressure 50 to 100 kPa to work properly; garden hoses need about 200 kPa, but they are at ground level. Mains pressure plumbing is limited to buildings no higher than five stories, to allow for loss of pressure in the mains due to heavy use and for frictional pressure drop in the building's own pipework. This height limit can vary according to local conditions; for example, in a hilly city the pressure is greater in the valleys than on the hilltops (Figs. 2.3.1 and 2.3.2).

Taller buildings require a water tank on the roof. For a building of eight stories it may be possible to fill this at night, when the mains pressure is high. Taller buildings require a pump to get the water to the top.

In a *downfeed system*, the water is pumped to a storage tank on the roof and allowed to feed down by gravity. The tank must be elevated above the roof, to give enough pressure to the top story. In a tall building the pressure for the lower floors would become so great that ordinary pipes and faucets would not be able to cope with it. This problem is overcome by using staging tanks at every tenth story in the building, or by installing pressure-reducing valves.

If the building is tall enough to require a water storage tank at the top, it also needs water storage for firefighting. The storage tank is made big enough for both domestic and firefighting purposes. The domestic water outlet is located partway up the side of the tank (Fig. 2.3.3). The bottom half is reserved for fire purposes and cannot be used for other purposes, but any water above that level can be used for firefighting if necessary. Furthermore, any silt suspended in the water settles into the bottom section.

Alternatively, a building can be supplied with water by means of a pumped *upfeed system*. Instead of pumping all the water to the top and allowing it to come down again, it is pumped up to maintain a constant pressure (Fig. 2.3.4). This can be done with a tank in the basement half full of water and half of compressed air. As the water is used, a pump cuts in to supply more water to the tank. A more elaborate system uses one or more variable-speed pumps to maintain the pressure.

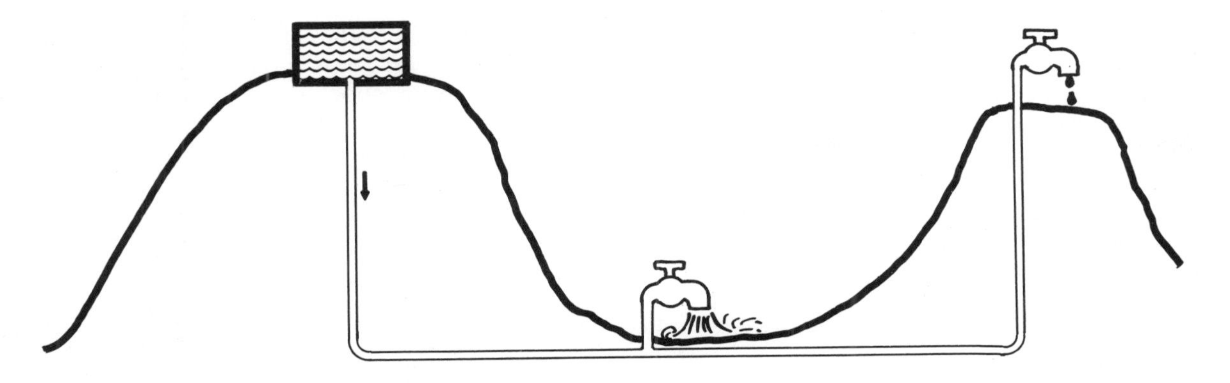

Fig. 2.3.1. The water pressure depends on the height of the reservoir above the water faucet.

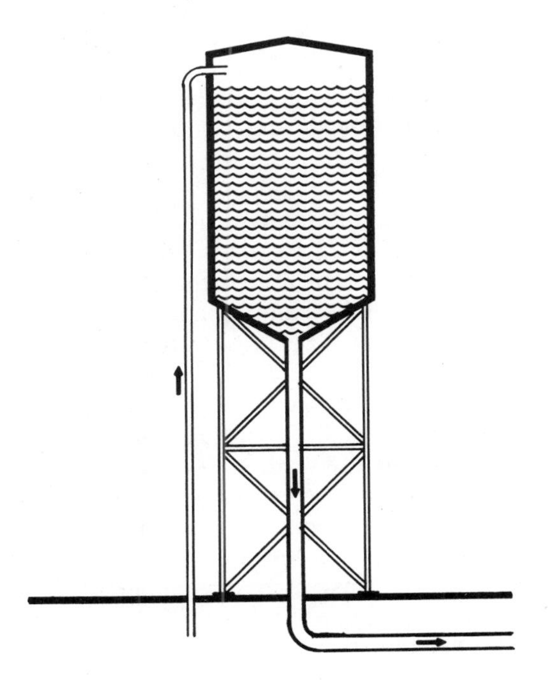

Fig. 2.3.2. If there is no high ground on which to build the reservoir, it can be elevated on a tower to provide pressure in the water mains.

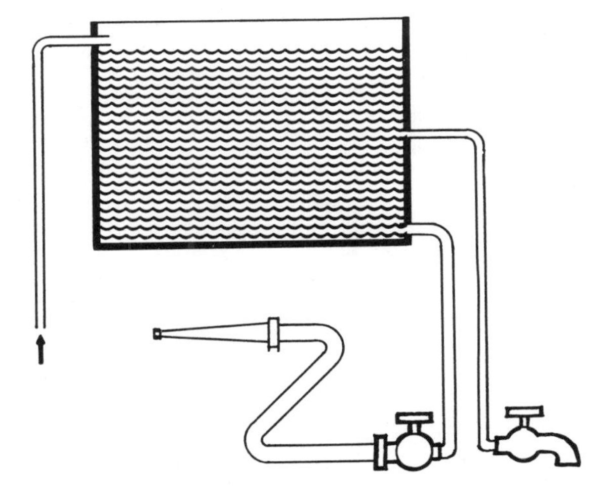

Fig. 2.3.3. In a tall building, a storage tank is provided on the roof. To ensure that there is always enough water in it for firefighting, the domestic water outlet is located partway up the side of the tank.

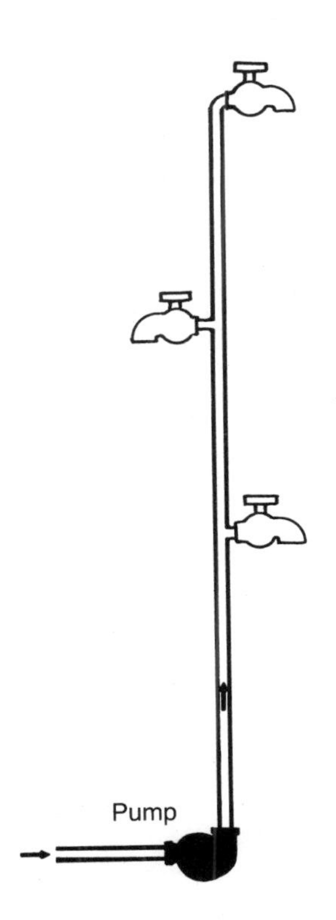

Fig. 2.3.4. In an upfeed system, a pump maintains sufficient pressure to overcome gravity and supply water to all floors.

We noted that the mains pressure varies according to the amount of usage of the whole system. The mains pressure drops if there is accidental breakage in a pipe or if part of the pipework has to be shut off for maintenance or extension work. To meet these eventualities, it is necessary to store some water on-site for essential purposes, and it is also necessary to prevent water from the fixtures from siphoning back into the empty mains.

When a building is supplied by a downfeed system, the storage tank at the top of the building takes care of the need for on-site storage. When the building has a pumped upfeed system, any required storage must be provided in a storage tank in the basement. In this case it is essential that the pumps not fail in an emergency. Several pumps operating in parallel reduce the risk of a total mechanical failure. A standby diesel engine is used to ensure that water is available for firefighting even if the electricity supply is cut off.

Although building fires and the breakdown of the city water supply are both events of low probability, it is possible that earthquakes, hurricane winds, or explosions (from accident, terrorist attack, or military action) could cause fires, breakage of water mains, and failure of the electricity supply all at the same time. In such an event, a major building should have water available in on-site storage, with an independent, engine-driven pump to supply pressure until the fire department can connect its pumps. Insurance companies set minimum standards for the size of this water storage.

City water departments protect the quality of their water supply from back-siphonage from a customer's premises into a temporarily empty main. In the case of sinks and bathtubs, the bottom of the faucet must be above the top of the fixture bowl. A similar air gap is possible when a storage tank is fed through a ball valve. When a tank is connected directly to the pipework, as in a mains-pressure hot water tank, a check valve must be fitted in the connection to the tank (Fig. 2.3.5).

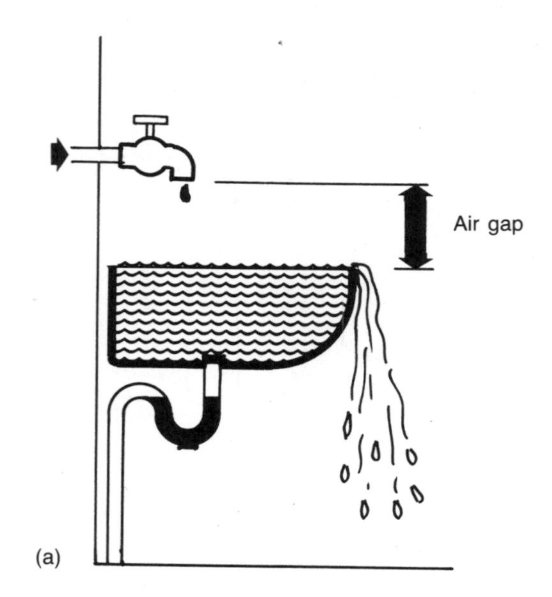

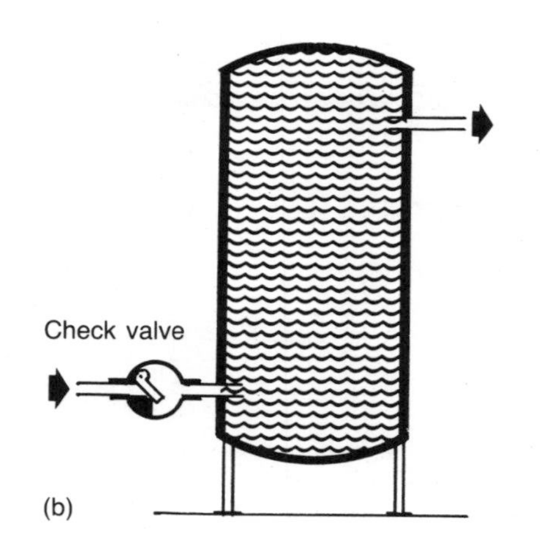

Fig. 2.3.5. Protection from back-siphonage.
(a) The air gap between the rim of a sink and the lowest point of the faucet above it ensures that water in the sink cannot be siphoned back into a temporarily empty main.
(b) When an air gap is not possible, a check valve prevents back-siphonage.

2.4 Supply of Hot Water

In an apartment or hotel building, nearly half the water used is likely to be hot water. In other buildings the proportions change depending on the amount of hot water needed for washing, or cold water for outdoor purposes. However, the hot water component is always significant. The temperature at which the water is used varies from about 60°C for dishwashing to about 40°C for personal hygiene. Therefore hot water can be stored at between 65°C and 75°C and mixed with enough cold water to give the right temperature for each use. Alternatively, the storage can be kept as hot as 95°C, just below the boiling point, so that a tank of a given size produces more hot water at the desired temperature. However, the hotter the tank is, the greater the heat loss from it and the hot pipes [see also Section 13.3]. As another alternative, hot water can be stored at about 45°C* to serve the bathrooms, and a separate system at a higher temperature can supply the kitchen and laundry.

We have come to expect hot water to be available at the turn of a faucet. This was not always so; a few generations ago a hot bath started with the chopping of wood to heat the water. Instant hot water can be supplied from a storage tank, as described above, or by means of an instantaneous or *tankless heater* which starts working as soon as there is a flow of water through it (Fig. 2.4.1.a). An instantaneous heater must be capable of supplying a large quantity of heat to the water rapidly, while the heater in a storage

* Temperatures below about 50°C may not be sufficient to prevent the growth of certain microorganisms in the stored water.

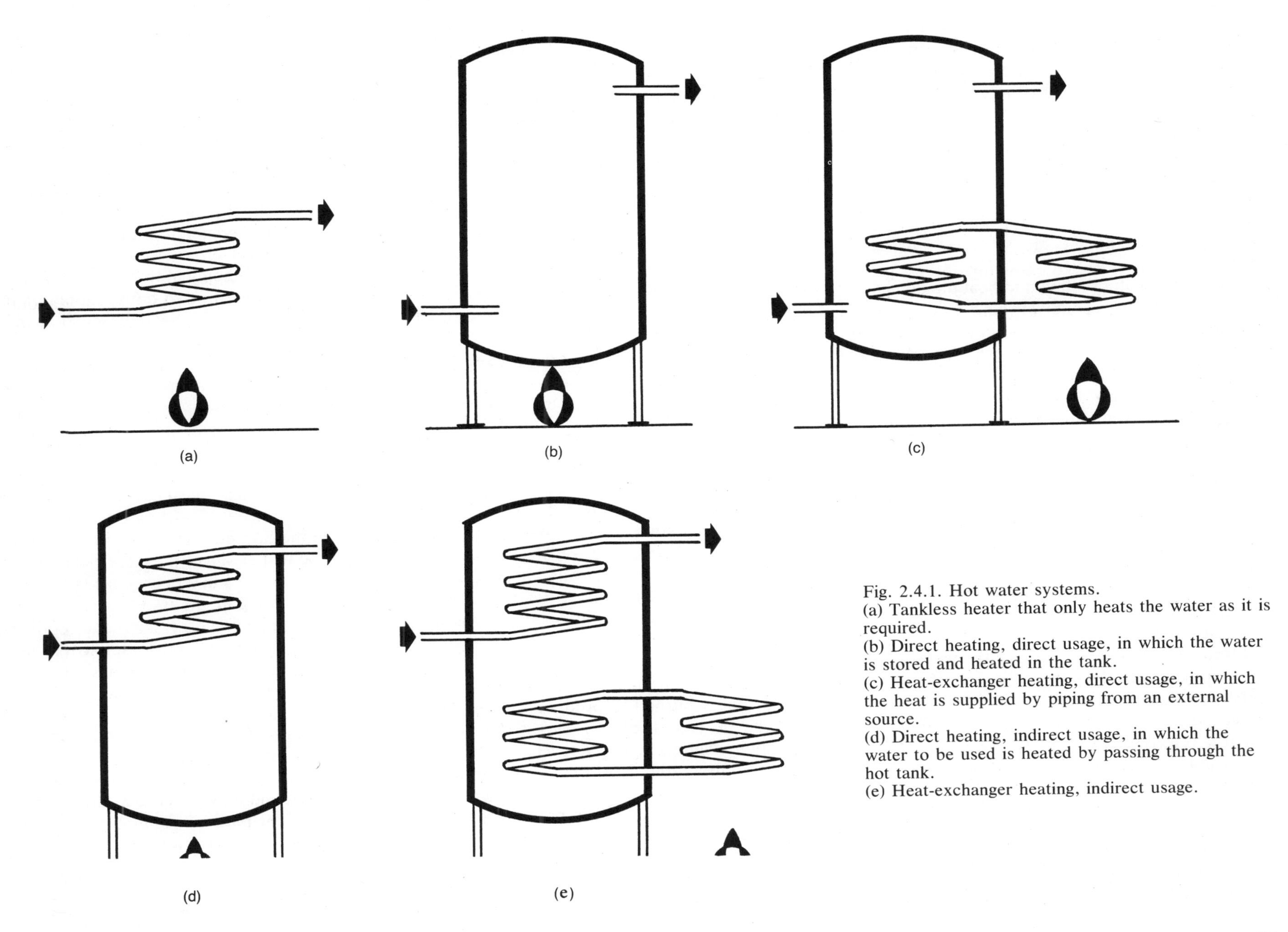

Fig. 2.4.1. Hot water systems.
(a) Tankless heater that only heats the water as it is required.
(b) Direct heating, direct usage, in which the water is stored and heated in the tank.
(c) Heat-exchanger heating, direct usage, in which the heat is supplied by piping from an external source.
(d) Direct heating, indirect usage, in which the water to be used is heated by passing through the hot tank.
(e) Heat-exchanger heating, indirect usage.

tank can work at a slower rate for many hours to heat the same amount of water. Instantaneous heaters can use gas or electricity, although the large, short-duration load is better suited to gas. They are more energy-efficient when hot water is used infrequently or in small quantities, because it is not necessary to maintain a tank at a high temperature all the time.

If the building has a central heating system, or an air conditioning system that includes hot water for warming the air, or a steam-raising plant for any purpose, we should consider using the same hot water or steam to heat the domestic hot water. A large boiler is more efficient than a small one, and the domestic hot water can be heated at times when the boiler is working at less than full capacity for its other loads. However, a long pipe run may be required to serve the hot water system from the boiler location, and in summer the boiler may have to be fired merely to supply hot water. The last objection can be overcome by installing an electric immersion element to operate the hot water system at those times when the main boiler is not in use.

The temperature required for hot water is relatively low, and almost any heat source can be used, such as gas, oil, or solid-fuel burners, electricity, heat pumps, solar energy, or waste heat from any process.

Storage hot water systems are classified as direct or indirect. In a direct system the stored hot water is used and replaced with cold water; in an indirect system it remains in the tank and exchanges its heat with cold water passing through a coil immersed in it. Their control is simple: A thermostat is required to turn off the source of energy, and this does not need to be very accurate.

The *direct system* allows most of the hot water to be drawn off the top of the tank, while cold water replaces it from the bottom, with very little mixing (see also Fig. 2.4.5). It has two disadvantages: If mains pressure is used, the whole storage tank must be made strong enough to resist the pressure; furthermore, if the water is hard, scale is deposited on the hot parts of the heating element, reducing the efficiency of heat transfer to the water (Fig. 2.4.1.b, c).

In an *indirect system* the heated water acts as a store of energy. The water to be used passes through a coil of copper pipe within this hot water and is heated by conduction through the pipe walls. Thus the stored water can be at atmospheric pressure with a vent at the top, while the water for use can be at mains pressure so long as the coil of pipe is strong enough to resist that pressure. Scale from hard water develops inside the coil rather than in the storage tank, but as the temperatures are lower than at the heater element surfaces, the rate of scale deposition is less (Fig. 2.4.1.d, e).

With either system the heat can be supplied directly to the stored water (Fig. 2.4.1.b, d), or by means of another heat exchange coil in which water, steam, or any suitable fluid brings heat from an external source (Fig. 2.4.1.c, e). Since every heat-exchange surface results in some decrease in the overall efficiency and in an increase in the overall cost of the plant, heat exchangers are only introduced where direct contact is not appropriate.

Direct immersion in the stored water is the best solution for an electric resistance element. When a fuel is burned to produce the heat, the water can be contained in a jacket around the combustion chamber. However, when the burner is not operating, the flow of air up the stack results in great loss of heat from the tank, and for all but small systems it is preferable to use indirect heating. A thickly bunched coil of piping is placed in the combustion chamber, and water is circulated through it and through the water storage tank.

When solar energy is the source of heat, the *solar collector* must be located on the roof or in some other suitable location exposed to the sun. Water or some other fluid is circulated between the collector and the storage tank. The storage tank usually needs to be larger for a solar system than for other heat sources, because there is no solar radiation during the night, and little on cloudy days. In all but the most favorable locations, a gas or electric booster is required to carry the system through a prolonged period of cloud or rain. It is not economical to build a storage tank for more than one to two days' supply unless it is integrated with space heating [see Section 6.7].

The intensity of solar radiation is low [Section 3.1], seldom exceeding 1 kW/m^2 in summer. The energy required to heat water is greatest in winter, because the feedwater temperature is low, standing losses to the cold air are higher, and the hours of available sunlight are short. Collectors are therefore designed to make the best use of the winter sun. A basic flat-plate collector consists of a large area of black material facing south (north in the Southern Hemisphere) and inclined so that the winter sun strikes it at close to normal incidence. As the sun's energy heats up the surface, the working fluid passes over it in tubes to take away the heat energy. To reduce convective heat losses to the air and losses to the sky by long-wave radiation, one or more glass covers are placed over the plate (Fig. 2.4.2). To reduce losses through the back of the collector, it is insulated with 50 to 100 mm of mineral wool or equivalent insulating material.

Losses occur in spite of the insulation and glass covers, and they increase as the temperature of the collector plate rises. Flat plate collectors can be made to produce water at 90°C in summer conditions, but at this temperature the efficiency is very low. For efficient performance in winter they should not be expected to produce water hotter than 45 to 55°C. These figures, of course, will vary with the quality of the collector and the severity of the climate.

It is desirable to avoid heat exchangers and use a direct system, so that the same water is stored in the tank, circulated through the collector, and drawn off the outlets.

In a freezing climate the collectors must be drained whenever the temperature of the water in them drops close to freezing point, or they may by heated by a small backflow of hot water. The first of these precautions is subject to human or mechanical failure, while the second wastes some of the solar energy collected.

Alternatively, a nonfreezing fluid is placed in the collector, with a *heat exchanger* to transfer the heat energy to the hot water that is to be used. This fluid can be a mixture of antifreeze and water, paraffin oil, or air. Antifreeze solutions such as ethylene glycol are toxic, and a double heat exchanger (that is, glycol to stored water and stored water to water for use) should be used so that an accidental leak does not poison the hot water system. Paraffin has a lower heat capacity than water, but this can be overcome by increasing the flow rate. It allows cheaper materials to be used in the collector because it is noncorrosive. Air has a very low heat capacity and relatively poor air-to-surface heat transfer characteristics. To transfer heat from air to water, a finned coil is needed, similar to an automobile radiator. Relatively cheap materials, such as galvanized steel, can be used for air collectors, and workmanship is less critical because a small air leak does not cause damage inside the building, unlike a small water leak. However, air collectors are used more for space heating [see Section 6.4] than for hot water.

The heat transfer fluid must circulate through the collector and the storage tank when heat is being collected, but it must stop when the collector is cooler than the tank; otherwise all the collected heat would be dissipated every night. A simple thermosyphon circulation provides enough movement for a small installation. It has no devices that can fail, and it uses no extra energy (Fig. 2.4.3). It is essential that the tank be located above the collector, because the hot water is lighter than the cold water. This can be done inside a steep roof. For mild climates, an integral collector and tank can be placed on the outside of the roof (Fig. 2.4.4). This arrangement has the lowest plumbing costs, but it may require strengthening of the roof timbers, and it subjects the insulated storage tank to the full effects of the weather instead of having it sheltered inside the roof space.

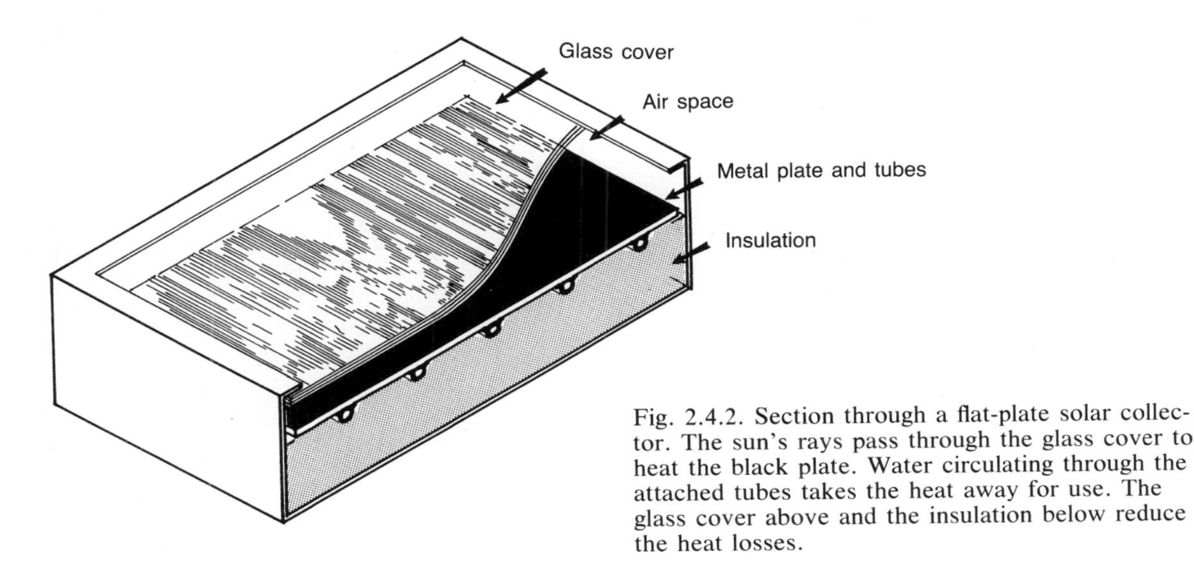

Fig. 2.4.2. Section through a flat-plate solar collector. The sun's rays pass through the glass cover to heat the black plate. Water circulating through the attached tubes takes the heat away for use. The glass cover above and the insulation below reduce the heat losses.

If the tank must be located below the collectors, or if for any other reason the thermosyphon principle cannot be used, a circulating pump is required. In addition we need a differential thermostat to detect when the water in the collector is hotter than that in the storage tank, and control gear to turn on the pump when this condition occurs.

As mentioned before, a solar hot water system needs a *booster* to make it reliable in all weather conditions. This should heat only a small amount of water when required, to ensure that it does not do most of the work. This can be achieved by placing a gas or electric instantaneous heater at the output of the storage tank, to add only enough energy to raise the temperature of the water to the minimum required for use. Alternatively an electric immersion element can be located near the top of the storage tank, to maintain a small part of it at the minimum temperature required.

The use of solar energy for space heating systems is discussed in Section 6.4.

In a small house or an apartment with its own hot water system, it is sufficient to keep the hot water in an insulated tank (Fig. 2.4.5). Cold water from the mains comes in at the bottom, and the hot water is drawn off at the top when needed.

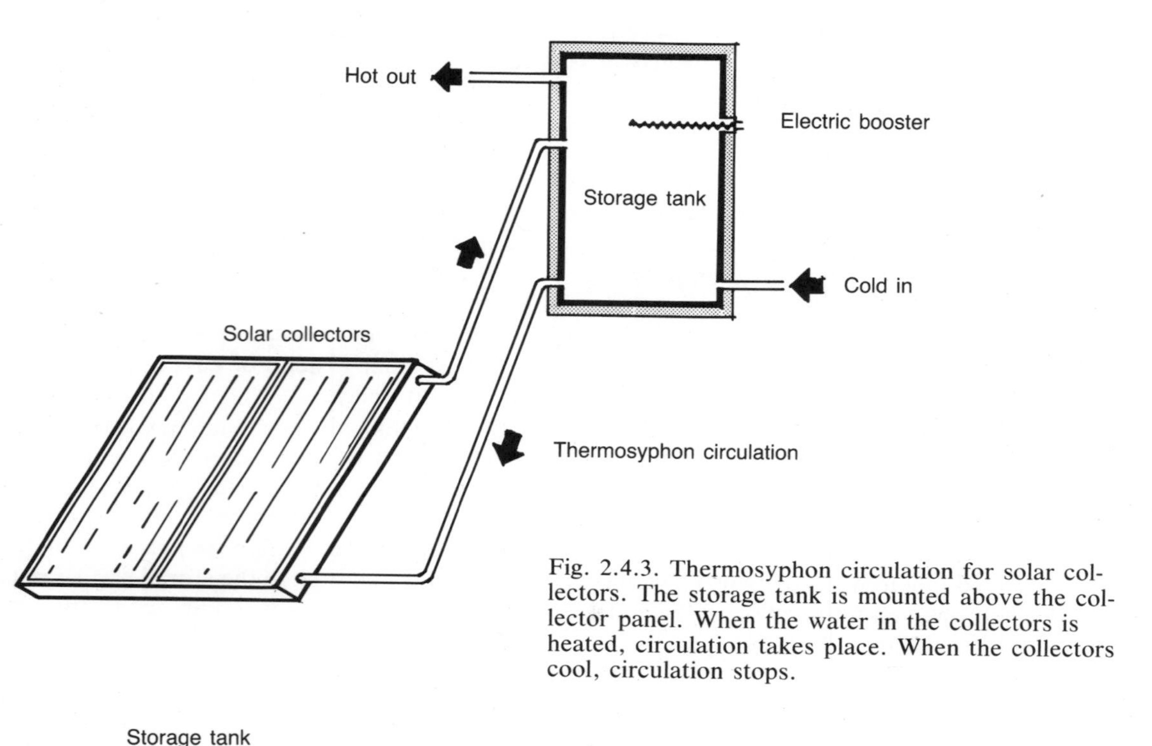

Fig. 2.4.3. Thermosyphon circulation for solar collectors. The storage tank is mounted above the collector panel. When the water in the collectors is heated, circulation takes place. When the collectors cool, circulation stops.

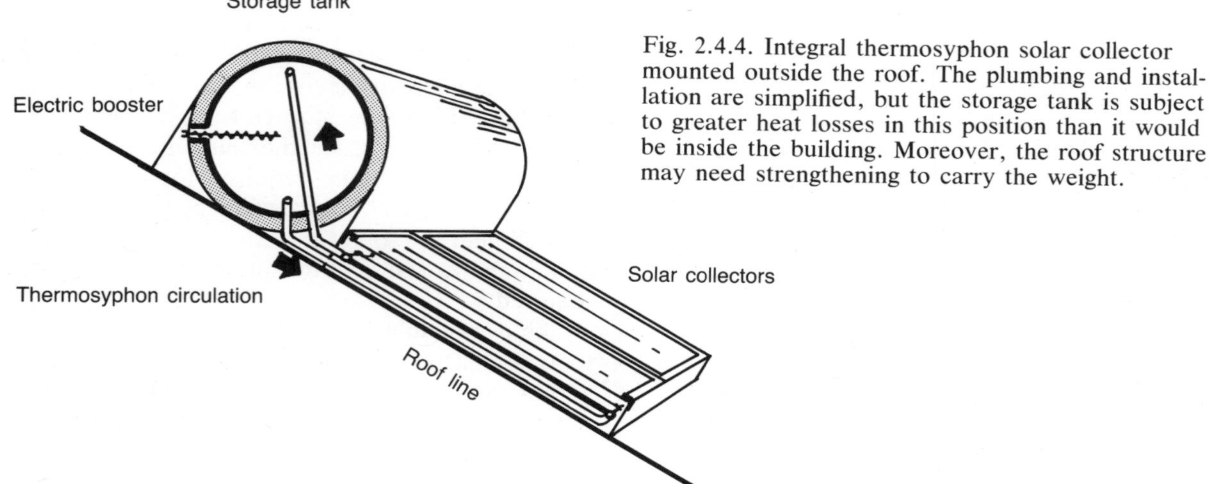

Fig. 2.4.4. Integral thermosyphon solar collector mounted outside the roof. The plumbing and installation are simplified, but the storage tank is subject to greater heat losses in this position than it would be inside the building. Moreover, the roof structure may need strengthening to carry the weight.

Although the hot water pipes are insulated, the small amount of water in them quickly cools to ambient temperature. When we turn on a hot water faucet somewhere, this cold water runs out of the pipe before the hot comes through. If we are running a bath, the few seconds of delay and the waste of energy represented by the hot water that has cooled are not important by comparison with the time and quantity of water needed to fill the bathtub. On the other hand, the delay can be annoying if we only want enough hot water to make a cup of coffee, and the amount of energy wasted can be more than the energy actually used in the cupful of hot water.

In a large building, we cannot allow the hot water standing in the pipework to get cold when it is not being used. Therefore, the hot water is circulated through a system of flow-and-return pipes, which have to pass through or close to every point for hot water in the building. Since the water in these pipes is always hot, they must be well insulated. Even so, the energy lost from this system is greater than that wasted in the simple, one-way flow system. In cold weather this energy is not entirely lost; it contributes to the heating of the building.

The circulation is usually maintained by a small pump, although it is possible to arrange the pipework so that the natural thermosyphon circulation causes enough flow in the loop (Fig. 2.4.6).

Since water expands when heated, a hot water tank must have provision for expansion. The most foolproof method is to supply cold water to the system from an open tank at the top of the building, and to have an expansion pipe connected to the hot water loop that can spill out into the open tank (Fig. 2.4.7). When the hot water tank is fed by mains pressure, a pressure-relief valve or safety valve allows excess pressure to discharge to waste.

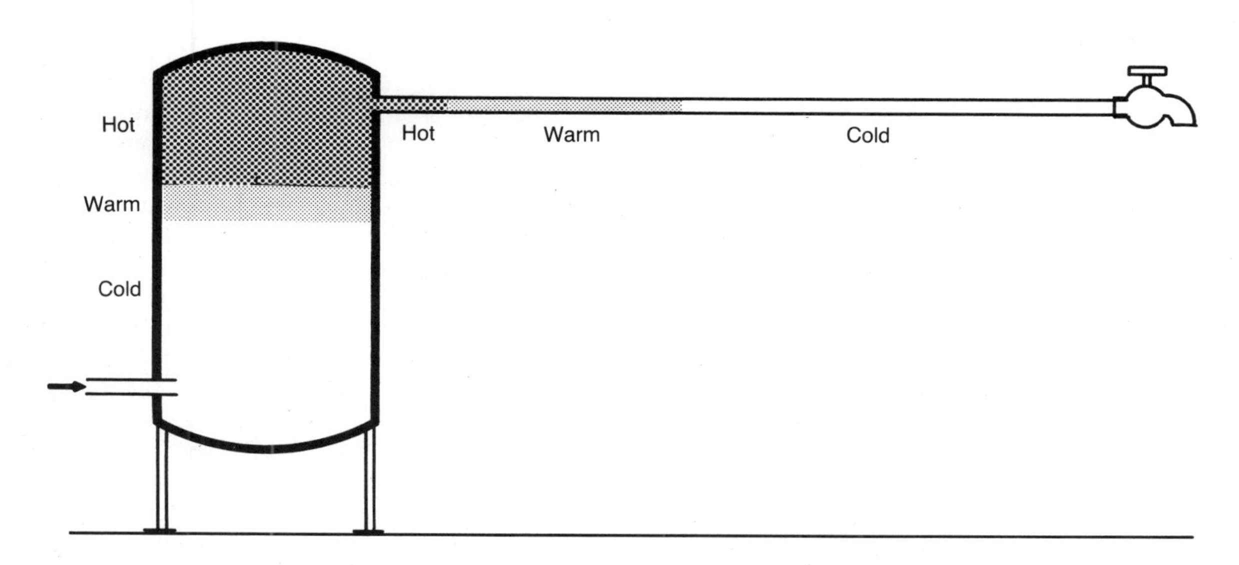

Fig. 2.4.5. Hot water system for a small house or apartment. The water in the outlet pipes cools when it is not being used. As the hot water is used and replaced by cold, there is little mixing because the cold water is denser than the hot.

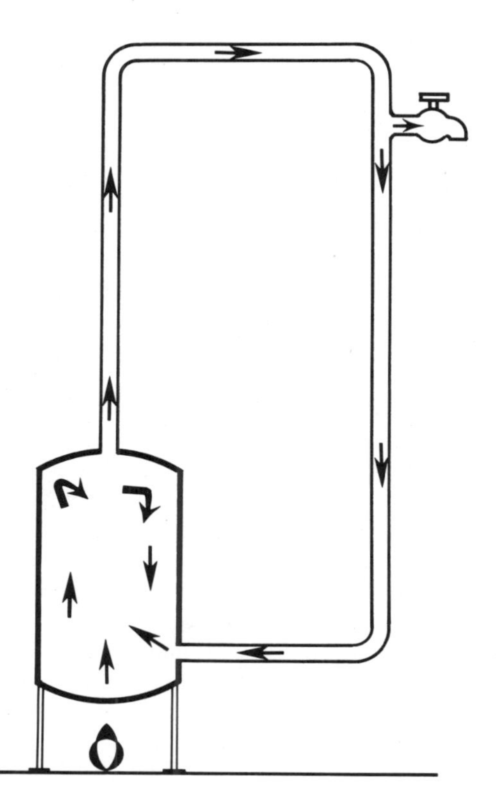

Fig. 2.4.6. Thermosyphon circulation. Hot water rises because it has a lower density than cool water. As it passes through the pipes, it loses some of its heat to the surroundings, and the slightly cooler water sinks back to the tank because of its higher density.

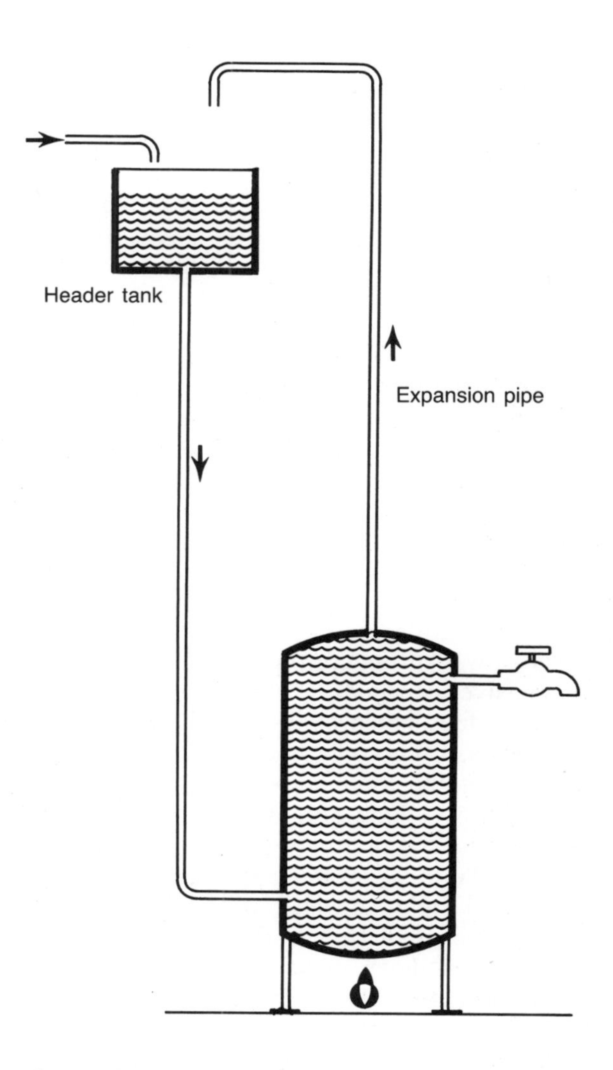

Fig. 2.4.7. Because water expands when heated, provision must be made for this expansion. If the system is fed from a header tank in the roof of the building, an expansion pipe can be provided.

2.5 Sanitary Drainage and Disposal of Liquid Waste

The disposal of human waste is a problem that quickly becomes more difficult as the size of a community increases. It can be broken into three operations: the collection, transport, and final disposal of the waste. All three are combined in the dry-pit privy built over a pit in the ground; the material filters away into the surrounding ground, and when the pit finally fills up, the privy is moved to a new location. In a crowded city, however, there is not enough ground to keep moving the privy, and it has to be emptied by some of the less fortunate municipal employees and transported to a disposal area.

The Romans had a better system, whereby the waste was carried away by running water, into a river. It was good to live in the city, but not so good to live downstream from it.

Piped sewerage systems began to reappear in major cities only in the late eighteenth century, and their extension beyond the inner city areas was slow. An underground sewerage system is expensive, especially if the natural slope of the ground does not provide the fall necessary for the pipes. It can only be undertaken by a prosperous, stable community that is prepared to be taxed over a long period of time to pay for a utility whose value may only be apparent after it comes into use. Water-borne sewerage requires a plentiful supply of water, and city water supplies had to be upgraded when sewers were introduced.

The breakdown of sewage occurs mainly by bacterial action. It results in the production of gases, including methane, which is explosive in air, and hydrogen sulfide, which has an offensive odor. The introduction of piped sewerage solved the problem of transporting the effluent, but the odors and the explosion risk were not brought under control until the late nineteenth century, through the invention of the system of venting and water-seal traps that feature prominently in our drainage regulations today. Only then could the collection end of the system, the toilet and the washbasin, be brought into the house.

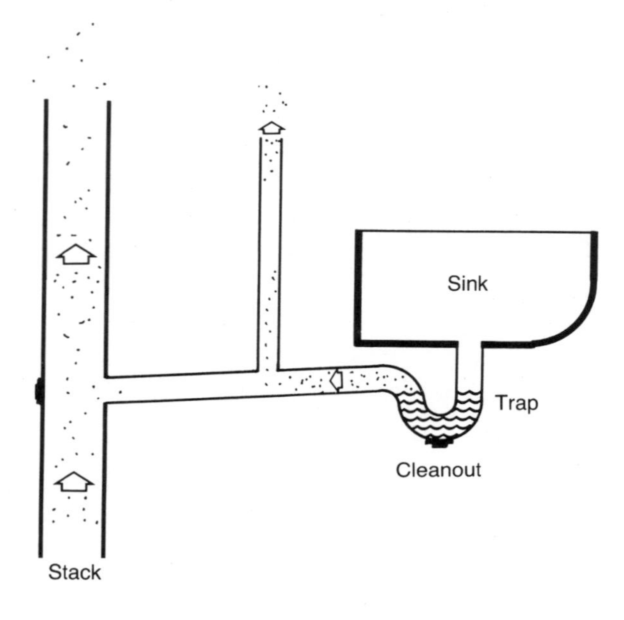

Fig. 2.5.1. The U-shaped trap holds water and prevents gas from the waste pipe from passing through the outlet of a fixture and into the room.

A trap is a U-shaped pipe that always holds enough water to prevent gas from one side passing through to the other (Fig. 2.5.1). There is always a trap in the drain as it leaves the building, so that gas in the public sewer cannot enter the drains inside the building. On the house side of this trap is a fresh air inlet, and a vent stack extends through the roof to form an air outlet (Fig. 2.5.2). All the sanitary drainage in the building is therefore ventilated. Where each toilet, lavatory, and kitchen sink connects to the drainage system there is also a trap, so that the air in the pipework cannot flow back through the fixture.

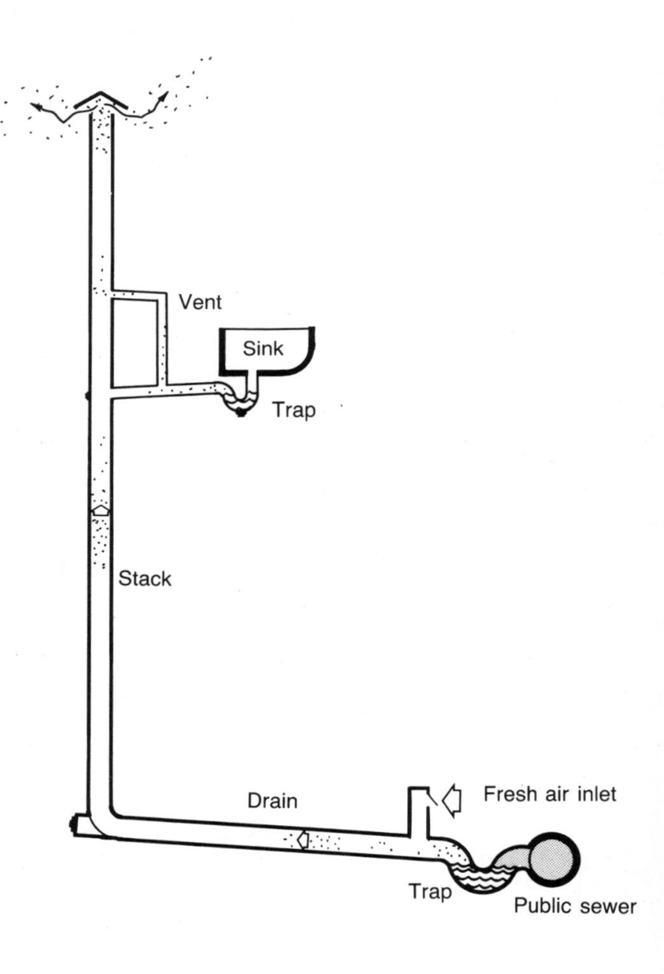

Fig. 2.5.2. The complete drainage system is vented by a fresh air inlet near the property boundary and an outlet above the roof. Methane gas, being lighter than air, escapes through the vent stack, avoiding the danger of a gas explosion.

Even with these precautions, it is possible that the rush of water from one fixture may cause the air pressure in the pipework to increase momentarily or decrease sufficiently to draw the water out of the trap of another fixture. Therefore, in certain locations close to a trap, an extra connection is made to the vent stack to allow the pressure to equalize. The precise requirements for venting the traps of individual fixtures are set by local plumbing codes. It often results in vent pipes being required in positions that are difficult to conceal in the architectural details. The provision of duct space, false ceilings, and in large amenities blocks (cafeteria, locker rooms, washrooms), false walls to contain the plumbing, should be considered early in the planning of the building, as we usually wish to conceal the pipework.

If the plumbing installation is of good-quality materials and the workmanship is good, leaks should be rare. However, blockages can happen from many causes beyond the plumber's control, and easy access to cleanouts is essential. Cleanouts are placed at the foot of each stack, just outside the trap in the house drain, and preferably at all points where horizontal runs enter stacks, particularly from toilets. It is also necessary to be able to remove fixture traps or have access to the cleaning openings at their base, to rescue items dropped down sinks and basins, either because such items are valuable or because they cause blockages.

2.6 Stormwater Drainage

Although rainwater must be carried away from the building to avoid waterlogging the ground or flooding the basement, it is essentially clean water that can be discharged into streams without fear of pollution. Wherever the water ran before the building was built, it will try to go afterwards. However, natural ground absorbs much of the rain that falls on it, and the vegetation cover holds back some of the water so that it runs off slowly. Water falling on the roof and driveway areas of a building runs off almost instantaneously. The rate of runoff from a built-up area, in a short, heavy storm, can be five to ten times that from a natural landscape. The disposal of rainwater, both from the roof and then from the drains into the city drainage system, therefore presents a serious problem.

If there is a sewerage system connected to the building, the temptation is to discharge the stormwater into the sewer. This was permitted in many cities in the early days of sewerage. The large underground sewers of many European cities were brick or stone-lined tunnels that drained sewage and stormwater directly into the nearby river, without treatment. In dry weather the flow was very small, but in heavy rain they ran with a considerable flow. This served to flush out the sewers.

When sewage treatment was introduced, to break down and purify the waste before it was discharged, the great variation in flow rates caused by adding stormwater to the sewage upset the operation of the treatment plant. Therefore most cities do not now allow stormwater in the sewerage system and have two sets of drains. Stormwater drains require no vents or traps. Many are simply open-top channels, discharging into the nearest natural stream. Sewers, on the other hand, are more expensive to build, may have to run farther to treatment works, but have to handle a relatively smaller flow rate than a stormwater system. Many U.S. cities are trying to eliminate their old "combined" sewers, but the cost is high in a long-established city.

In a tall building the stormwater stacks are usually run inside the building in the normal plumbing ducts. They cause little trouble as long as they are made of noncorrosive material and have no sharp bends to trap foreign objects. The outlets from the roof should be well guarded to keep debris out of the stacks. Because rainwater stacks are often made from lighter-gauge material than waste stacks, running water can produce a lot of noise. In a domestic-scale building the rainwater pipes are more usually placed outside.

The sizing of the stormwater system depends on the local history of intense, short-duration storms. In any area where very heavy rainfall can be expected (especially in hurricane areas, where leaves and other wind-borne debris are likely to block up the outlets) it is best to design the stormwater system on the fail-safe principle, that is, to make the pipes large enough to cope with the expected rainfall but to plan overflow routes to spill the water safely off the building before it can get inside and cause damage.

2.7 Disposal of Solid Waste

Garbage is an increasing problem in an affluent society. As with sewage, it can be considered in three stages: collection, transport, and disposal. The ultimate disposal is outside the scope of this book. There are now many proposals for waste recycling that require some sorting at source. This influences the provisions made in a building for collecting and storing wastes. We do not really dispose of waste; we "relocate" it.

In a single house, the garbage bin can be put outside the house or in the garage until it is collected, and this has only a minor influence on the planning of the building. A large apartment block, an office building, a hotel, or a hospital has more complex requirements for the handling of garbage.

On-site incineration was a convenient way of reducing the volume of garbage in many buildings, but clean-air legislation has made it difficult to operate an incinerator except on a large scale with all necessary controls to ensure complete combustion. With good sorting of materials before burning, an incinerator can produce some useful heat as it consumes the more easily burnt items. With less control over the input, it is usually necessary to add some fuel, such as gas, to get acceptable combustion.

Whenever garbage is stored on site awaiting collection, the building design must provide a suitable storage space with easy truck access. The store must be ventilated to keep odors out of the rest of the building, and if food wastes are included it may be necessary to refrigerate the storage to prevent it from rotting before collection.

The bulk and the manual handling of solid wastes are responsible for the cost of an operation that produces no useful product. Various means of mechanizing the process have been devised. A slurry can be made by grinding the solids in the presence of water. The transport of solids in slurry form is common in industry, which handles coal and various minerals in this way. Slurried waste is usually transported within the building in pipes, but then dewatered and compressed at a central position prior to being trucked away. The compressed material occupies much less volume than the original waste and can use the full load-carrying capacity of a truck. The slurry pipe system could easily be extended to serve an industrial estate, or a whole city if all the buildings could be converted to it, and thus eliminate the cost of the regular garbage service. However, the cost of the individual machines at each point of collection outweighs this saving in most buildings.

Solids can be conveyed along a pipe by air pressure as well as by water. Central-station vacuum-cleaning systems are often installed in larger houses and in multistory buildings to allow the cleaning of dust into a central collection filter. There are also experimental systems where garbage placed in plastic bags is conveyed to a central station by a large-diameter "vacuum-cleaner" pipe network under the streets in a whole suburb.

References

2.1 I.M. WINSLOW: *A Libation to the Gods*. Hodder and Stoughton, London, 1963. 191 pp.

2.2 *Report of the United Nations Conference on Human Settlements* (Habitat, Vancover, 1976). United Nations, New York, 1976, E.76. iv.7, A/CONS.70/15. 183 pp.

Suggestions for Further Reading

2.3 J.W. CLARK, W. VIESSMAN, and M.J. HAMMER: *Water Supply and Pollution Control*. IEP-Dun-Donnelley, New York, 1977. 857 pp.

2.4 W.A. BECKMAN, S.A. KLEIN, and J.R. DUFFIE: *Solar Heating Design by the f-Chart Method*. Wiley, New York, 1977. 200 pp.

2.5 J.A. SALVATO: *Environmental Engineering and Sanitation*. Wiley-Interscience, New York, 1972. 919 pp.

Chapter 3 The Sun

The sun is the source of most of the energy used in buildings, and we could use it to greater advantage than we do at present. Solar energy is free and plentiful, but it is thinly distributed over the earth's surface, and the solar collectors must be paid for. Thus solar energy for buildings frequently costs more than conventional energy.

The sun's apparent motion around the earth is complicated because its orbit is elliptical, not circular, and because the earth's equator is inclined to the plane of this orbit. To design sunshades, we can use models on a sun machine called a heliodon; or we can use sunpath charts with transparent shadow angle overlays; or we can use computer programs.

3.1 The Sun as a Source of Energy

All forms of energy currently used on earth, with the exception of atomic and geothermal energy (see Glossary), are derived from the sun. The most important of these is photosynthesis, the conversion by green plants of water and carbon dioxide into organic compounds. This chemical reaction requires energy, which is provided by solar radiation.

Firewood is produced directly by photosynthesis. Coal, natural gas, and oil are derived from plants that grew millions of years ago. Hydroelectric power is an indirect form of solar energy, for the sun evaporates the water that forms the clouds which eventually fall as rain to fill the reservoirs. Wind is partly produced by solar radiation and partly by the rotation of the earth. Most important of all, the energy expended by people and by animals is replenished by eating plants or meat.

Except for the fossil fuels, these forms of energy are renewable. We can maintain and even increase the production of food and timber. Present hydroelectric power stations can be maintained indefinitely, but the potential for further development is limited because the best sites are already in use. In theory far more power could be produced by windmills, but economical methods for large-scale exploitation have not been produced at the time of writing, although small windmills are efficient in remote areas for generating electricity and for pumping water from the ground.

Coal and oil have been known since antiquity, but they were little used until recent times. The tar content of coal created problems both in braziers for heating and in metallurgical processes, and it was difficult to burn oil with existing equipment until the distillation process was invented in the mid-nineteenth century. The principal fuel was charcoal made from wood. The demand for coal for heating increased after the sixteenth century, but it did not become an important material until the Industrial Revolution in the late eighteenth century. Oil was needed in large quantities only after the invention of the automobile in the early twentieth century. When oil became cheaper than coal for ships, railroads, and in many locations also for electric power stations, the demand for oil increased rapidly. In 1973 as a result of a war in the Middle East oil supplies were curtailed, and the price of oil increased rapidly, causing rises in the cost of other sources of energy. Since then there have been further steep increases in price, and restrictions of supplies for political reasons.

This is of greater immediate significance than the impending depletion of oil supplies. Natural gas, which can be used in place of oil for many purposes, will probably last beyond the year 2000, and coal supplies are sufficient for several centuries at the present rate of consumption. However, all fossil fuels must be exhausted in time, and it took several million years to produce them.

There are a number of energy sources that can be substituted for fossil fuels. Alcohol made from wood and from field crops is now used in automobiles in Brazil. Atomic energy is now used in several countries to generate electricity. Some regions could make greater use of hydroelectric or geothermal power. Pilot plants exist for the production of electricity from tidal currents in river estuaries and from wave action. However, the energy so produced may be more expensive.

Power can also be produced directly from the sun. The total amount of solar radiation falling on the earth *each year* is 2.7×10^{24} J, or 2.6×10^{21} Btu (see Appendix B for definition of units). This is equivalent to about 450 million million barrels of oil, or about one third of the *total proved* oil reserves in all countries (Ref. 3.1, p. 19). However, this huge quantity of energy is thinly distributed over a very large area.

Solar radiation can be concentrated by reflecting it from a large number of mirrors onto a boiler. The steam produced is used in a steam engine or turbine like steam generated by burning oil or coal. This method has been used since the late nineteenth century in pilot plants, and it is now economical for power plants in small communities remote from ports, railroads, and electricity supply lines.

An even more direct method employs photovoltaic cells that convert solar radiation into electricity. These supply the power for most earth satellites and for telecommunications links in remote areas (for example, between Alice Springs and Tennant Creek in Australia, a distance of 450 km, or 280 miles). However, the power required for these applications is small, so that the unit cost of producing the electricity is not a major consideration. The price of producing electricity by photovoltaics has decreased sharply, and further substantial reductions are anticipated; but at the time of writing (1982), it is still about 100 times that for electricity generated from locally produced coal (that is, it costs 10 000% more).

At present it does not seem likely that solar electricity will affect the design of buildings significantly during the twentieth century. However, two applications of solar energy are immediately useful. One is a more sophisticated version of the design principles known since Ancient Greek times [Section 1.2]; this is discussed in Sections 3.4, 5.15, and 5.16.

In the other, solar collectors are employed to heat water or some other fluid to a comparatively low temperature. The fluid from these collectors can be used for heating swimming pools, for heating water for washrooms, bathrooms, and kitchens, and for hot water for industrial processes [Section 2.4]. These applications are economical now where there is sufficient sunshine. Similar solar collectors can be used for the space heating of homes in winter and for the heat pumps of solar air conditioning systems [Section 6.7], but their economy is still open to question.

All these applications require expensive equipment that has to be replaced from time to time. Although the solar radiation is free, its collection can cost much more than energy produced from oil.

3.2 The Solar Spectrum

The effective temperature of the sun is about 5900 K (degrees Kelvin measured from absolute zero; see Section 4.1), which is equal to about 5600°C. The sun emits energy approximately as does a black body (that is, a body with a full absorptive power and no reflecting power), and the energy reaches the earth in that form. The distribution of radiation [see Section 4.5] outside the earth's atmosphere is shown in the upper curve in Fig. 3.2.1. The mean value of this radiation is 1395 W/m^2, and this is called the *solar constant*. Some of the solar radiation is absorbed by the earth's atmosphere, notably by ozone, water vapor, and carbon dioxide. The lower curve shows the distribution to be expected on a sunny day at sea level (Fig. 3.2.1).

The solar spectrum ranges approximately from 300 to 3000 nm (nanometers, or meters $\times$ 10^{-9}). The spectrum of visible light ranges from 390 nm (violet) to 760 nm (red). At one end beyond the visible spectrum is the ultraviolet range, and on the other side the infrared range. The infrared solar band is 6 times as wide as the visible spectrum, but the amount of radiation received from it is relatively low. The ultraviolet solar band is narrow. However, it is of particular importance when solar radiation is converted to electricity by photovoltaic cells, because only radiation with a wavelength of less than 1000 nm is effective when silicon cells are used (Ref. 3.1, p. 285).

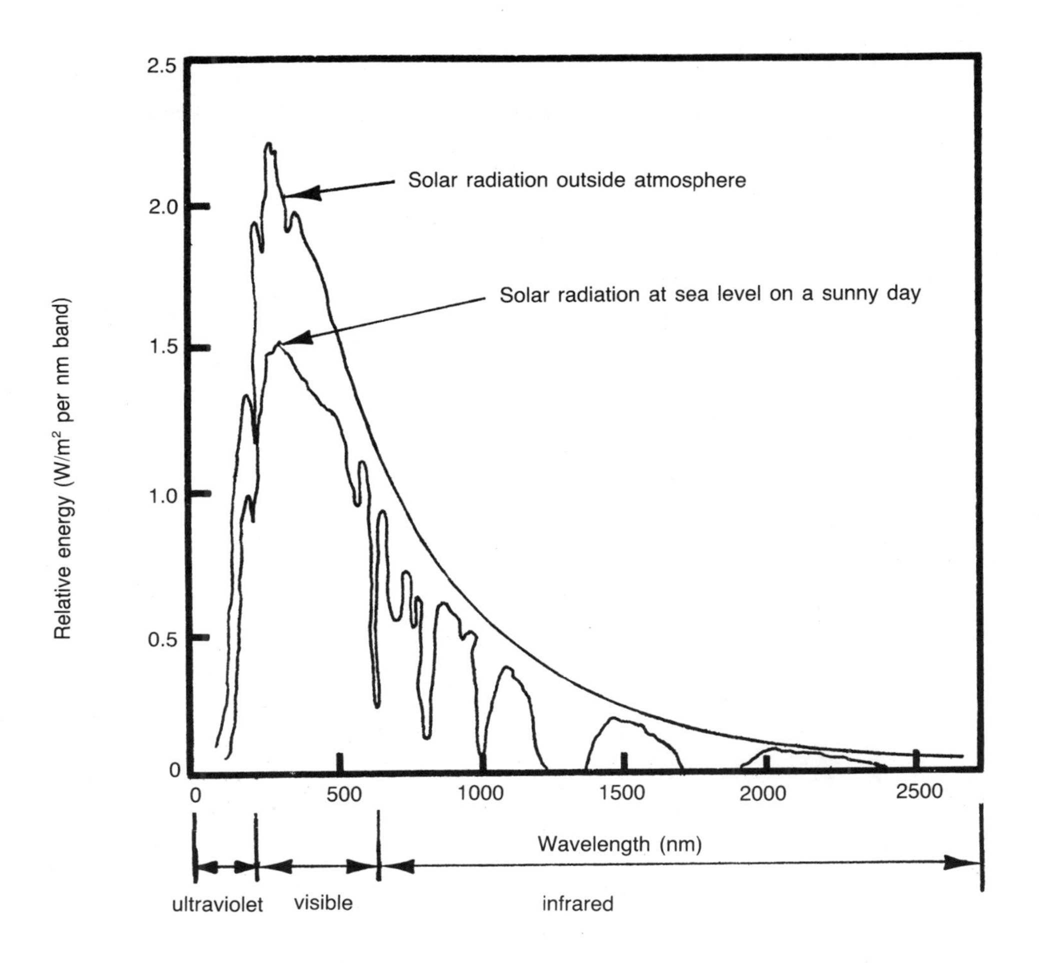

Fig. 3.2.1. Variation of solar radiation with the wavelength of the radiation. The upper curve shows the solar radiation outside the earth's atmosphere, and the lower curve the radiation received at sea level on a sunny day. Constituents of the atmosphere, such as ozone, carbon dioxide, and water vapor, absorb radiation at particular wavelengths.

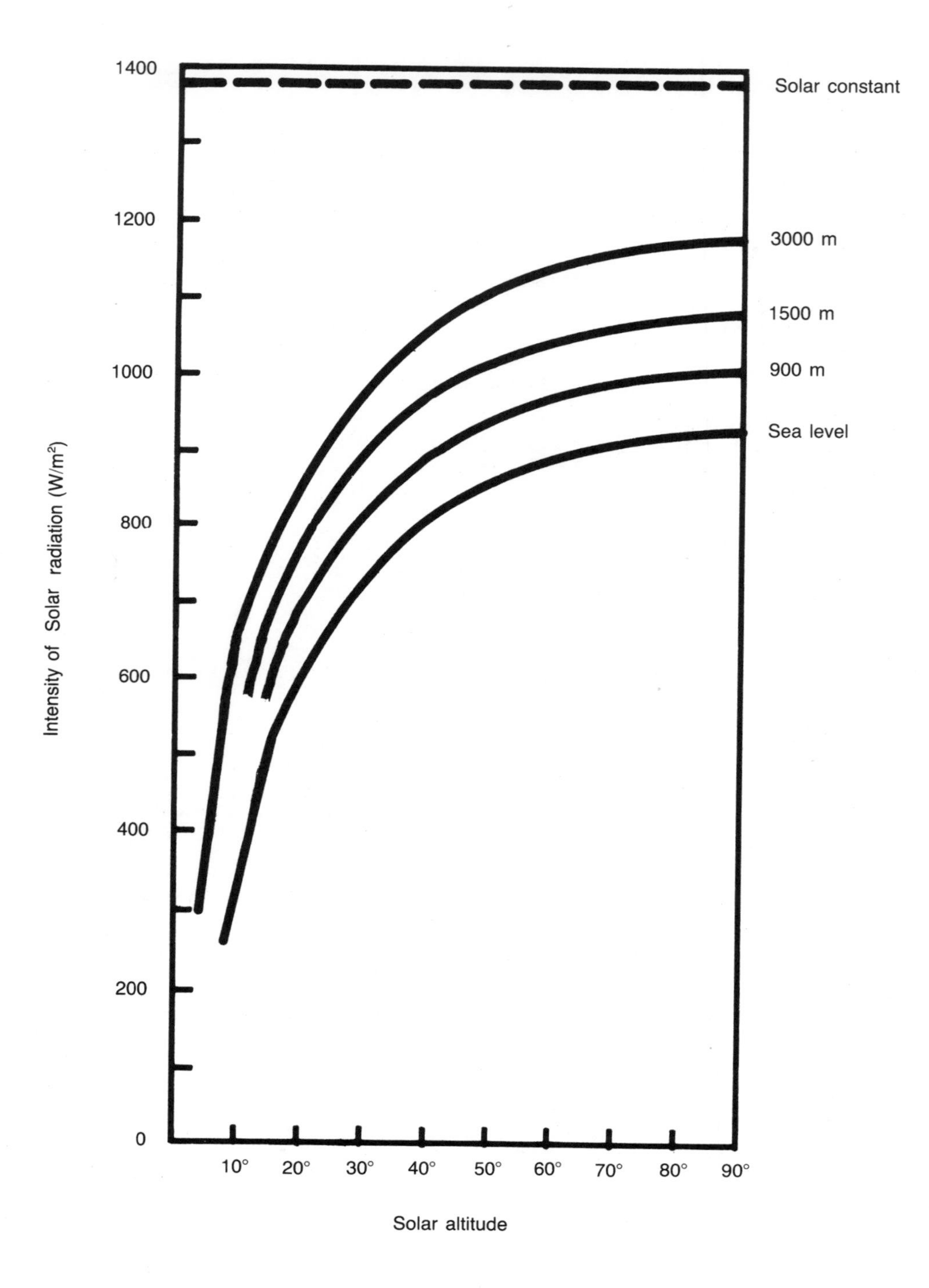

The amount of solar radiation varies with the altitude of the sun and with the height above sea level. The curves in Fig. 3.2.1 are based on the assumption that the sun is at the zenith, immediately overhead. The lower curve is at sea level, and the upper curve at a great height where the solar radiation is equal to the solar constant of 1395 W/m^2. Intermediate heights produce intermediate curves. As the solar altitude decreases, the solar radiation is reduced, and it becomes zero at an altitude of 0° when the sun drops below the horizon. These variations are shown in Fig. 3.2.2.

The sun never reaches an altitude of 90° outside the tropics, so that the intensity of the solar radiation varies with latitude. Solar radiation is further reduced by cloud cover, air pollution, and water vapor. The geographical variation is shown in Fig. 3.2.3. The greatest amount of radiation is received in the Sahara Desert, Saudi Arabia, and central Australia.

For solar design we need information on this radiation throughout the year. It is common practice to use published data, where these are available. The U.S. Weather Bureau collects recordings of solar radiation intensity from all countries in the world, and a number of other national organizations provide this information.

In the absence of data [Section 4.5], the daily solar radiation can be calculated approximately from the solar constant (Fig. 3.2.2) and the hours of sunshine. One such equation is quoted in Ref. 3.2, p. 283, and Ref. 3.3, p. 8. The hours of sunshine are easily measured with a sunshine recorder; this consists of a magnifying glass that burns holes into a graduated chart when the sun is shining.

Fig. 3.2.2. Variation of solar radiation with solar altitude and with height above sea level.

3.3 The Annual and Daily Movements of the Sun

The earth rotates once each day about an axis joining the north and south poles, and it revolves around the sun once each year. Until the seventeenth century it was generally believed that the earth was stationary and the sun moved around it; for the purpose of solar design it is convenient to retain this concept, and to consider the apparent path in the sky described by the sun each day.

This differs from day to day, because the earth's orbit around the sun is not circular but elliptical, and because the earth's equator is inclined to the plane of this orbit at an angle of 23°26′ (Fig. 3.3.1).

The rotation of the earth alone would cause the sun to rise daily in the east at the equator, reach an altitude of 90° at noon, and set in the west. At other locations, say with latitude λ, the highest point of the sun's movement would be $90° - \lambda$. The inclination of the earth's equator to the plane of the earth's orbit causes the seasons. The two planes intersect on March 21 and September 23. On these dates the length of the day

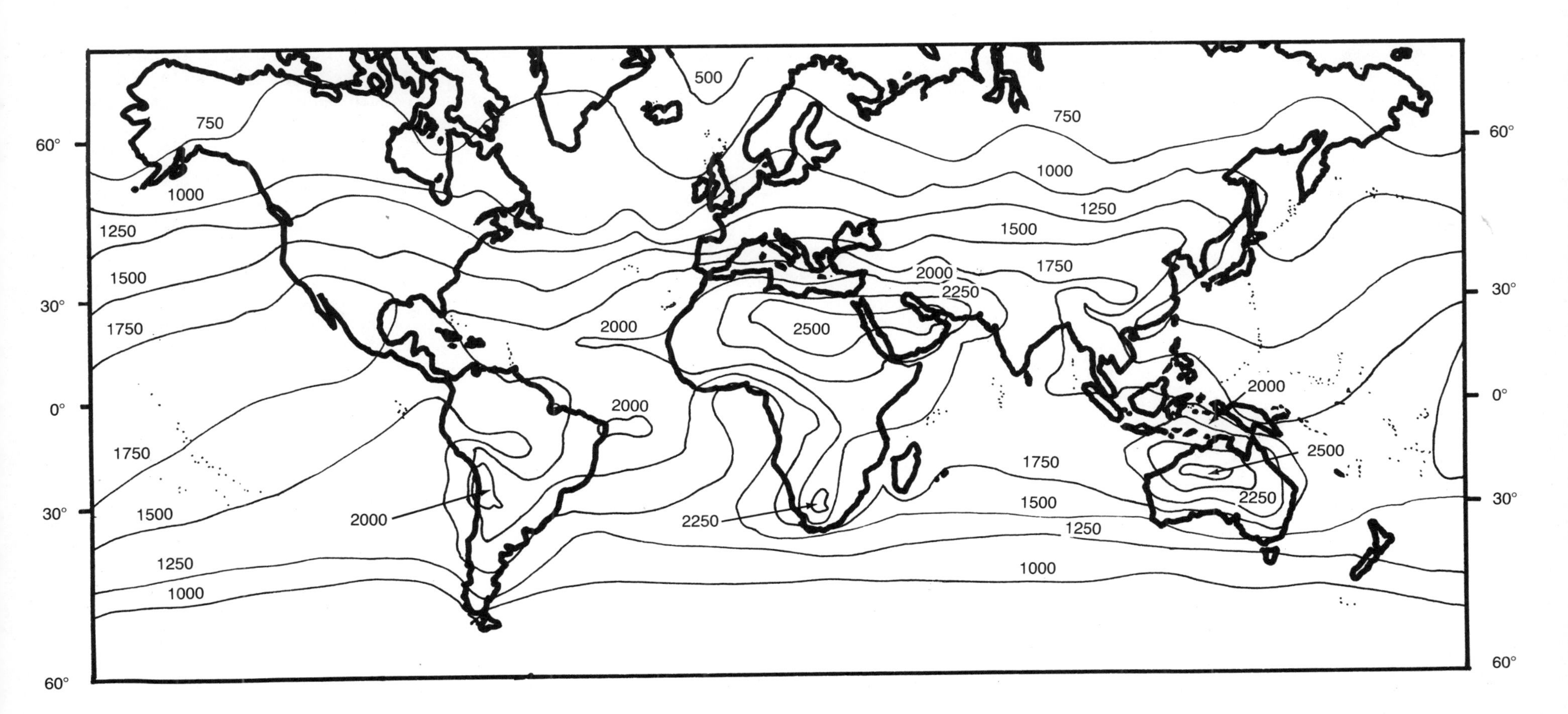

Fig. 3.2.3. Geographical variation of total annual solar radiation in kWh/m².

and the length of the night are equal everywhere on the earth's surface, and these days are therefore called the *equinoxes*. The greatest difference between the length of the day and the night occurs on the *solstices*, which are on June 21 and December 22. In the Northern Hemisphere the winter solstice is in December, and in the Southern Hemisphere it is in June.

Between March and June the sun appears to move northward as far as the *Tropic of Cancer*, 23°26′ north of the equator. It then appears to move back to the equator until September, when it appears to move south as far as the *Tropic of Capricorn*, 23°26′ south of the equator, and so on. The tropics are the region between latitudes

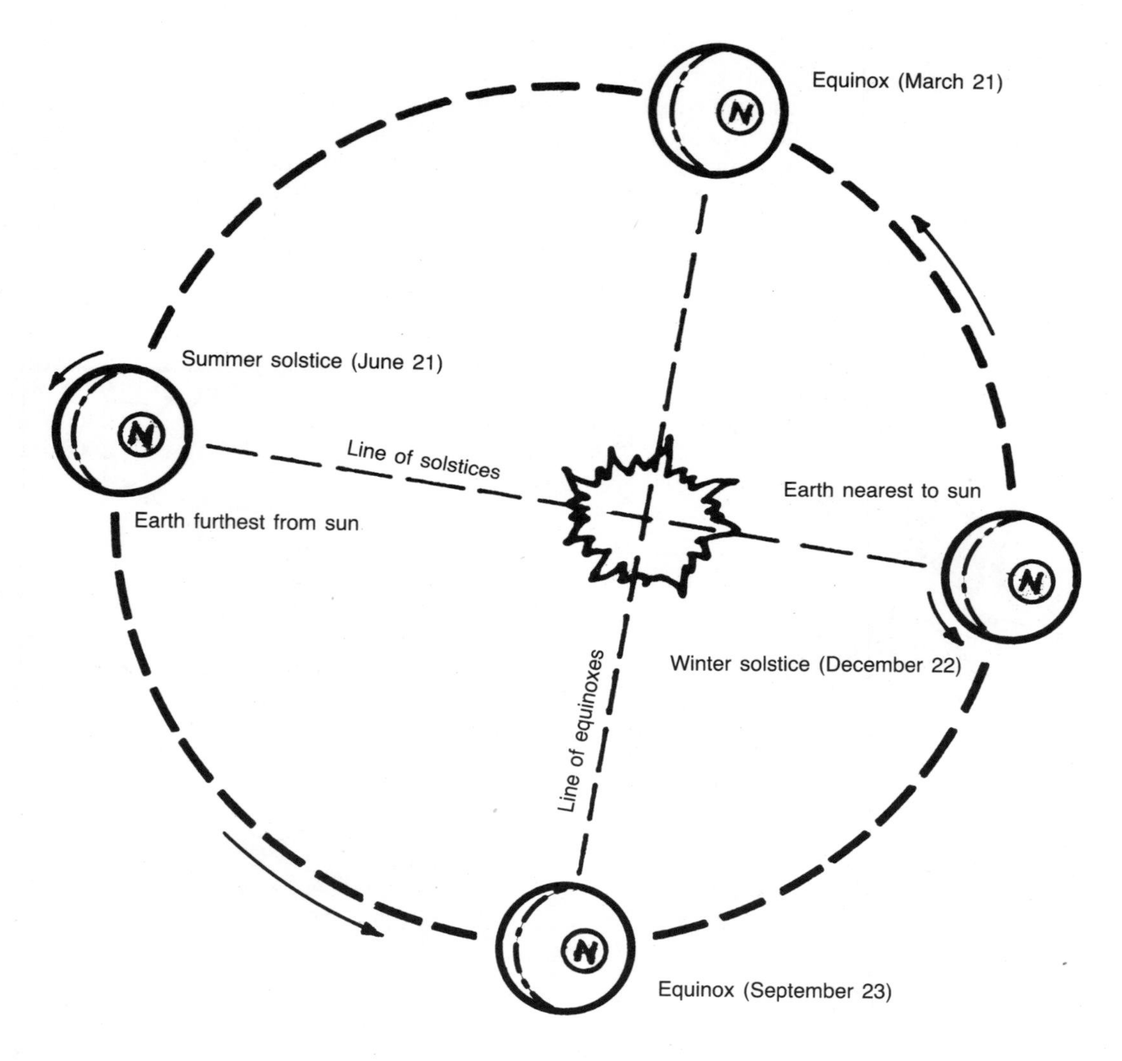

Fig. 3.3.1. Earth's orbit around the sun, and seasons in the northern hemisphere. The earth rotates once daily about an axis joining the north and south poles. The axis of rotation is inclined to the plane of the earth's orbit around the sun at an angle of 23°26′. On the equinoxes (March 21 and September 23, when day and night are of equal length) the line joining the centers of the earth and the sun is in the plane of the earth's equator, shown in the equinox positions. On the solstices (June 21 and December 22, respectively the longest and shortest days of the year), the line joining the centers of the earth and the sun is perpendicular to a line joining the earth's arctic and antarctic polar circles, shown in the solstice positions. The earth is farthest from the sun in early July and nearest to the sun in early January.

Fig. 3.3.2. Buildings in the tropics receive solar radiation from both the north and the south at different times of the year. In buildings in the subtropics and the temperate zone either the northern or the southern facade is permanently in the shade. If it is possible to orient a building with its long facades facing north and south, and to use windowless short walls east and west, only one facade requires sunshading.

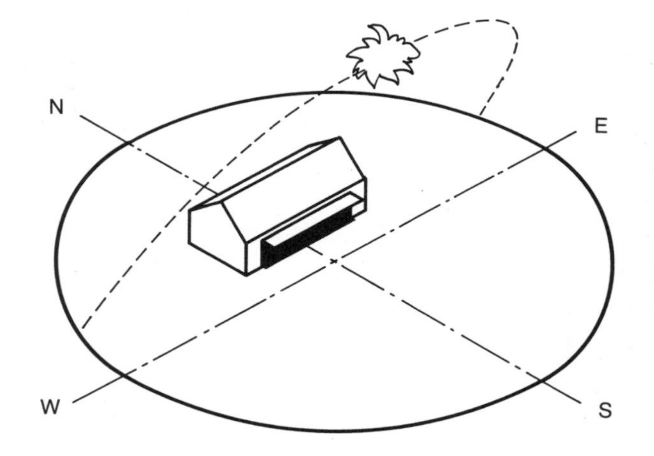

Table 3.1

Equation of Time

Date	Equation of Time (minutes)
January 15	+ 9
February 15	+ 14½
March 15	+ 9½
April 15	+ 0½
May 15	− 4
June 15	0
July 15	+ 5½
August 15	+ 4½
September 15	− 4½
October 15	− 14
November 15	− 15½
December 15	− 5

23° N and 23° S; in the tropical zone the sun appears directly overhead at some time. Outside the tropics it never reaches an altitude of 90°. Furthermore, in the tropics the sun may shine on both the south and the north face of a building. Outside the tropics a building facing north and south has one face permanently in the shade, except for brief periods in the morning and the evening during summer (Fig. 3.3.2); however, the angle of the sun shining on the south face (in the Northern Hemisphere) gets lower as one moves away from the equator, so that sunshades must project more from the facade.

The ellipticity of the sun's orbit causes a variation in the distance between the sun and the earth of about 2%, which is not significant, and a variation in the earth's actual movement around the sun, which ranges from −14½ minutes to +15½ minutes (Table 3.1). Watches and clocks are designed to operate at a uniform rate, and they indicate *mean solar time*. To obtain the *apparent solar time* indicated by the sun's motion, it must be corrected by the *equation of time*. Similarly, sun dials, which indicate apparent time, must be corrected by the equation of time to give the mean time shown by a watch:

$$\text{Equation of Time} = \text{Apparent Solar Time} - \text{Mean Solar Time} \quad (3.1)$$

Furthermore, the watch time must be corrected to local time. In 1884 it was agreed that the world's standard time should be that corresponding to the longitude of Greenwich Observatory, near London. The time throughout the world differs generally by a multiple of 1 hour from that in Greenwich. Thus Central Standard Time used in Chicago is 6 hours behind Greenwich Standard Time, but the longitude of Chicago is 87°37′ W. The world's circumference is divided into 360° of longitude and 24 hourly time zones, so that 1 hour corresponds to 15° of longitude (Fig. 3.3.3).

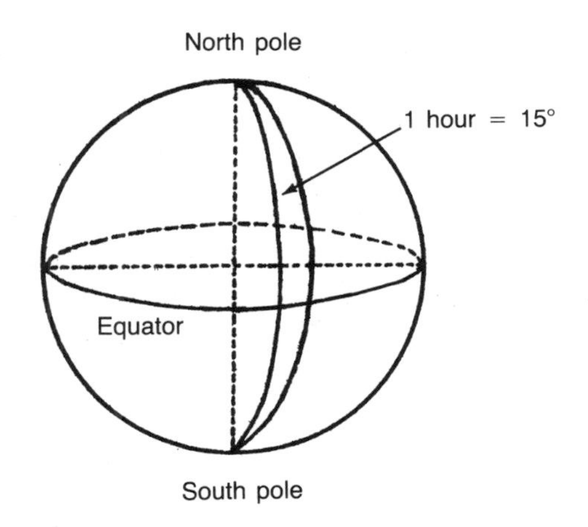

Fig. 3.3.3. The earth's surface is divided into 360° of longitude and into 24 hourly time zones. Therefore 1 hour corresponds to 15° of longitude.

If the time indicated by a watch in Chicago on November 5 is 3 hours 45 minutes, we need to correct the watch time by (90 − 87.6) × 60/15 = +10 minutes to obtain local mean time, and by the equation of time (= −15 minutes) to obtain local apparent time. Thus the local apparent time is 3 h 45 m + 10 m − 15 m = 3 h 40 m.

3.4 Devices for Sun Control

The designer must determine what is to be achieved with a sunshade. Xenophon [Section 1.2] 2300 years ago favored the exclusion of the summer sun and the admission of the winter sun. In most parts of the world this is a desirable objective, but if the admission of winter sun is of marginal benefit it may not be worth the extra expense. As with all other applications of solar energy, we must balance the extra capital cost against the energy savings or the improved performance that will result, and also consider the appearance of the sunshading device. Thus we will not necessarily choose the optimal solution from the scientific point of view; we may even decide not to use sunshading at all.

If the plan of a building can be arranged so that all facades face east, north, west, and south, the northern facade is always in the shade (in the Northern Hemisphere), except in the tropics. It is relatively easy to shade the southern facade when the sun is high in the sky. Sunshades for eastern and western exposures are more difficult, and the western sun in particular can cause distressing overheating even in the temperate zone. The most favorable plan has short, solid walls east and west and long walls with windows facing north and south, with the south face sun-shaded (Fig. 3.3.2).

There are many ways in which the sunshading can be provided, for example, by means of a projecting reinforced concrete floor slab that forms a balcony. Let us assume that we wish sunlight penetration to commence at noon on one equinox and to terminate at noon on the other equinox (Fig. 3.4.1), so that sunshine is wholly excluded in midsummer, wholly admitted in midwinter, and partially admitted in the spring and the fall. On a wall facing south in the Northern Hemisphere, the vertical shadow angle, ϕ, is

$$\phi = 90^\circ - \lambda \tag{3.2}$$

If the distance from the soffit of the projecting floor slab or balcony to the window sill is H, the required projection of the floor slab or balcony to prevent the sun shining into the window is

$$L = H \tan (90^\circ - \phi) = H \tan \lambda \tag{3.3}$$

Example 3.1 *Assuming that the distance* H *from the soffit of the balcony or projecting slab is 1.5 m (Fig. 3.4.2), find the projection* L *required at a latitude of 34° for a building facing south (north in the Southern Hemisphere). This is the latitude of Los Angeles, Atlanta, Casablanca, Beirut, Sydney, Capetown, and Buenos Aires.*

Solution *From Eq. (3.3)*

$$L = H \tan 34^\circ = 1.5 \times 0.6745 = 1.012 \text{ m}$$

A projection of 1 m would be appropriate.

In American units, a distance H of 5 ft would require a distance L of 3.37 ft, that is, 3 ft 4 in.

We can assure more accurately that the sun is excluded in hot weather and is admitted in cool weather by determining the days when the temperature exceeds, say, 21°C (70°F). This overheated period is then marked on the *sunpath chart*, and the sunshades can be made of a size to fit this period. This is further discussed in Section 3.5 and Fig. 3.5.9.

Fig. 3.4.1. Vertical (ϕ) and horizontal (η) shadow angles. At the equinoxes (March 21 and September 23) the sun's path crosses the equator, and therefore its altitude at noon is 90° − λ, where λ is the latitude of the place. When the facade faces south (north in the Southern Hemisphere), the vertical shadow angle is equal to the complement of the latitude.

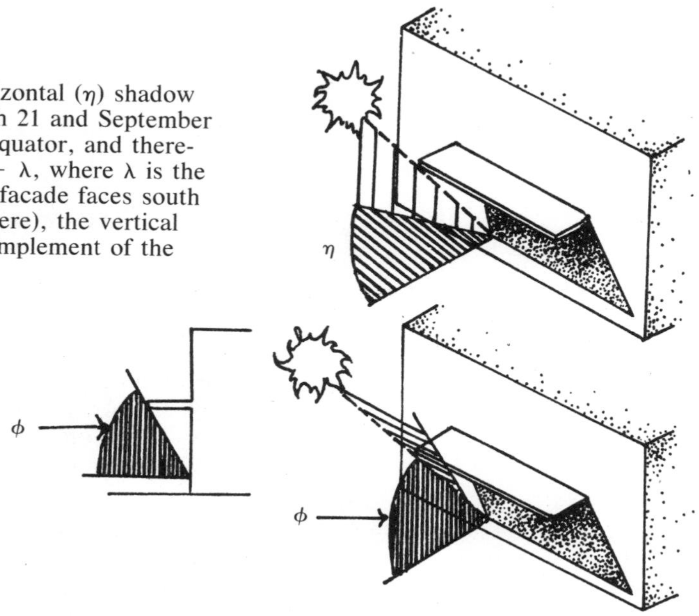

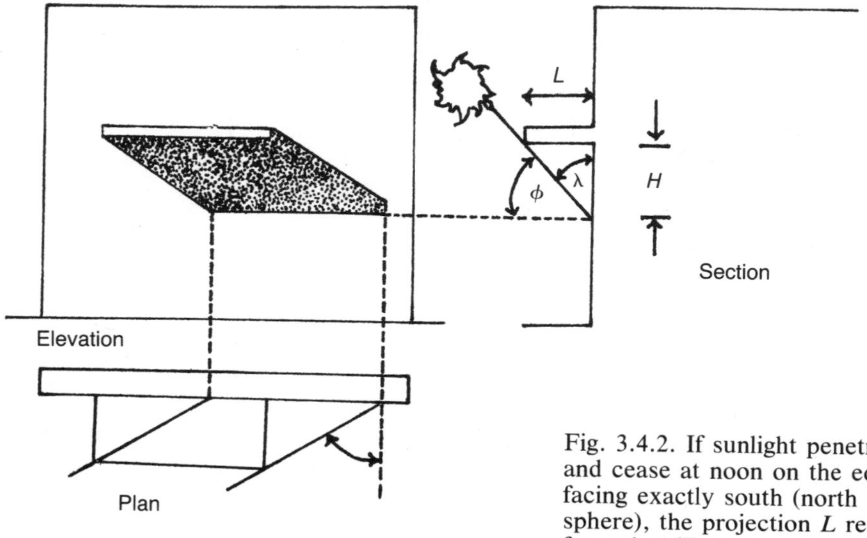

Fig. 3.4.2. If sunlight penetration is to commence and cease at noon on the equinoxes for a facade facing exactly south (north in the Southern Hemisphere), the projection L required for a height H from the sill is $L = H \tan \lambda$.

Fig. 3.4.3. Some sunshading devices.
(a) Mathematics Building, University of Sydney, latitude 34° S, northern facade. A northern facade in the Southern Hemisphere presents a relatively simple problem. The shade of inclined aluminum slats is precisely designed to provide complete shading in midsummer and allow unimpeded entry of solar radiation in midwinter. The slats allow free passage for natural ventilation, required during the humid summer in a building that is not air-conditioned.
(b) Banco de Credito del Peru, Iquitos, Peru, latitude 4° S, northeastern and southeastern facades. Complete sunshading on southeastern and northeastern facades requires deep louvers, even in the tropics. This has been provided for the upper floors, but on the ground floor mirrored reflective glass [see Section 5.9] has been used to provide a view from the banking chamber. (*Photograph by courtesy of Mr. E.L. Harkness, University of Newcastle, N.S.W.*)
(c) Union Club Building, Durban, South Africa, latitude 30° S, western facade. Sunshading on a western facade is particularly difficult because of the low angle of incidence of the sun and the high temperatures that may occur on a summer afternoon. The vertical louvers curved in the horizontal plane permit a view toward the southwest but prevent sunlight penetration until 2:30 P.M. in midsummer. The inside of each louver reflects the daylight, which results in good natural lighting. With a stronger horizontal curvature of the louvers complete sunshading could have been achieved, but at the cost of obstructing the view. These sunshades are nonstructural, but similar shades could be utilized as load-bearing members. (*Photograph by courtesy of Mr. E.L. Harkness, University of Newcastle, N.S.W.*)
(d) Greater Pacific House, an office building in North Sydney, latitude 34° S. The loadbearing reinforced concrete walls are built from precast units with integral sunshades, identical for all facades. This is a compromise solution that results in excessive shading in winter on the northern facade and insufficient shading during summer on the western facade.

(a) (c) (d)

(b)

The projection required is not excessive even if the latitude is increased.

Example 3.2 *Assuming that the distance* H *is 1.5 meters, as in Example 3.1, what is the projection* L *required at a latitude of 46° for a building facing south (north in the Southern Hemisphere)? This is the latitude of Montreal, Portland (Oregon), Milan, Harbin (China), and Dunedin (New Zealand).*

Solution *From Eq. (3.3)*

$$L = H \tan 46° = 1.5 \times 1.036 = 1.553 \text{ m}$$

A projection of 1.5 m would be appropriate.
In American units, a distance H of 5 ft would require an equal distance L of 5 ft.

However, as the wall is rotated east or west, the required width of the projection increases rapidly. It then becomes necessary to use vertical sunshades as well. The design of a wall facing other than south or north is more complicated and is discussed in Section 3.5.

During the cold part of the year the sun can be utilized for heating by allowing solar radiation to enter the building. Sunlight penetrates through glass, which is mainly transparent to the radiation of the solar spectrum [see also Section 5.9]. However, glass is mainly opaque to the longer-wave radiation emitted by the walls and the floors, so that the heat is retained in the room. This is known as the *greenhouse effect*. Thus windows act as solar collectors of relatively low efficiency, but also of negligible cost.

The thermal inertia of a reinforced concrete floor can be utilized if a hard, heat-absorbing floor surface, such as marble paving, tile, or terrazzo is used. The concrete then acts as a heat store [see Section 5.15], and the heat collected by it during the day is released during the night, thus increasing the efficiency of the solar collector. Carpet is an insulator, and it greatly reduces the usefulness of the floor as a heat store.

Whether a floor is given a hard finish or a carpet is largely a matter of custom. In many hotels and residential and commercial buildings in Italy marble, tiles, or terrazzo are regarded as desirable finishes. In the southern United States, in a similar climate and with a similar type of occupancy, a carpet is usually preferred. It is doubtful whether the saving in energy resulting from the use of a hard surface would compensate people who prefer a carpet; but the choice is at least worthy of consideration.

In climates where winter temperatures are not low enough to warrant solar heating, it is appropriate to design the sunshades to prevent sunlight penetration throughout the year. This is also simpler, as there is only one limit to be considered, but it may reduce the amount of daylight admitted by the sunshades unless this is checked [Section 9.7]. If artificial lighting has to be used unnecessarily, it increases the energy consumption not only directly but also indirectly by raising the air conditioning load. This is a common occurrence when venetian blinds are used for sun control without external sunshading devices.

Evidently sunshading is not worthwhile in places where high summer temperatures occur only occasionally.

There are many other economical sunshading devices (Ref. 3.9), some of which are shown in Fig. 3.4.3.

3.5 Determination of Sunlight Penetration and Design of Sunshading Devices

Even if no sunshading device is used, or if the device is a simple horizontal or vertical slab, the determination of sunlight penetration is very complex because of the variation in the annual and daily movement of the sun, unless the building faces precisely north and south.

The problem can be solved experimentally using a sun machine, called a *heliodon* or *solarscope*. There are two types. In one the model is placed on a horizontal table, and a lamp representing the sun is set in the correct position for the latitude of the place, the time of the year, and the time of the day (Fig. 3.5.1). The shadows thrown by the lamp are then measured or photographed. This machine is read directly, but as the lamp must be connected to the table through an arm, the distance at which it can be placed is limited, so that the rays of light striking the model are not parallel as in sunlight. This introduces an error. In the other type the lamp is placed at the far end of the room; in a long room approximately parallel light rays can be achieved. The model

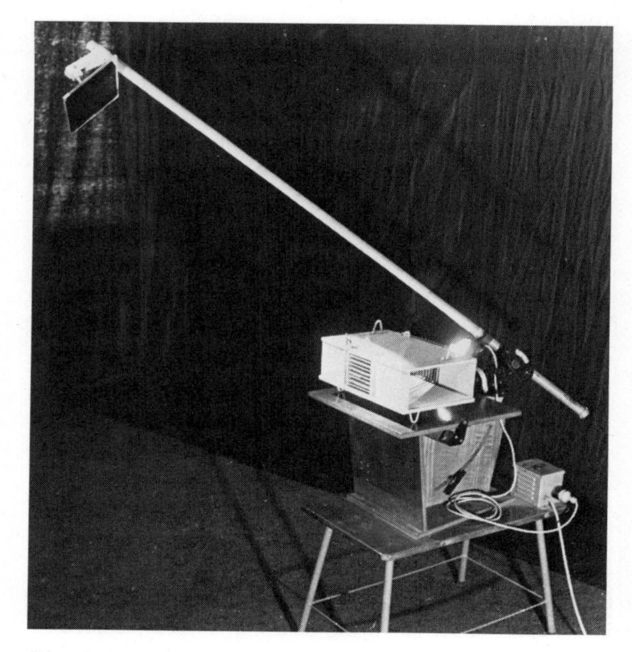

Fig. 3.5.1. Heliodon with fixed model table and movable light. This instrument is less accurate than that shown in Fig. 3.5.2, but because it is read directly it is more convenient to use.

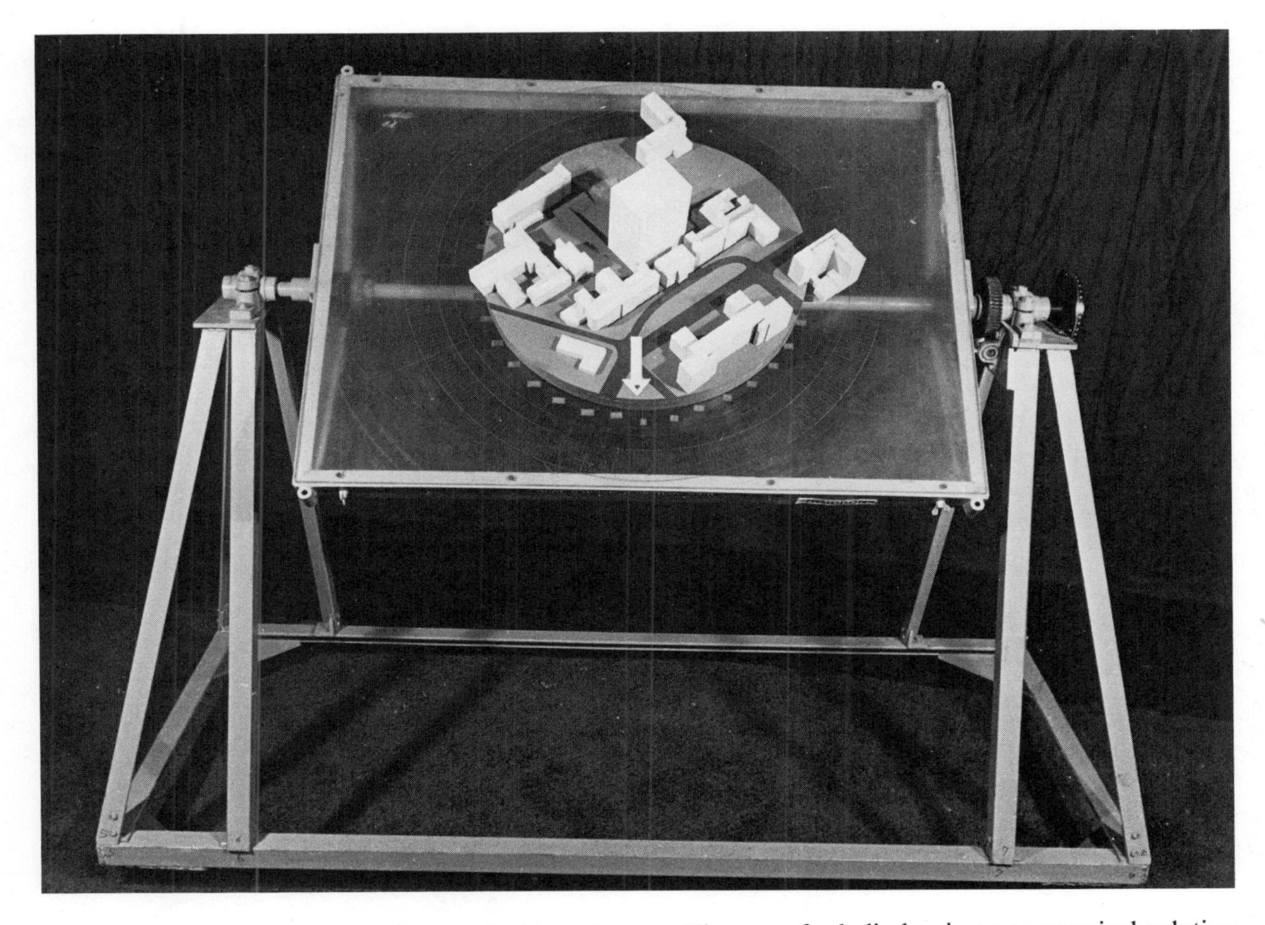

Fig. 3.5.2. Heliodon with movable model table and fixed light. This is an accurate instrument, but the azimuth (horizontal angle) and altitude of the sun must be translated into the time of year and the time of the day for the given latitude in order to interpret the results.

table is rotated, and the time of the year and of the day are calculated from the altitude and azimuth (see Glossary) of the model (Fig. 3.5.2). Thus the answer cannot be read directly from the machine.

The use of a heliodon is an economical solution if a suitable architectural model that fits on the table of the heliodon is already in existence. If a model must first be made, it is generally quicker to use sunpath charts with transparent overlays or to employ computer graphics.

A sunpath chart shows the three-dimensional movement of the sun during the year on a flat piece of paper. There are more than a dozen projections for this purpose. The stereographic projection, probably devised by Hipparchus in the second century B.C., is the one most frequently used; it is explained in specialized books (for example, Ref. 3.6) and also in books on solid geometry or astronomy.

A sample stereographic sunpath chart is shown in Fig. 3.5.3. It shows the altitude of the sun measured as an angle with the horizon, and the azimuth, or horizontal angle made by the sun with the north. The movement of the sun is marked by two sets of curved lines. One set shows the variation during the year, generally by six lines at monthly intervals; each line can be used twice, as the movement of the sun is symmetrical about the solstices. The other set of lines shows the movement of the sun during the hours of daylight. The intersection of the two sets gives the horizontal and vertical angles made by the sun for each hour of the year.

A different chart is needed for each latitude, and separate dates and graduations are required for the Northern and Southern Hemispheres. Sets of these charts have been published in several books, for example Ref. 3.2, pp. 292–303, and Ref. 3.4, pp. 219–36. Computer programs are also available for drawing the charts.

To determine sunlight penetration, we place a transparent shadow angle protractor (Fig. 3.5.4) over the sunpath chart; the protractor shows the horizontal shadow angles (η) by radial lines and the vertical shadow angles (ϕ) by curved lines. A transparent shadow angle protractor is included with Ref. 3.2. Reference 3.4 gives only the diagram, but a transparency can be made with a copying machine. It is also a simple matter to draw a shadow angle protractor with a compass.

As discussed in Section 3.4, we determine the times when sunlight penetration is to be prevented and when it is to be permitted (if at all). This sets the limits on the sunpath chart. We then translate these limits into shadow angles, using the transparent shadow protractor. These enable us to design suitable sunshading devices.

For example, Fig. 3.5.5.a shows a horizontal sunshade. To prevent sunlight penetration, a vertical shadow angle $\phi = 45°$ is required. We take the curved line marked 45° on the protractor (Fig. 3.5.5.b) and mark its intersection on the sunpath chart for our latitude. This indicates the time of year when sunlight penetration is completely prevented, and the time of year when sunlight penetrates for part of the day. The hours when this occurs can also be read off the sunpath

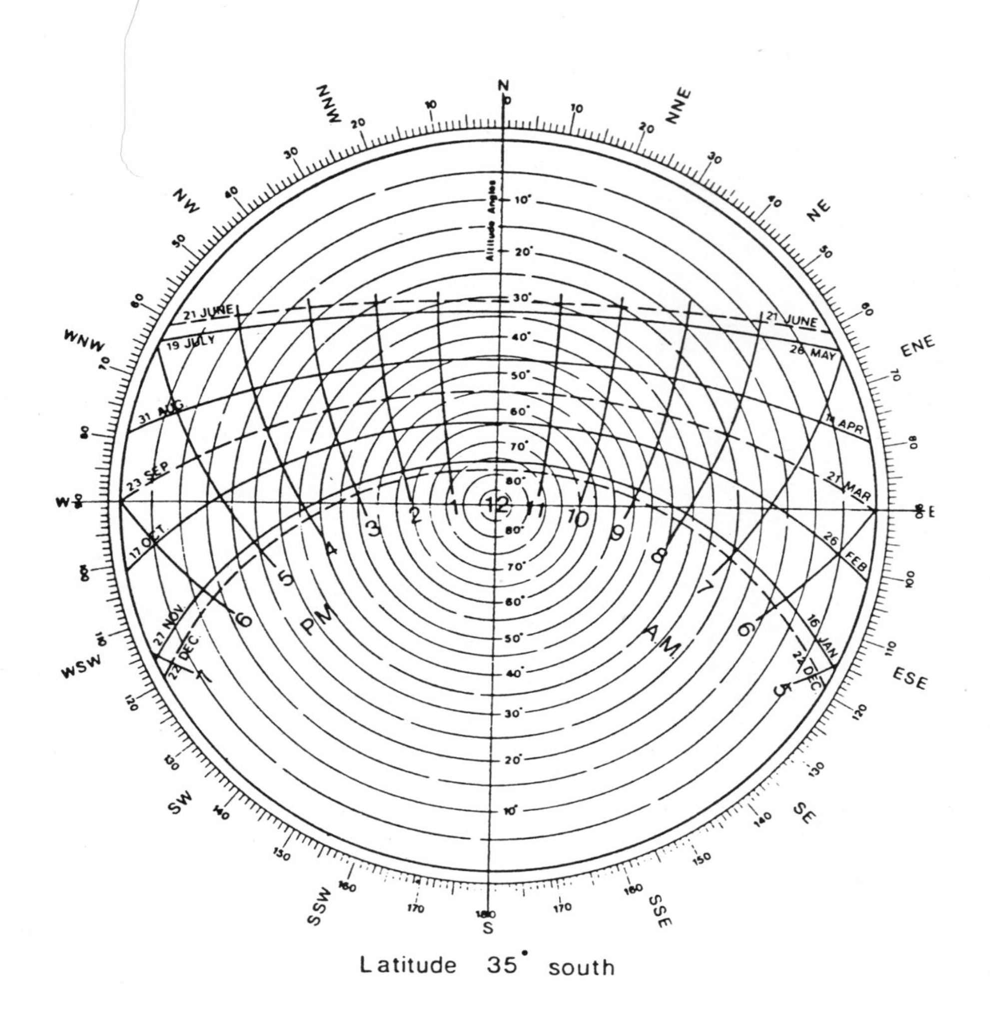

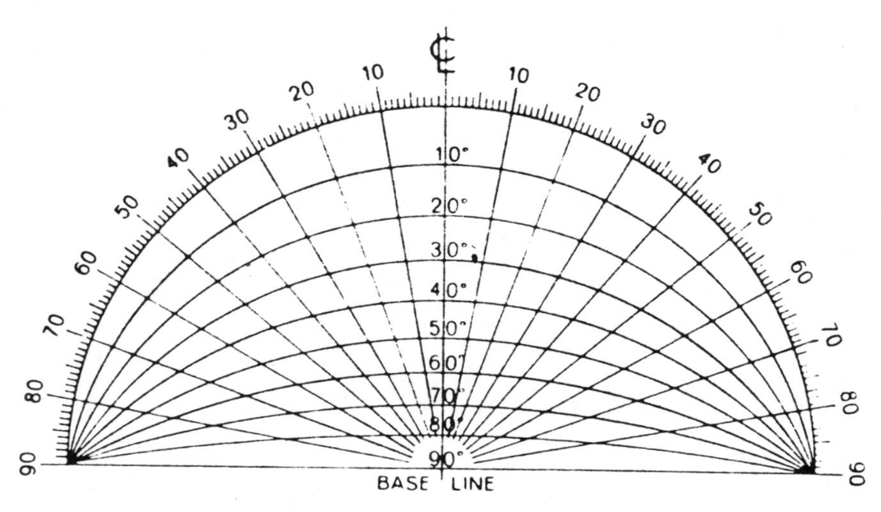

Fig. 3.5.3. Sunpath chart for latitude 35° S. It can be used for latitudes from 33° S to 37° S; for example, Sydney, Australia, 34° S: Canberra, 35° S; Auckland, New Zealand, 37° S; Cape Town, 34° S; and Buenos Aires, 34° S. The chart shows the altitude of the sun (its vertical angle from the horizon) and its azimuth (its horizontal angle from the north-point) for each month of the year and for each hour of daylight. Separate sunpath charts are required for each latitude, but it is sufficiently accurate to use a chart for a latitude within 5° of that required. It is also necessary to have different scales for the Northern and Southern Hemispheres, although the chart itself is the same for the same latitude.

Fig. 3.5.4. Shadow angle protractor. If a transparency of this protractor, to the same scale, which can be made with a copying machine, is placed over the sunpath chart, the times during the year when sunlight penetrates into the interior of the building can be determined. The radial lines on the protractor represent horizontal shadow angles (η) and the curved lines vertical shadow angles (ϕ).

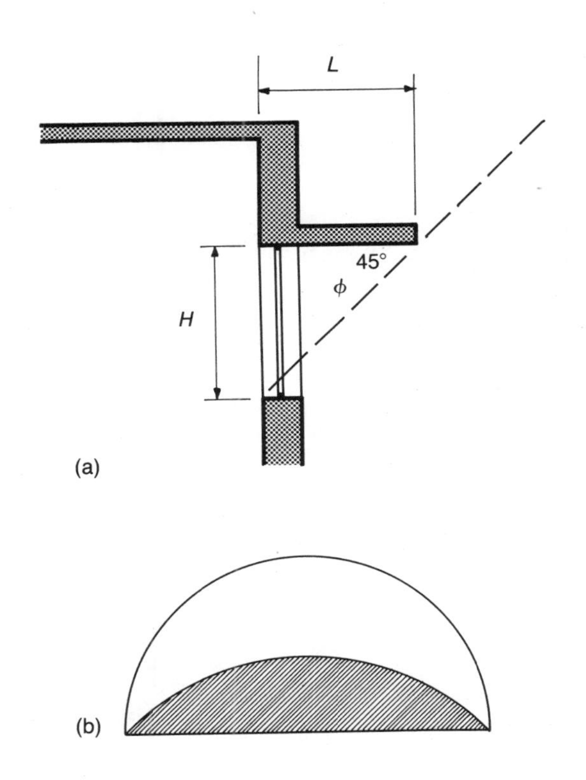

Fig. 3.5.5. (a) Section of a horizontal sunshade that prevents sunlight penetration at a vertical shadow angle $\phi = 45°$.
(b) Vertical shadow angle $\phi = 45°$ marked on the shadow angle protractor shown in Fig. 3.5.4. The cross-hatched area indicates the extent of the screening provided by the horizontal shading device shown in (a).

Instead of using a transparent protractor, a mask shaped like the hatched area can be used. The intersection between the curved line and the lines on the sunpath chart (see Fig. 3.5.3) indicates the time of year when sunlight penetration is completely prevented and the time of year when sunlight penetrates for part of the day. The hours when this occurs can also be read from the sunpath chart by intersection with the curved 45° line. (*After Harkness and Mehta, Ref. 3.9.*)

chart covered either with a transparent protractor or with a shading mask cut from cardboard along the 45° curve.

Figure 3.5.6.a shows a vertical shading device perpendicular to the facade. To prevent sunlight penetration, horizontal shadow angles of $\eta = \pm 45°$ are required. We take the radial lines marked 45° on the protractor (Fig. 3.5.6.b), and the same procedure applies otherwise.

Figure 3.5.7 shows inclined vertical sunshades that produce vertical shading angles of $-30°$ and $+60°$. We therefore take the radial lines 30° to the left and 60° to the right of the center line of the transparent protractor.

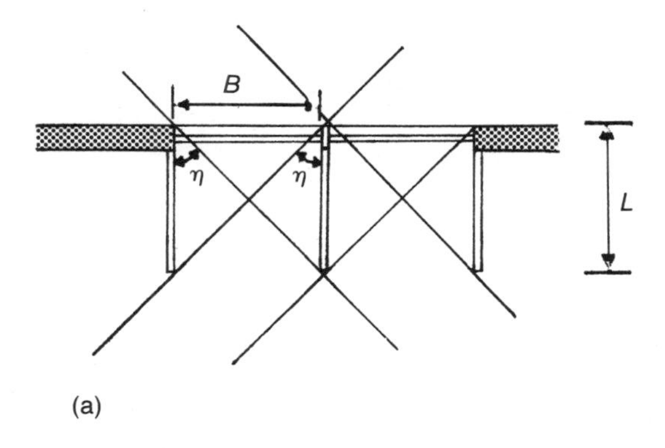

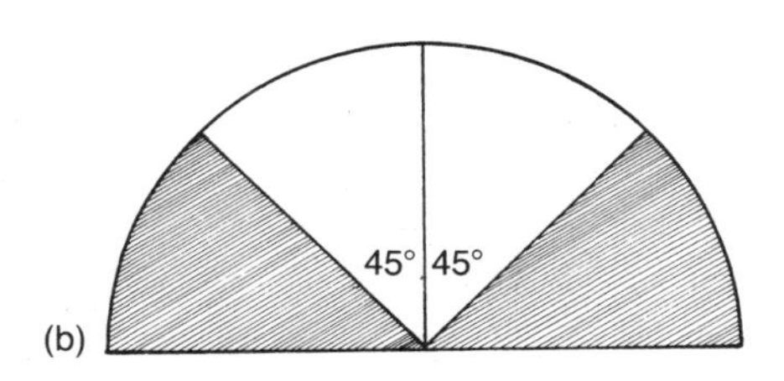

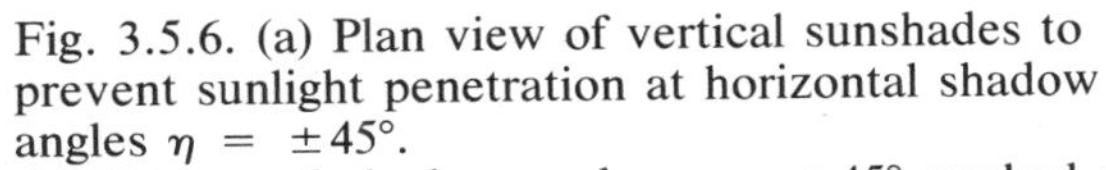

Fig. 3.5.6. (a) Plan view of vertical sunshades to prevent sunlight penetration at horizontal shadow angles $\eta = \pm 45°$.
(b) Horizontal shadow angles $\eta = \pm 45°$ marked on shadow angle protractor. The cross-hatched area indicates the extent of screening provided by the vertical sunshades shown in (a). A mask of the shape of the hatched area may be used instead. The procedure is the same as for Fig. 3.5.5. (*After Harkness and Mehta, Ref. 3.9.*)

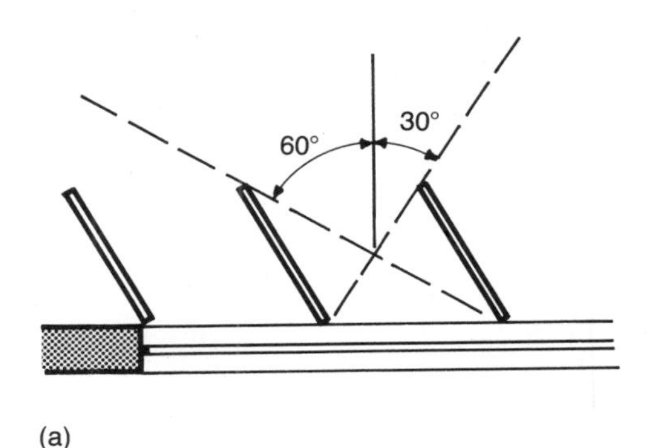

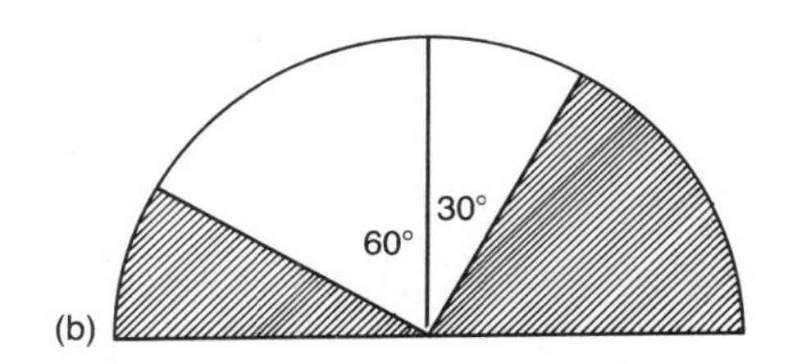

Fig. 3.5.7. (a) Plan view of inclined vertical sunshades, and
(b) Shadow angle protractor or shading mask for these sunshades. The procedure is the same as for Fig. 3.5.5. (*After Harkness and Mehta, Ref. 3.9.*)

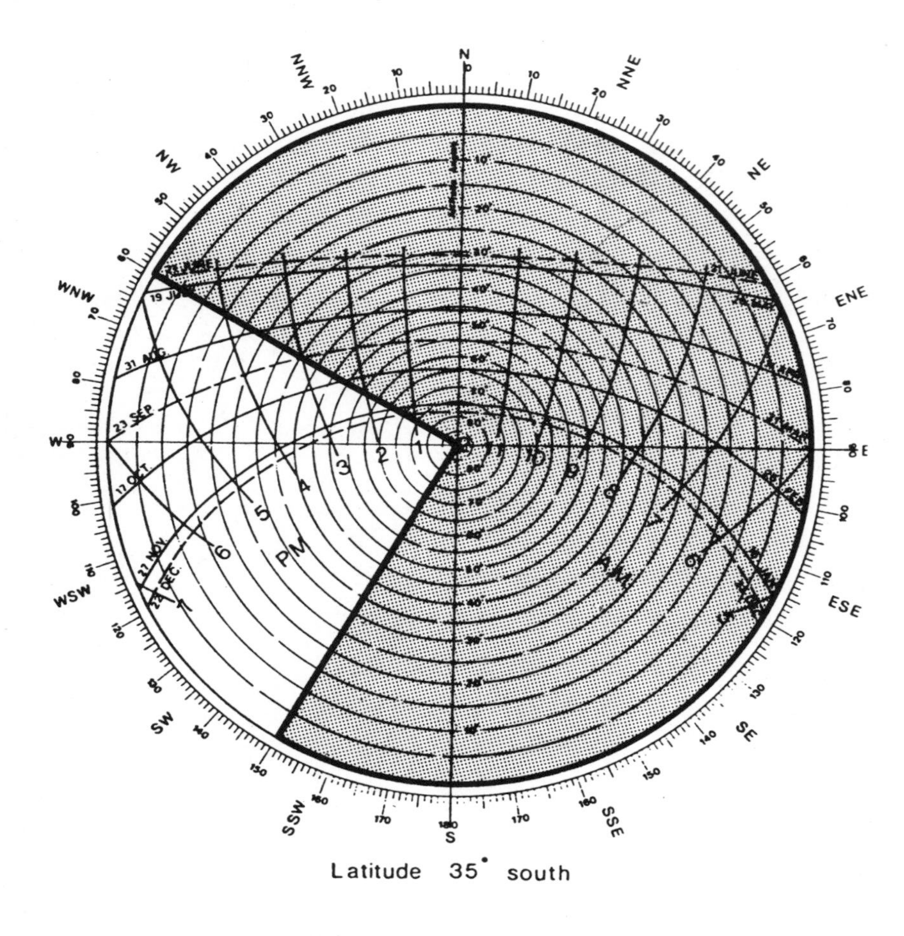

Fig. 3.5.8. Shading mask for the inclined louvers shown in Fig. 3.5.7 superimposed on the sunpath diagram for latitude 34° S for a facade facing precisely west (*Example 3.3*).

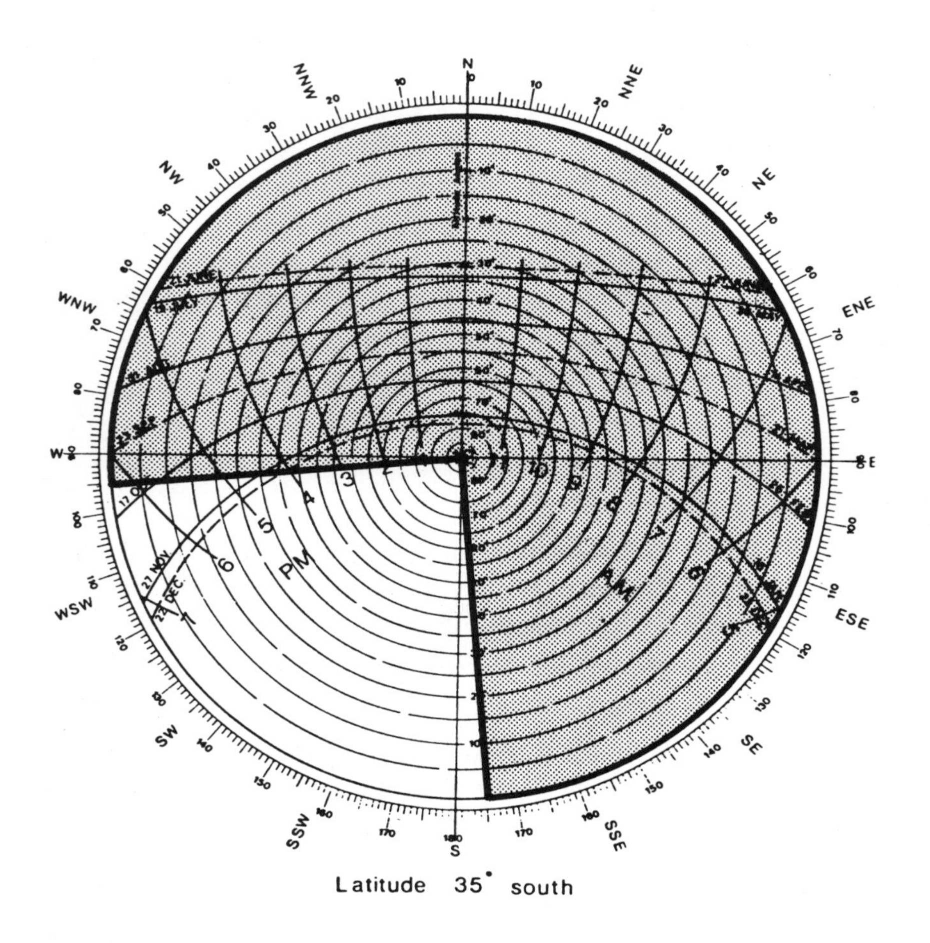

Fig. 3.5.9. Shading mask for the inclined louvers shown in Fig. 3.5.7 superimposed on the sunpath diagram for latitude 35° S for a facade facing 35° south of west (*Example 3.4*).

Example 3.3 *Using the sunpath chart and the transparent shading mask (Figs. 3.5.3 and 3.5.4), determine the hours of sunlight penetration permitted by the shading device shown in Fig. 3.5.7 on a western facade in Sydney, Australia (34° S).*

Solution *We will use the 35° S chart shown in Fig. 3.5.3. The superposition of the shading mask, oriented west, on that chart is shown in Fig. 3.5.8.*

The shading device is useless for a western facade. On December 22 sunlight penetration commences at 1:20 P.M. On October 17 and on February 26 sunlight penetration starts at 2:20 P.M. Evidently the louvers need to be curved or, if straight, placed at a sharper angle or closer together.

Example 3.4 *Determine the hours of sunlight penetration permitted by the shading device shown in Fig. 3.5.7 on a facade oriented 35° south of west in Sydney (34° S).*

Solution *The superposition of the shading mask, oriented 35° S of W, on the 35° S chart is shown in Fig. 3.5.9.*

The shading device does not provide complete shading for that orientation. However, sunlight penetration on the longest day, December 22, does not start until 4 P.M., and by October 17 and February 26 it does not commence until 5:40 P.M., about three quarters of an hour before sunset.

If we use both horizontal and vertical sunshades, for example, in an egg-crate shading device, we need both the vertical and the horizontal shadow angles on the protractor (Fig. 3.5.10).

We noted in Section 3.4 that sunlight penetration for the winter months could be designed to occur at a certain time of the year, such as the equinox, or it could be designed for a certain temperature, say, 21°C (70°F). This requirement can be incorporated in the sunpath chart. The days when the temperature exceeds 21°C, and the times of the day when this occurs, are known for most cities as an average over several years. This information is plotted, in simplified form, on the sunpath chart. By marking the intersection of the borders of the shaded portion in this figure with the shadow angle of the transparent protractor (Fig. 3.5.11) we can determine the shadow angles needed throughout the year to prevent sunlight penetration during the "overheated period." Evidently, different angles may be needed at different times of the year. For most buildings it is best to adopt a compromise solution and use a fixed shading device of pleasing appearance.

Fig. 3.5.10. Horizontal shadow angle $\eta = \pm 45°$ and vertical shadow angle $\phi = 45°$ marked on shadow angle protractor. The cross-hatched area indicates the extent of the screening provided by both vertical and horizontal sunshades to form an "egg crate." A mask in the shape of the hatched area can be used instead. Evidently, an egg-crate sunshade is more efficient than either horizontal or vertical sunshades used by themselves, but it also interferes more with natural ventilation, which is important for buildings without air conditioning in warm-humid climates.

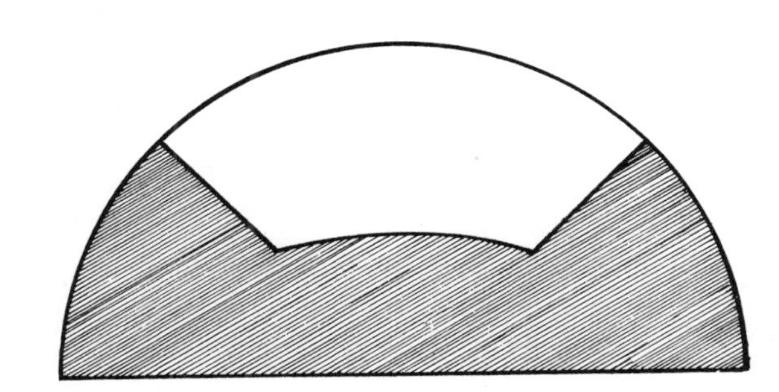

Adjustable shading devices can be used, but many have become immobile after a few years through corrosion or dirt; even when they can be moved, the occupants of the building rarely remember to make the adjustment on the right date. It would be possible to install electric motors operated by a computer to make the necessary seasonal adjustments to the shadow angles, but this is unlikely to be economical.

Example 3.5 *A sunshade on a wall facing north consists of a slab projecting horizontally for a distance of 1 m at a height 1.5 m above the window, as shown in Fig. 3.4.2. Consider its effect on the thermal performance of a building in Sydney.*

Solution *This is a more accurate version of Example 3.1, which assumed that sunlight penetration would commence at noon on the equinox in spring and cease at noon on the equinox in fall. The vertical shadow angle is*

$$\phi = tan^{-1}\frac{1}{1.5} = 56.3°$$

which is the complementary angle for the latitude:

$$\lambda = 90° - 56° = 34°$$

We are using the sunpath chart for latitude 35° S (Fig. 3.5.3) and the curved line on the shadow angle protractor for a shadow angle of 90° − 35° = 55°. The result is shown in Fig. 3.5.12.

This simple sunshade, designed by an approximate method, provides quite a good design in this instance. Some overheating occurs, particularly between 1 and 2 P.M. in March and April, but this is acceptable and provides passive solar energy [see Section 5.15] if it can be stored. However, solar radiation would be helpful in the early hours of the morning, when this sunshade excludes it from September 23 to March 21.

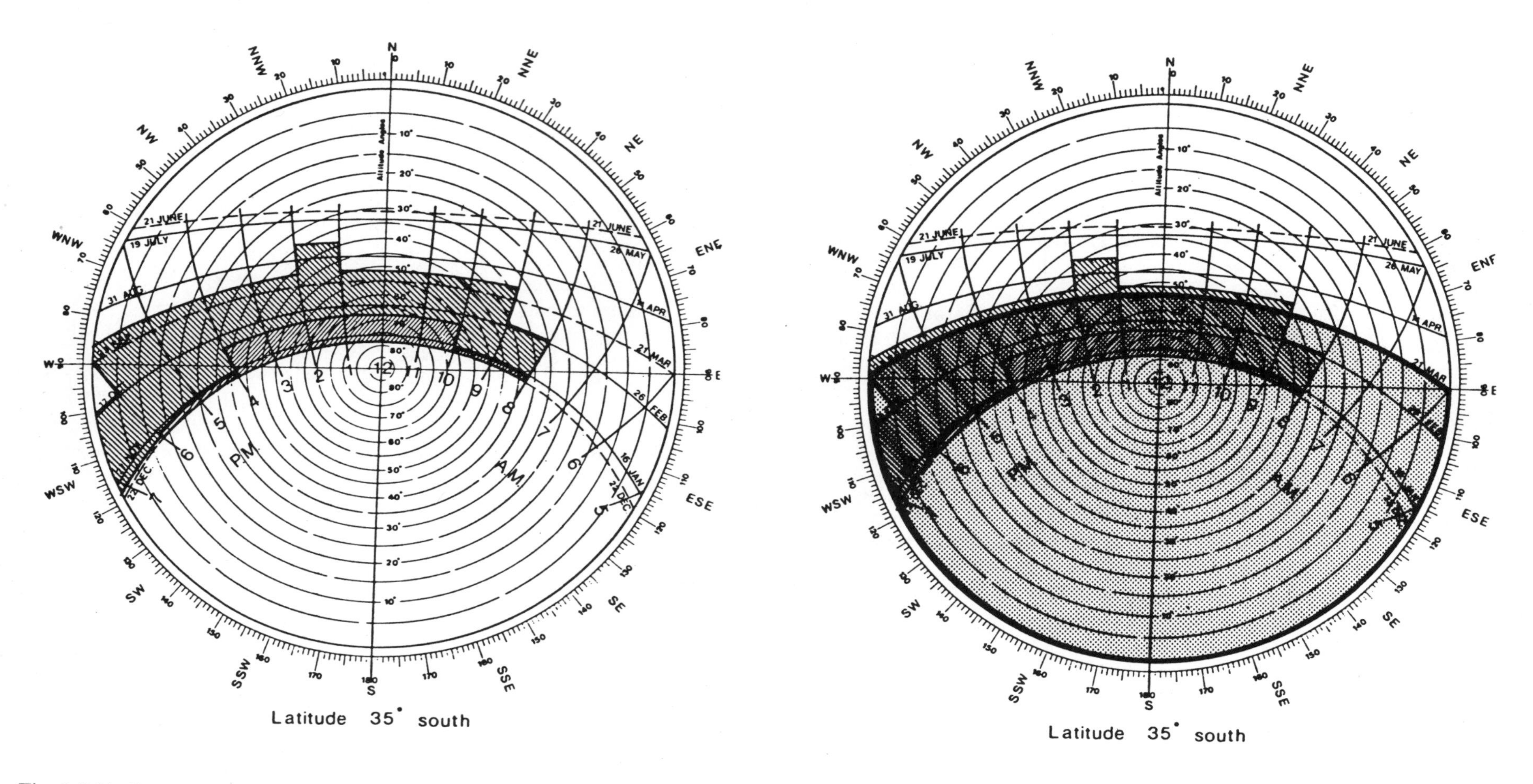

Fig. 3.5.11. Sunpath chart for latitude 35° S, on which the times have been marked when the temperature in Sydney exceeds 21°C (70°F). Dark hatching is used for the period from June to December, and light hatching from December to June. This indicates the period when overheating would occur if the windows were not suitably shaded.

Fig. 3.5.12. Shading mask for a horizontal projection with a shadow angle of 35° superimposed on a sunpath chart for latitude 35° S for a facade facing precisely north. The overheated period in Sydney has been marked on the sunpath chart (*Example 3.5*).

A sunpath chart showing the local climate gives a more accurate design for the sunshading than one on which this information does not appear. Evidently, it must be prepared especially for each city. The chart is otherwise used in the same way in conjunction with a shadow protractor or mask of the type shown in Figs. 3.5.4, 3.5.5, 3.5.6, 3.5.7, and 3.5.10. (*Reproduced from Ref. 3.5, p. 87.*)

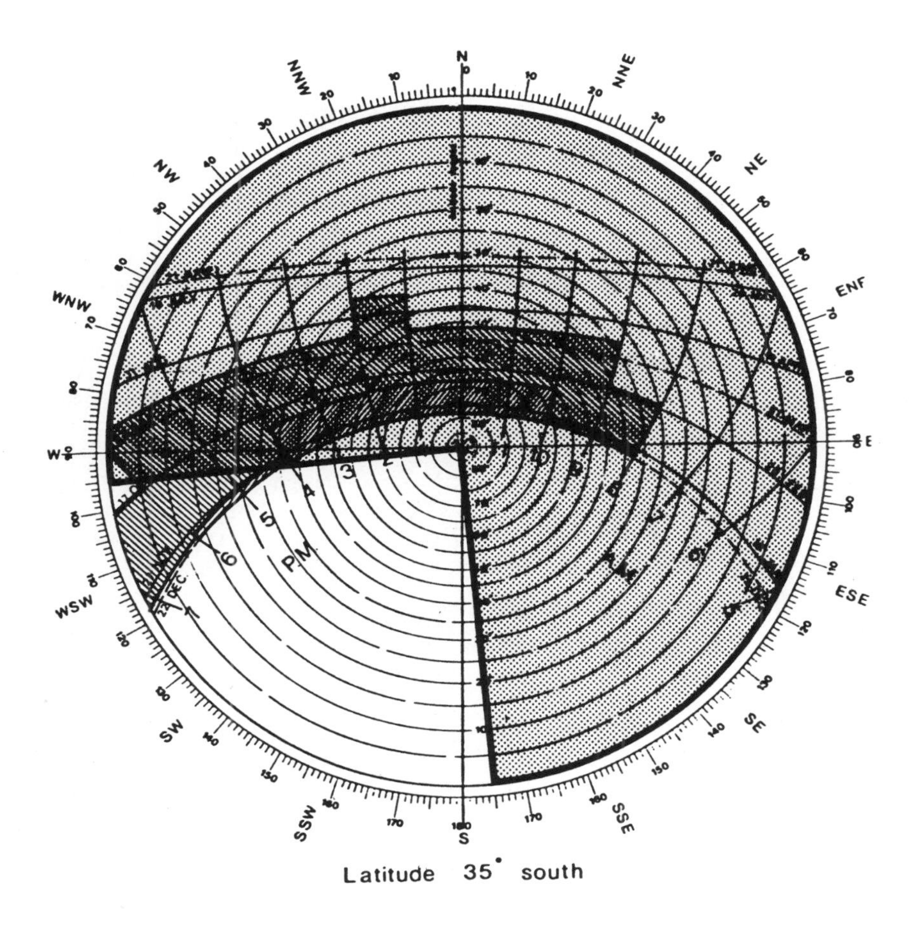

Fig. 3.5.13. Shading mask for the inclined louvers shown in Fig. 3.5.7 superimposed on the sunpath diagram for latitude 35° S for a facade facing 35° south of west. The overheated period for Sydney has been marked on the sunpath chart (*Example 3.6*).

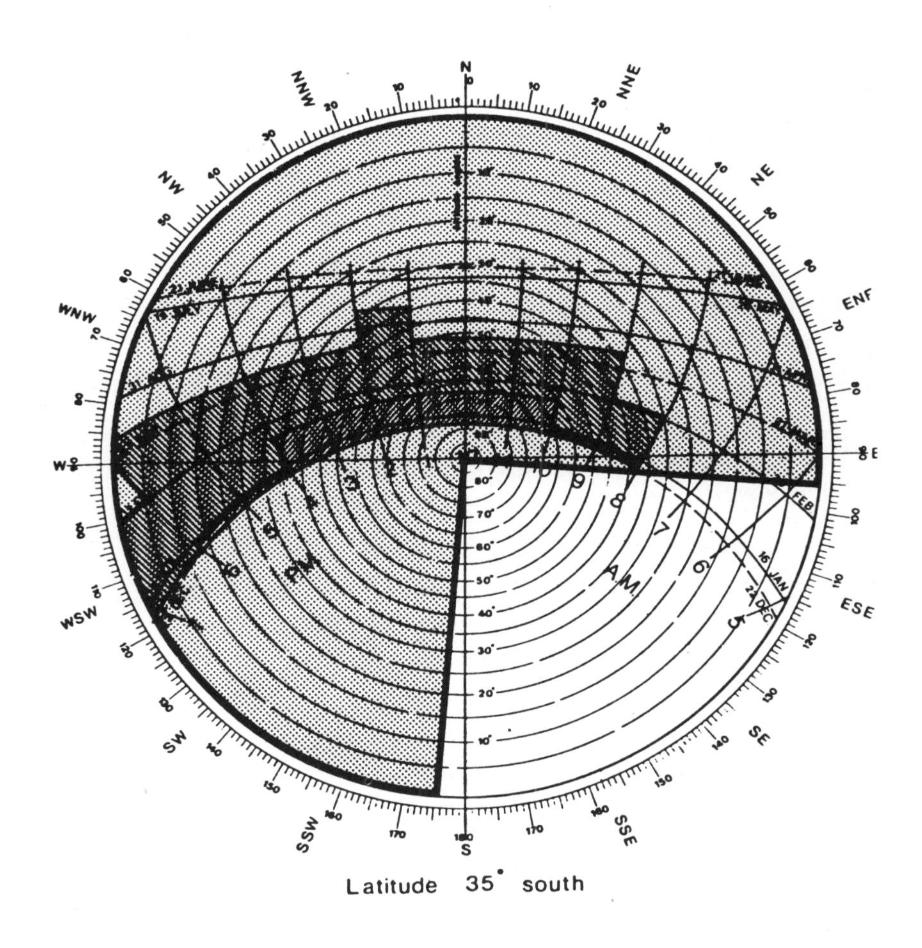

Fig. 3.5.14. Shading mask for inclined louvers similar to those shown in Fig. 3.5.7, but tilted in the opposite direction, superimposed on the sunpath diagram for latitude 35° S for a facade facing 35° south of east. The overheated period for Sydney has been marked on the sunpath chart (*Example 3.7*).

Example 3.6 *We will repeat Example 3.5, using the sunpath chart of Fig. 3.5.11, on which the times have been marked at which the temperature exceeds 21°C (70°F). This is more accurate than the use of the sunpath chart in Fig. 3.5.3, on which the overheated period has not been marked.*

Solution *We are examining the effectiveness of the shading device shown in Fig. 3.5.7 on a facade oriented 35° south of west in Sydney (34° S). The superposition of the shading mask on the temperature-marked sunpath chart is shown in Fig. 3.5.13. In this instance the result is almost the same as for Example 3.4 (Fig. 3.5.9), because temperatures above 21°C occur mainly in the afternoon. The louvers provide some excessive shading in November and December between 5 and 6 P.M., when temperature records are considered, but this is not a major disadvantage.*

Example 3.7 *We will examine the effect of the same shading device on a wall facing 35° south of east in Sydney (34° S).*

Solution *The superposition of the shading mask on the temperature-marked sunpath chart (Fig. 3.5.11) is shown in Fig. 3.5.14. From a geometric point of view, a shade oriented 35° south of east is as effective as one oriented 35° south of west, provided the louvers are tilted towards the north in each case. However, overheating before 8 A.M. does not occur at any time in Sydney, and before 9 A.M. it is limited to two months in the year. The louvers provide excessive shading when the temperature record is considered.*

Additional and more complex examples of the use of sunpath charts are given in specialized books (for example, Refs. 3.2, 3.4, 3.5, and 3.9).

Sunlight penetration and the design of sunshading devices can also be investigated by computer graphics (Fig. 3.5.15). A number of programs are available (Ref. 3.5, pp. 75–79), and their versatility is improving rapidly.

The effect of sunshading on daylight is discussed in Chapter 9.

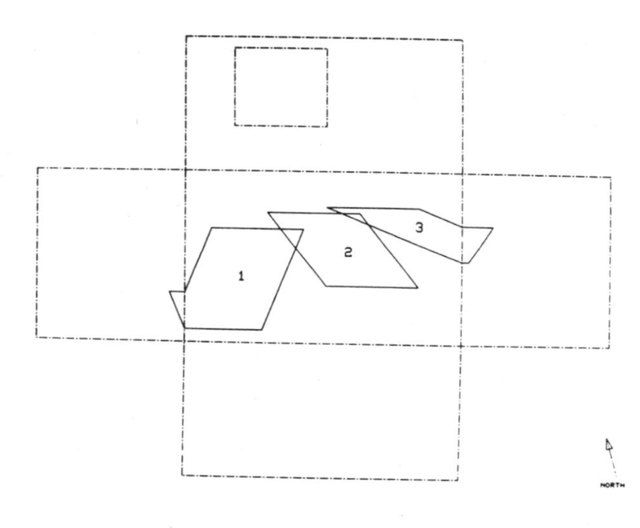

Fig. 3.5.15. Graphic output of sunlight penetration computer program for a rectangular room with an off-center window facing 15° east of north. The drawing shows the plan of the room and the elevations of the four walls. These can be cut out and folded along the dotted lines to produce a three-dimensional model. The polygons 1, 2, and 3 show the sunlight penetration through the unshaded window at 10 A.M., 1 P.M., and 3 P.M. on April 22 in Sydney, Australia (latitude 33.92° S, longitude 151.17° E; time 10 hours later than Greenwich Mean Time).

References

3.1 H. MESSEL AND S.T. BUTLER (Eds.): *Solar Energy*. Pergamon, Oxford, 1975. 340 pp.

3.2 O.H. KOENIGSBERGER, T.G. INGERSOLL, A. MAYHEW, and S.V. SZOKOLAY: *Manual of Tropical Housing and Building, Part One, Climatic Design*. Longman, London, 1974. 320 pp.

3.3 S.V. SZOKOLAY: *Solar Energy and Building*. Second Edition. Architectural Press, London, 1975. 174 pp.

3.4 T.A. MARKUS and E.N. MORRIS: *Buildings, Climate and Energy*. Pitman, London, 1980. 540 pp.

3.5 A.M. SALEH: "Design of Sunshading Devices," in H.J. COWAN (Ed.): *Solar Energy Applications in the Design of Buildings*. Applied Science Publishers, London, 1980. pp. 33–94.

3.6 F.W. SOHON: *The Stereographic Projection*. Chemical Publishing Co., New York, 1941. 210 pp.

Suggestions for Further Reading

References 3.1 and 3.3.

3.7 FARINGTON DANIELS: *Direct Use of the Sun's Energy*. Yale University Press, New Haven, 1964. 374 pp.

3.8 OFFICE OF TECHNOLOGY ASSESSMENT, CONGRESS OF THE UNITED STATES: *Application of Solar Technology to Today's Energy Needs*. Volume 1. U.S. Government Printing Office, Washington, D.C., 1978. 525 pp.

3.9 E.L. HARKNESS and M.L. MEHTA: *Solar Radiation Control in Buildings*. Applied Science Publishers, London, 1978. 271 pp.

Chapter 4 The Thermal Environment

Heat may be transfered by conduction, convection, and radiation. This results in changes of the temperature and humidity of the air. Their measurement is described, and the assembly of the data in graphical form on a psychrometric chart is explained.

Air movement and radiation also affect human comfort, although to a lesser extent; their measurement is discussed. In addition, thermal comfort greatly depends on the amount of clothing worn and the amount of work performed; two tables give comparative figures for these factors. All the data on thermal comfort in buildings are then assembled to define the comfort zone. The temperature and humidity range of this zone can be drawn on a psychrometric chart.

4.1 Definitions and Thermal Units

Heat is a form of energy. This was demonstrated in 1843 by the English physicist James Prescott Joule, when he measured the rise in temperature of a quantity of water resulting from the expenditure of mechanical work. The fact that there is a constant relation between the amount of heat gained and the amount of energy lost, and vice versa, is now known as the *First Law of Thermodynamics*.

The *joule* (J) is the unit of mechanical energy both in the old metric system and in the SI metric system. In the SI metric system the joule is also used as the unit of heat energy. In the old metric units, heat is measured in calories. (In American units, the old British units, mechanical energy is measured in foot-pounds, and heat in British thermal units, abbreviated Btu).

The joule is a small unit, and in practice we measure energy in kilojoules or megajoules.

$$1 \text{ kJ} = 1000 \text{ kJ}$$
$$1 \text{ MJ} = 1\,000\,000 \text{ J} = 1000 \text{ kJ}$$

The flow of energy, or energy per unit time (that is, per second), is measured in *watts* (W), where

$$1 \text{ W} = 1 \text{ J/s}$$

Larger units of energy flow are measured in kW or MW. It has become common practice to give the consumption of electrical energy in kWh, that is, 1 kilowatt used for a period of 1 hour. Since there are 3600 seconds in 1 hour,

$$1 \text{ kWh} = 3.6 \text{ MJ}$$

In SI metric units the watt is used to measure both the heat energy per unit time and the mechanical energy per unit time. In the old metric units and in American units there is no special unit for heat energy per unit time, and calories per hour or Btu per hour are employed.

Conversions between the various units are given in Appendix B.

There are basically two forms of heat. *Sensible heat* can be detected by the human senses. Addition of sensible heat raises the temperature; removal of sensible heat reduces the temperature. *Latent heat* is the heat required or released by a change of state without a change of temperature. Thus the melting of a solid and the evaporation of a liquid require latent heat. The condensation of a vapor and the freezing of a liquid release latent heat.

A change of temperture is measured in the metric system in *degrees Celsius* (°C) or *degrees Kelvin* (K). The Celsius and Kelvin scales are determined by the boiling point (100°C) and the freezing point (0°C) of water (Fig. 4.1.1). (The term "degree centigrade" was formerly used as an alternative to "degree Celsius." In American units temperature is measured in degrees Fahrenheit.)

Absolute zero is the lowest temperature attainable, in theory, when a perfect gas ceases to exert any pressure. This is zero degrees on the Kelvin scale, and −273°C on the Celsius scale. Thus a temperature of 30°C is equal to 303 K. Many thermal laws are related to absolute zero and are therefore more conveniently expressed in the Kelvin scale. The Celsius scale, on the other hand, gives a clearer impression of human reaction to temperature ("30°C is hot," "10°C is cold").

Temperature differences are the same, whether they are measured in degrees Celsius or degrees Kelvin. It is customary when using SI units to express temperature differences in degrees Kelvin (K).

4.2 Transfer of Heat by Conduction, Convection, and Radiation

Thermal conduction is the transfer of heat from one part of a body to another, or from one body to another body with which it is in contact. All substances consist of very small particles that are in continuous motion. This motion depends on the temperature. The particles on the warmer side of a body move more intensely than those on the colder side, and they transfer some of their energy to the colder particles by colliding with them. As their motion increases, that part of the body becomes warmer. The rate of heat transfer depends on the temperature difference and on the thermal conductivity of the material:

$$Q = \frac{k(T_1 - T_2)}{d} \tag{4.1}$$

where Q is the quantity of heat passing through a unit area of a body (for example, a wall or a floor) in unit time, measured in watts per square meter (W/m^2);

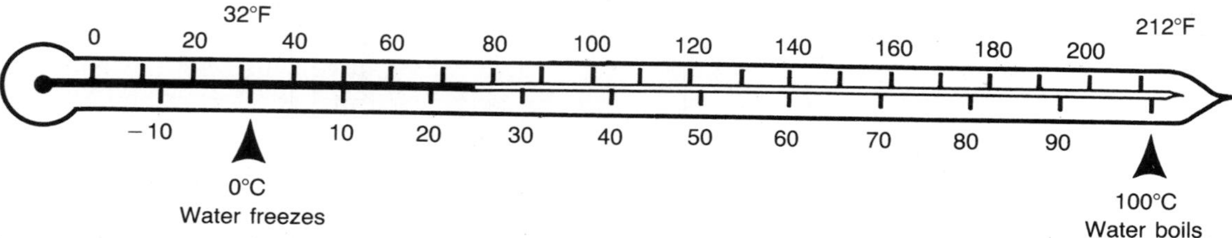

Fig. 4.1.1. The Celsius scale. Water freezes at 0°C and boils at 100°C.

k is the thermal conductivity of the material, measured in watts per meter per degree Kelvin (W/mK);

T_1 is the temperature on the warmer side of the body;

T_2 is the temperature on the cooler side of the body; the temperature difference $T_1 - T_2$ is measured in degrees Kelvin (K);

d is the distance between the two sides of the body in meters (m).

All materials, whether solid, liquid, or gaseous, conduct heat; but some are better conductors than others. Thus metals are excellent conductors of heat, while still air is a poor conductor. Porous materials that contain small pockets of air are thus good insulators.

In liquids and gases *convection* is generally a more effective method of heat transfer. The term is used to describe the transfer of heat by the mixing of one portion of a liquid or a gas with the remainder that is at a higher or a lower temperature. While conduction cannot be seen, it is possible to make thermal currents due to convection visible, for example, by introducing particles of dye into water that is heated by a burner.

Convection is due to the fact that most liquids and gases change their density as they change their temperature. Generally they get lighter as they get hotter. Thus if a pot of water is heated at its base by a burner, the warm particles of water rise to the top and the colder particles sink to the bottom, where they are heated in turn. Water between 0°C and 4°C is an exception; the density of water increases with cooling until it reaches 4°C but then decreases again until it freezes. This is an important property for life on earth; if it were otherwise, lakes would freeze solid from the bottom, instead of forming a skin of ice on top.

In buildings, convection is of particular importance because it increases the transfer of heat by air flowing over the surfaces of walls, floors, roofs, and pipes used for heating or cooling. However, air is a poor conductor of heat.

The quantity of heat transferred by convection is

$$Q = h(T_1 - T_2) \tag{4.2}$$

where Q is the quantity of heat transferred per unit area, measured in watts per square meter (W/m²);

h is the boundary layer heat transfer coefficient, measured in watts per square meter per degree Kelvin (W/m²K);

$T_1 - T_2$ is the temperature difference measured in degrees Kelvin (K).

The numerical value of h depends on the nature of the flow, the physical properties and velocity of the fluid, the shape and dimensions of the surface, and the temperature. Values required for use in buildings are given in Chapter 22 of the *ASHRAE Handbook—Fundamentals* (Ref. 4.1).

Thermal radiation is energy in the form of electromagnetic waves, that is, the same type of energy that transmits visible light, television, radio, and X-rays. Solar heat energy has a wavelength ranging from 0.3 to 2.4 μm (that is, 300 to 2400 nm; Fig. 3.2.1) Energy reradiated from the surfaces of the building has longer wavelengths, from 2.5 to 100 μm.

The quantity of heat radiated is governed (Fig. 4.2.1) by a law first proposed by Joseph Stefan and proved by Ludwig Boltzmann in 1884; it is known as the Stefan-Boltzmann law:

$$Q = \sigma \varepsilon T^4 \tag{4.3}$$

where Q is the quantity of heat emitted from a unit area of a body (for example, the surface of a wall or a floor), measured in watts per square meter (W/m²);

σ is the Stefan-Boltzmann radiation constant, which is 5.68×10^{-8} W/m² K⁴;

ε is the emissivity of the surface, a ratio that is 1.0 for a "black body," namely, a body that has no reflecting power but is capable of absorbing all the radiation it receives;
For ordinary building materials, such as concrete and brick, ε is about 0.9; for dull aluminum and for bright galvanized steel it is about 0.2, and for polished aluminum it is about 0.05.

T is the temperature, measured from absolute zero in degrees Kelvin (K).

Fig. 4.2.1. Thermal radiation. Energy is radiated from the surfaces of a building with a wavelength from 2.5 μm to 100 μm, a longer wavelength than that of solar radiation (see Fig. 3.2.1).

4.3 Temperature and Humidity, and Their Measurement

Thermal comfort is determined mainly by the *air temperature*. This is measured with a thermometer. The most common type consists of a glass tube, with an enlarged end, partly filled with mercury. The temperature is measured by the expansion of the mercury, which remains liquid at temperatures as low as −40°C and is still a liquid when heat-resistant glass softens and loses its shape. An alternative thermometer fluid for use in buildings is alcohol, which freezes at −80°C and boils at 70°C.

Remote-control temperature measurements are made with *thermocouples,* which consist of two electrically dissimilar metals joined to produce an electromotive force, measured in millivolts, when the temperature changes. The electric measurement is then converted to temperature. Thermocouples are frequently made by the junction of copper and constantan, an alloy of 40% nickel and 60% copper.

Thermocouples can also be used for *recording thermometers*. An alternative nonelectric method employs a *bimetallic strip* formed by two metals with different rates of thermal expansion. This strip bends as the temperature changes, and the deformation of the strip moves a needle vertically over a piece of paper. The paper is moved horizontally by a clockwork, so that a graph of the variation of temperature over, say, a week is produced.

Humidity is important for thermal comfort, particularly in hot-humid climates. It is most conveniently measured with a *wet-bulb thermometer;* this is a mercury thermometer whose bulb is covered by a damp wick or piece of cloth dipping into water. Next to the wet-bulb thermometer is an ordinary, or *dry-bulb,* thermometer that measures the air temperature (Fig. 4.3.1).

If the air is completely saturated with water vapor, the temperature recorded by the wet-bulb thermometer is the same as the dry-bulb temperature; but when, as is commonly the case, the air is unsaturated, the wet-bulb temperature is below the dry-bulb temperature. The drier the air, the more rapidly water is evaporated from the wet bulb, and the more that bulb is cooled. Hence at any given temperature the difference between the dry-bulb and wet-bulb thermometers increases as the humidity of the air decreases.

This difference can be translated into either absolute humidity or relative humidity, by means of a chart or a table. The *absolute humidity* (AH) is the mass of water vapor per mass of dry air. As the temperature decreases, the ability of the air to absorb water vapor decreases (see Fig. 4.4.1); therefore as the temperature is reduced, some of the water vapor in the air may be condensed into water. This accounts for the condensation on windows on a cold night.

The *relative humidity* (RH) is the ratio of the water vapor contained in the air to that contained in saturated air. As air is cooled, its relative humidity rises unless some of the moisture is removed from the air. Thus a relative humidity of 80% is quite common in cold weather and is not felt to be particularly unpleasant, whereas the same relative humidity in hot weather would strike most people as oppressive and it corresponds to a much higher absolute humidity.

An instrument for measuring humidity is called a hygrometer (from the Greek *hygros*, "moist"), and thus a wet-and-dry-bulb thermometer may also be referred to as a hygrometer. It is the simplest and most accurate instrument for this purpose, but it cannot *record* humidity. A recording hygrometer commonly uses a ribbon of hair that changes its length with a change in the humidity and thus causes a pointer to move up or down on a chart. The time element is provided by a clock.

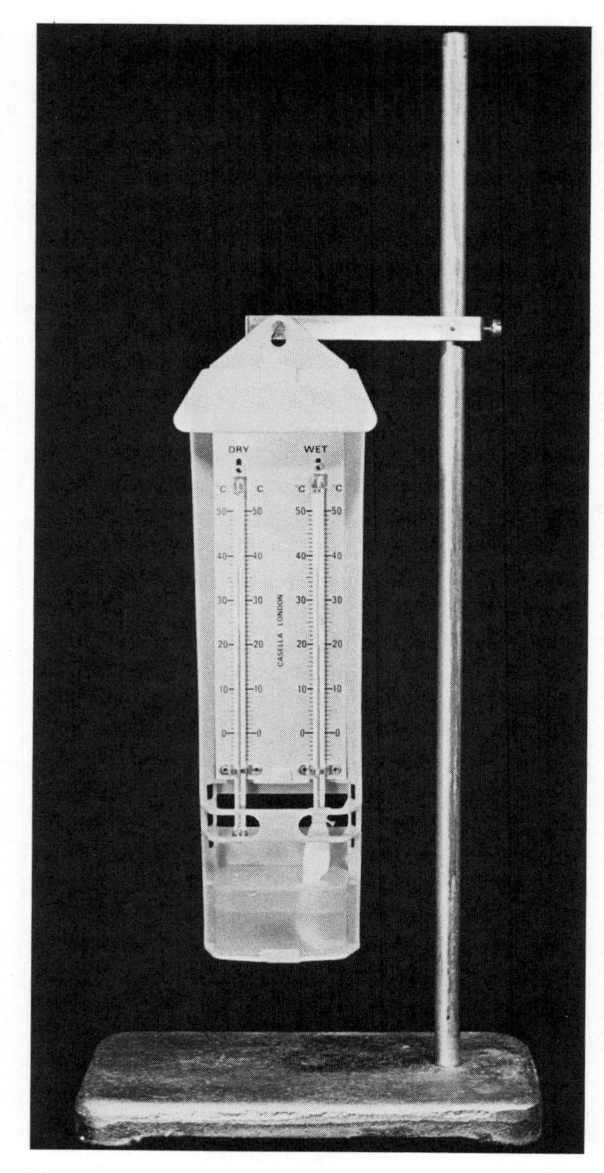

Fig. 4.3.1. Humidity is important for thermal comfort, particularly in hot-humid climates. It is most easily measured with a wet-and-dry-bulb thermometer.

4.4 The Psychrometric Chart

The temperature and humidity of the air, and the heat contained in the moist air, can be plotted on a psychrometric chart (from the Greek *psychros*, "cold"). This greatly aids the discussion of human comfort criteria and the determination of the heating or cooling required to produce a comfortable environment. It has the dry-bulb temperature (DBT) in degrees Celsius as its horizontal coordinate and the absolute humidity (in grams of water vapor per kilogram of dry air, or g/kg) as its vertical coordinate. Each value of the absolute humidity corresponds to a particular vapor pressure, measured in kilopascals (kPa), ranging from 0 kPa for zero absolute humidity to 4.0 kPa for an absolute humidity of 25 g/kg; this is shown by an additional vertical coordinate. A number of curved lines are plotted on these coordinates corresponding to the relative humidity (RH). When the relative humidity reaches 100%, water must be removed by condensation, and the 100% RH line therefore terminates the chart (Fig. 4.4.1).

On this chart the lines of equal wet-bulb temperature (WBT) form straight inclined lines (Fig. 4.4.2).

The density of air changes with temperature and humidity. For air conditioning calculations we need to know the volume of each kilogram of dry air to be handled, that is, the *specific volume* (SV) in cubic meters per kilogram (Fig. 4.4.3). This is the reciprocal of the density of the air.

Another set of straight lines is formed by the *enthalpy* (H); this term is derived from the Greek *thalpo,* "to make hot." The enthalpy is the heat content of the moist air, measured in kilojoules per kilogram of air (kJ/kg), and it is the sum of the sensible heat and the latent heat [Section 4.1]. The sensible heat is the amount of heat required or released during a change of temper-

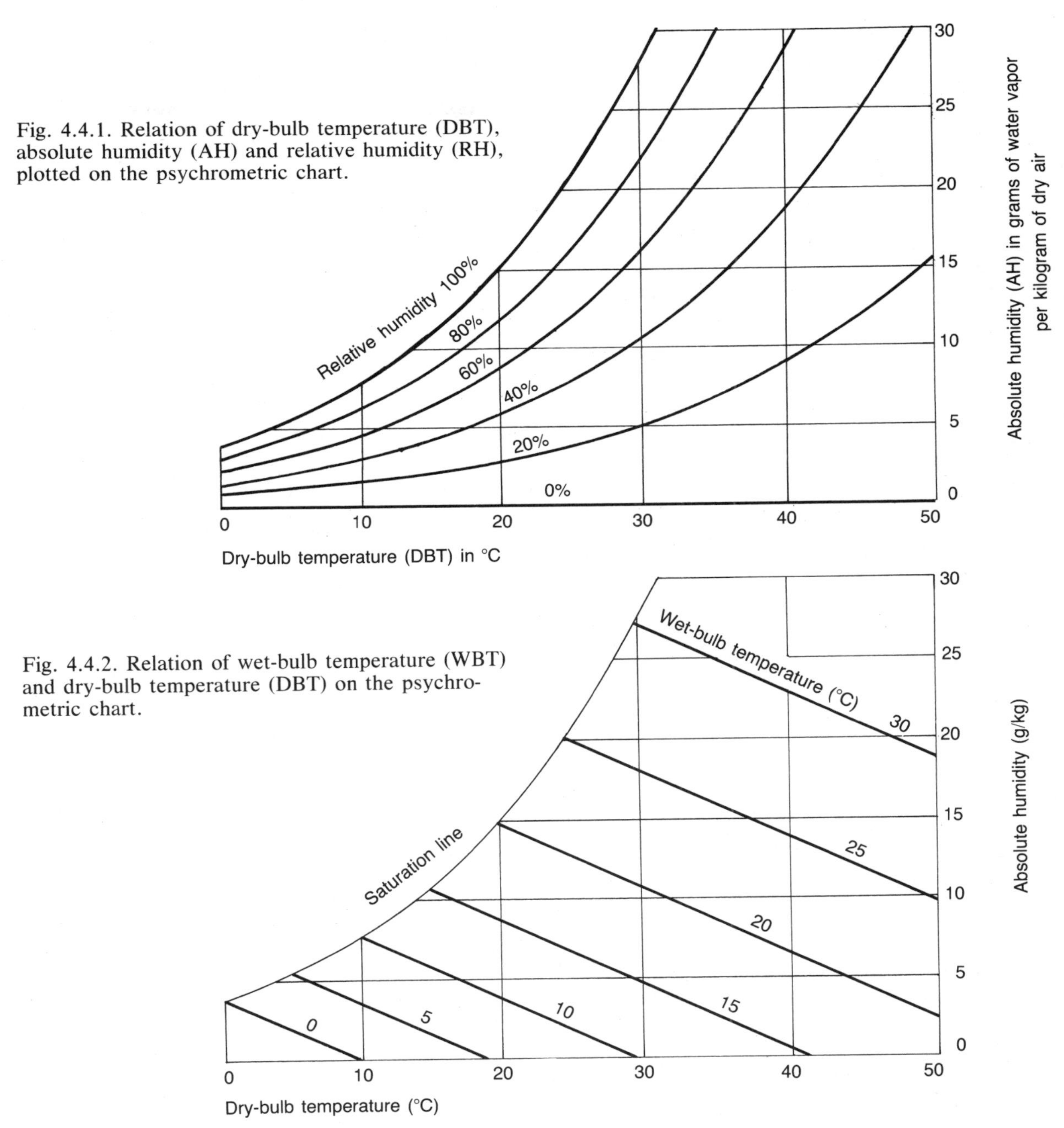

Fig. 4.4.1. Relation of dry-bulb temperature (DBT), absolute humidity (AH) and relative humidity (RH), plotted on the psychrometric chart.

Fig. 4.4.2. Relation of wet-bulb temperature (WBT) and dry-bulb temperature (DBT) on the psychrometric chart.

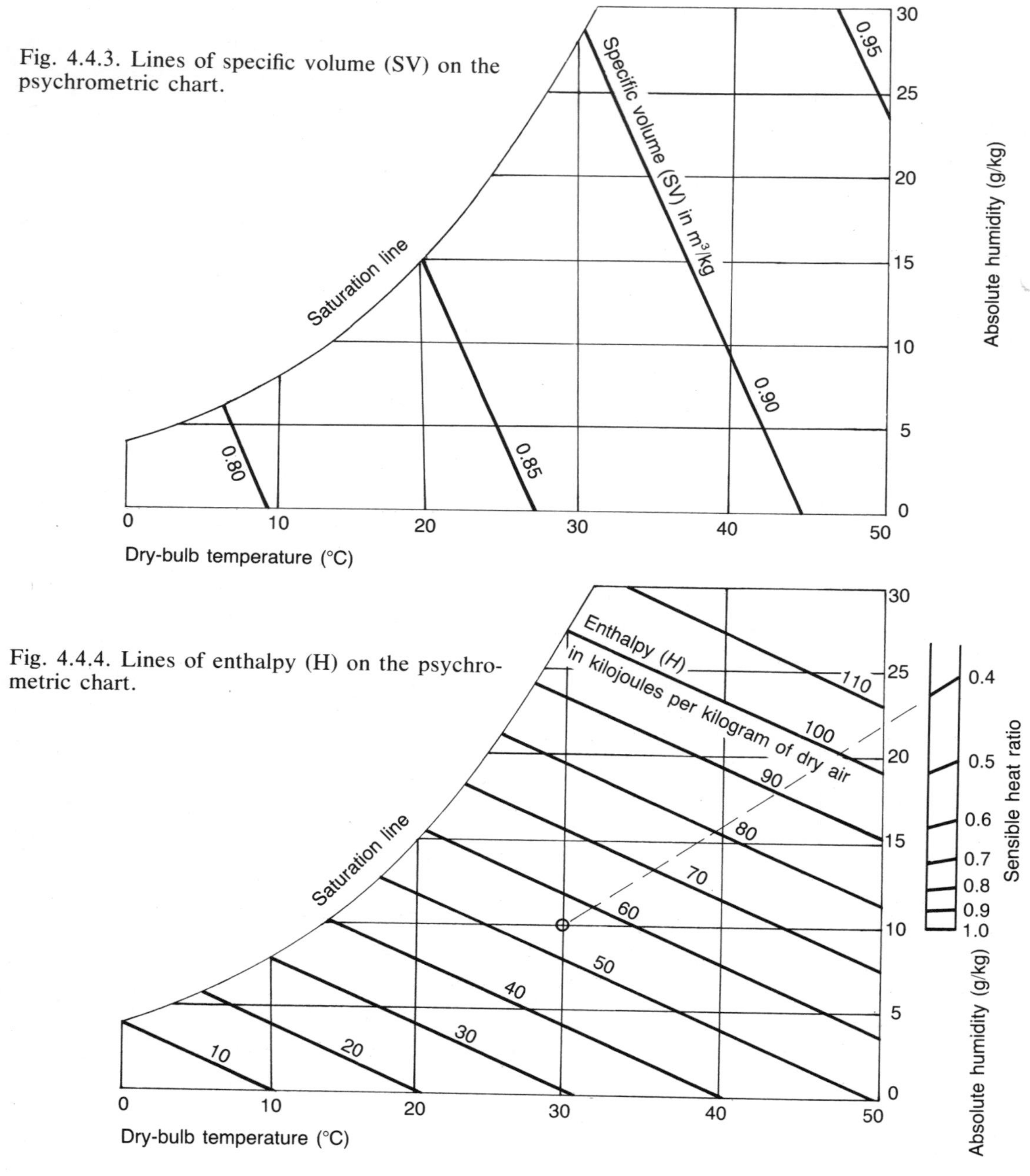

Fig. 4.4.3. Lines of specific volume (SV) on the psychrometric chart.

Fig. 4.4.4. Lines of enthalpy (H) on the psychrometric chart.

ature at a constant absolute humidity, for example, the heat needed to raise the dry-bulb temperature from 20°C to 25°C without changing the mass of water contained in the air. The latent heat is the amount of heat required or released when the absolute humidity is changed at a constant temperature, because we must evaporate or condense some of the moisture in the air. The lines of enthalpy are shown in Fig. 4.4.4. The ratio of sensible heat to total (sensible plus latent) heat is shown on the right-hand edge.

In Fig. 4.4.5 the vertical lines of dry-bulb temperature, the horizontal lines of absolute humidity, the curved lines of relative humidity, and inclined straight lines of wet-bulb temperature and of specific volume are combined to form the psychrometric chart. The inclined lines of enthalpy are shown only around the edges of the chart to make it easier to read, and they can be joined by placing a transparent ruler on the chart.

It is now possible to plot on the psychrometric chart the range of temperatures and humidities that are considered comfortable for a particular type of occupancy, and the temperature and humidity of the outside air. Hence it is possible to determine the change in temperature and humidity required to produce a comfortable environment, and the enthalpy, that is, the energy required to do so (see Fig. 4.7.1).

Example 4.1 *A dry-bulb thermometer reads 20°C, and an adjacent wet-bulb thermometer reads 15°C. What are the relative humidity and the absolute humidity of the air?*

Solution *From Fig. 4.4.5, the relative humidity is 59%, and the absolute humidity is 8.6 grams of water vapor per kilogram of dry air.*

Example 4.2 *Air with a dry-bulb (DB) temperature of 20°C and a wet-bulb (WB) temperature of 15°C is to be heated to a dry-bulb temperature (DBT) of 26°C without a change in its moisture content. What is the new wet-bulb temperature (WBT), and what is the change in the relative humidity?*

Solution *From Fig. 4.4.5, the wet-bulb temperature at a dry-bulb temperature of 26°C is 17°C. The relative humidity decreases from 59% to 41%.*

Example 4.3 *Air with a temperature of 20°C DB and 15°C WB is to be cooled to a dry-bulb temperature of 10°C. Will condensation occur?*

Solution *From Fig. 4.4.5, the relative humidity reaches 100% at a dry-bulb temperature of 11.5°C. The wet-bulb temperature is then also 11.5°C. If the temperature is to be further decreased, some of the water vapor must condense. At a temperature of 10°C and a relative humidity of 100% the absolute moisture content is 7.6 grams of water vapor per kilogram of dry air, a reduction of 12%.*

Example 4.4 *Air at a temperature of 0°C DB and a relative humidity of 80% is to be heated to 26°C DBT. What is its relative humidity after heating? If a relative humidity of at least 25% is required for comfort, how much water vapor must be added to the air?*

Solution *From Fig. 4.4.5, the moisture content of air at 0°C DBT and a relative humidity of 80% is 3 g/kg. At 26°C DBT, its relative humidity is reduced to 14%. The wet-bulb temperature is only 12°C. A relative humidity of 25% at 26°C DBT requires a moisture content of 5.3 g/kg. Therefore 2.3 grams of moisture must be added for each kilogram of air.*

Example 4.5 *What is the change in the enthalpy, that is, the heat content, of the moist air at a temperature of 20°C DB and 15°C WB, if it is heated without a change in its moisture content to 26°C?*

Solution *As pointed out in Example 4.1, the absolute humidity of the air at 20°C DB and 15°C is 8.6 g/kg. From Fig. 4.4.5, the enthalpy increases from 42 kilojoules per kilogram of air to 48 kJ/kg. The additional energy of 6 kJ/kg is required to heat the air.*

Example 4.6 *What is the change in enthalpy of air at a temperature of 20°C DB and 15°C WB, if it is cooled to 10°C?*

Solution *As pointed out in Example 4.3, some of the moisture condenses if the air is cooled to 10°C. The enthalpy at 10°C DB and 10°C WB is 29 kJ/kg, so that 13 kJ/kg of energy are released.*

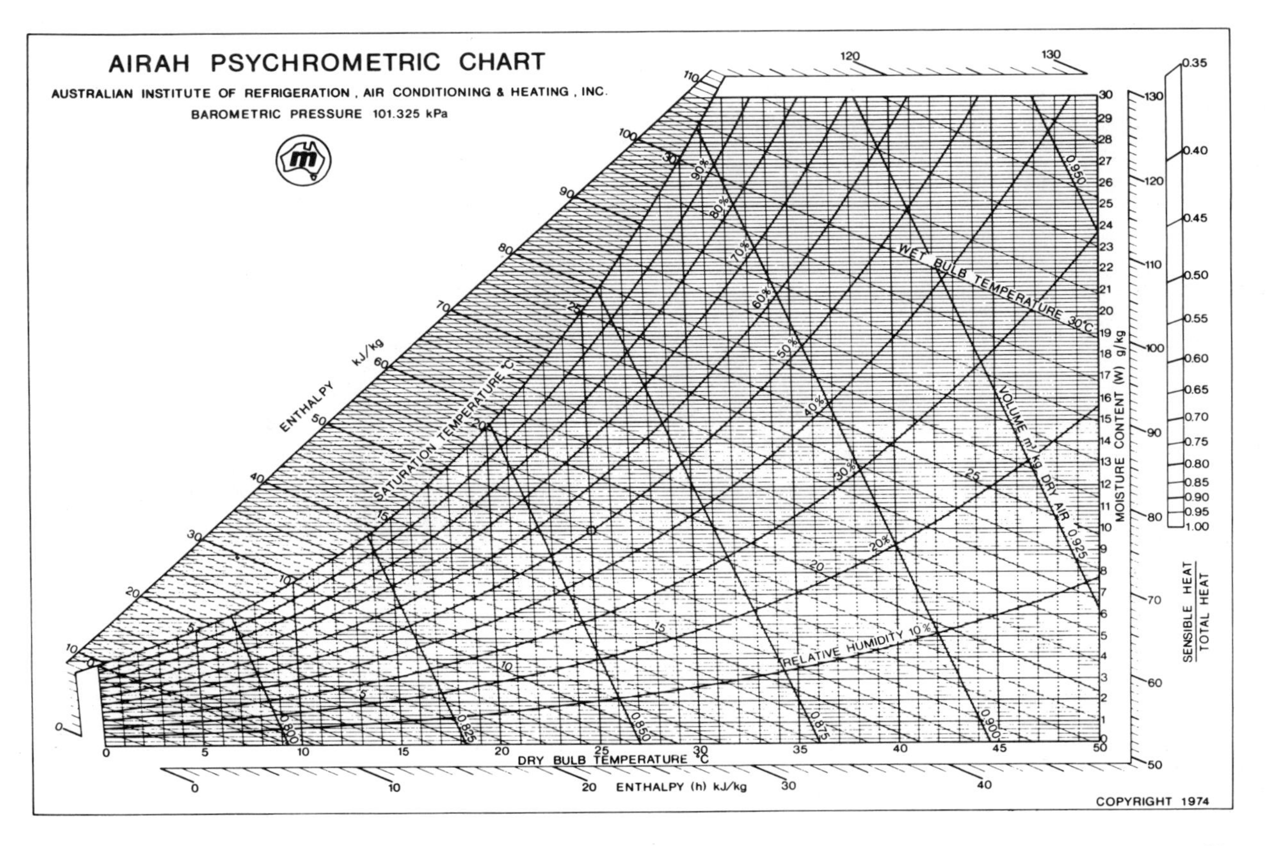

Fig. 4.4.5. The psychrometric chart that combines the lines from Figs. 4.4.1, 4.4.2, 4.4.3, and 4.4.4. (*Reproduced by courtesy of the Australian Institute of Refrigeration, Air Conditioning and Heating.*)

4.5 Air Velocity and Radiation, and Their Measurement

Air velocity may either reduce or increase human comfort. In hot-humid weather air movement provides relief, but in cold weather and in hot, dry weather it can be distressing. The velocity of wind is usually measured in meters per second (m/s) with a *cup anemometer* (from the Greek *anemos*, "wind"), which is mounted on a roof or on a tall mast so as to be clear of the ground, which through friction reduces the wind velocity. A *vane anemometer* is a small instrument for use indoors, employing the same principle (Fig. 4.5.1).

Air velocity can also be measured by a *Pitot tube anemometer*. This has a small hole at the end of a tube facing the air current to measure the dynamic wind pressure, and small holes at the sides of the tube to measure the static wind pressure. A pressure gauge measures the difference between the dynamic and the static wind pressure, and hence the air velocity.

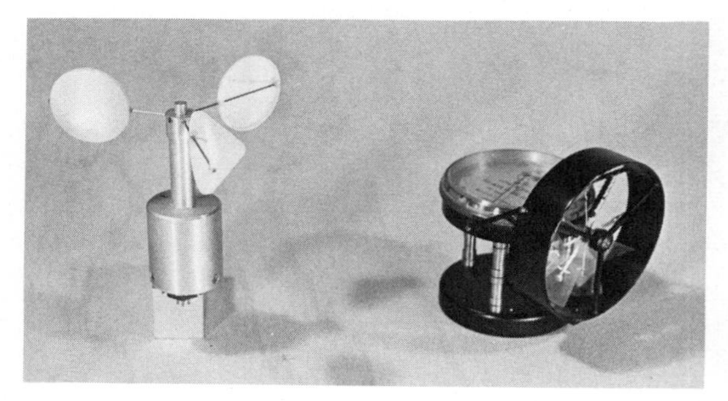

Fig. 4.5.1. Cup anemometer for measuring the wind velocity in the open air, and vane anemometer for measuring the velocity of air indoors. In each case the rotations of the vertical spindle per minute are counted, using a gearwheel, to measure the wind velocity in meters per second (m/s). Cup anemometers frequently incorporate a weather vane to indicate the direction of the wind; this can be included in the graphical record with the wind velocity.

The *kata thermometer* was devised in 1914 by Sir Leonard Hill, one of the pioneers of studies on thermal comfort (Ref. 4.2). It measures the cooling power of air by allowing a thermometer with a large bulb to cool through a standard temperature interval and measuring the time taken. The *hot-wire anemometer* is a more accurate instrument employing a similar principle (Fig. 4.5.2).

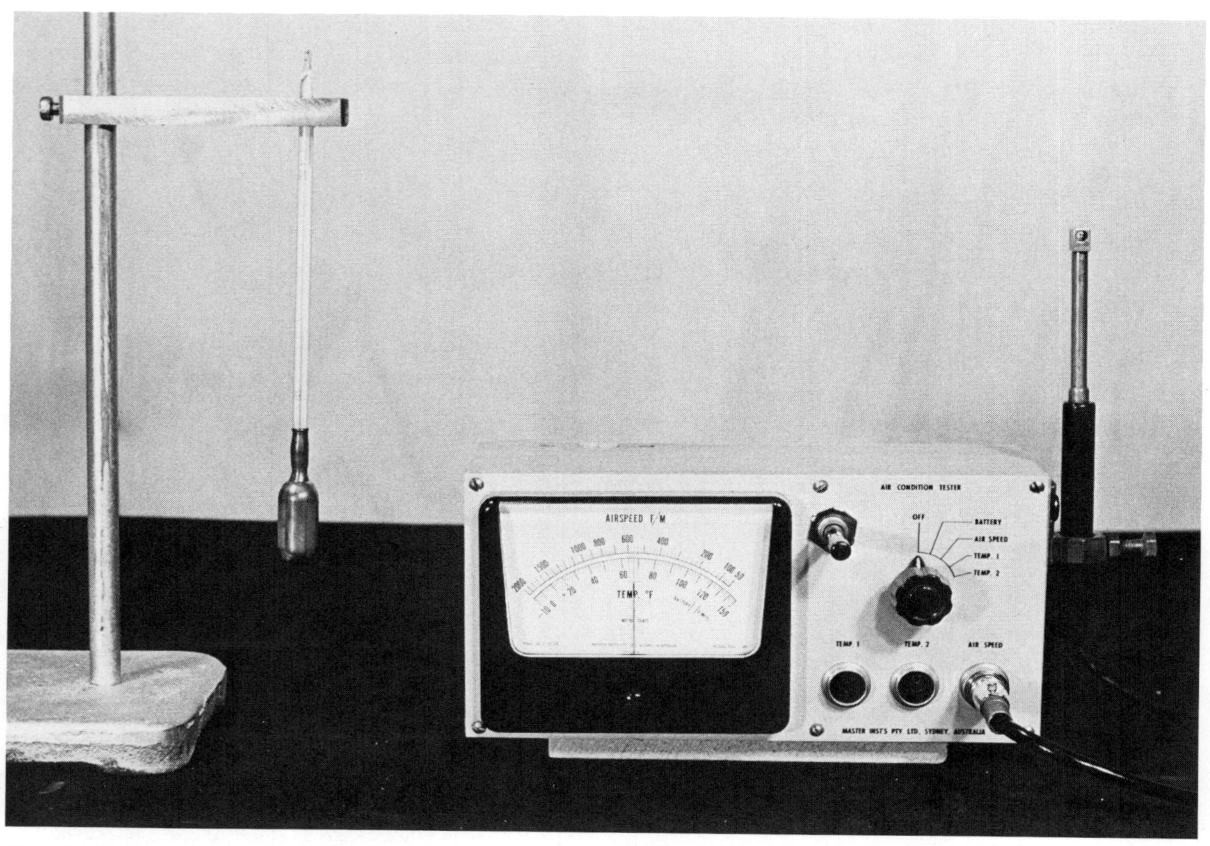

Fig. 4.5.2. Kata thermometer and hot-wire anemometer. In both instruments the cooling power of air is measured by the drop in temperature produced.

The kata thermometer is a thermometer with a large bulb and only two marks. In the one illustrated, the upper mark, 37.8°C, is a little above the temperature of the human body, and the lower mark, 35°C, is a little below body temperature. The kata thermometer is placed in hot water, and the time taken for it to cool in air from 37.8°C to 35°C is measured. The "cooling power" of the air is determined from a chart.

The hot-wire anemometer is a more precise instrument. A wire filament is heated by an electric current, and its temperature is measured with a thermocouple. By comparing the rate of cooling in still air (from a calibration) and the rate of cooling in a current of air, the velocity of the air can be determined.

The *total solar radiation* received at the earth's surface consists of the *direct solar radiation*, received directly from the sun, and the *diffuse solar radiation*. We noted in Section 3.2 that the earth's atmosphere interacts with solar radiation. Part of it is transmitted directly; part of it is absorbed or reflected back into the sky; another part is diffused and eventually transmitted to the earth's surface. Thus the sun provides illumination through diffused sunlight, even when the sun itself is hidden by clouds. Similarly, the sun provides diffused thermal radiation at the earth's surface.

Solar radiation is measured with a *solarimeter*. The simplest type measures the radiant temperature with a thermopile, that is, a number of thermocouples connected together. Photographic exposure meters cannot be employed, because their photovoltaic cells respond only to visible light, not to infrared radiation. Thermopiles measure the temperature due to the total solar radiation, but the diffuse solar radiation can be obtained by eliminating the direct solar radiation with a small sunshade. The direct solar radiation is then the difference between the total and the diffuse radiation.

From the radiant temperature recorded by the thermopile the *irradiance* (or flow or solar energy) can be determined in watts per square meter. Alternatively, the *irradiation* (or total energy over a period of time, such as an hour or a day) can be measured in joules per square meter.

Data are available for some, but not all, weather stations in North America and in other parts of the world from the U.S. Weather Bureau. Both total and diffuse solar radiation are available separately for some of these; for others only the total solar radiation is recorded.

Solar measurements are expensive, and in the absence of data, the total solar irradiation can be estimated from the solar constant [see Section 3.2] and the hours of sunshine, measured with a *sunshine recorder*. This is a simple instrument consisting of a spherical magnifying glass and a paper chart. The sun shining through the magnifying glass burns a hole in the chart. While other recording instruments require a clock mechanism, this is not needed for a sunshine recorder, because of the sun's motion across the sky.

An alternative and more accurate method requires the observation of the type of cloud in accordance with a standard specification (for example, stratus cloud, cirrus cloud), and the amount of cloud cover in tenths. From these data a ratio called the *cloud cover modifier* is determined, by means of a table. The approximate solar radiation actually received on a cloudy day is then taken as the product of the clear sky radiation and the cloud cover modifier.

The radiation from the surfaces of a room varies with the temperature of its surface. Our concern is with the *mean radiant temperature* (MRT) of all the surfaces, not with the irradiance of each individual surface. The MRT can be measured with a globe thermometer (Fig. 4.5.3), and the radiant heat gain or loss from the surroundings determined in watts per square meter.

Fig. 4.5.3. Globe thermometer. A completely nonreflecting surface, that is, a black surface, emits radiation at the maximum possible rate at any given temperature [see Section 4.2]. The globe thermometer consists of a copper globe of 150 mm diameter painted black, containing at the center of the sphere the bulb of an ordinary thermometer or the junction of a thermocouple.

If the temperature of the surroundings is higher than the air temperature, the globe thermometer absorbs long-wave radiation and records a temperature higher than the air temperature. Conversely, if the surroundings are cooler than the air, the globe thermometer records a temperature below air temperature. The globe thermometer is also influenced by air movement. The mean radiant temperature in degrees Celsius and the radiant heat gain or loss in watts per square meter (W/m^2) can be calculated from the temperature of the globe thermometer, the air temperature, and the velocity of the air, by means of a formula or a chart (Ref. 4.2, pp. 39–40).

4.6 The Physiological Basis of Human Comfort

The temperature of the central part of the human body is approximately constant at 37°C, although the temperature of the skin varies from place to place and from time to time. In conditions of thermal comfort the skin temperature is approximately 33°C, but in cold weather the skin temperature of the hands and feet may be much lower.

All the energy produced in the body by the oxidation of food is converted into heat, except that used for performing work. If the body is to maintain its temperature constant, this heat must be dissipated:

$$H = E_d + E_{sw} + E_{re} + L + K \quad (4.4)$$

where H is the heat dissipated by the body, measured, like all subsequent quantities, in watts;
E_d is the heat loss by vapor diffusion through the skin;
E_{sw} is the heat loss due to evaporation of sweat from the skin;
E_{re} is the latent respiration heat loss (see below);
L is the dry respiration heat loss (see below);
K is the heat lost by the body to its surroundings (see below);

Heat and water are absorbed by the inspired air by convection and evaporation from the respiratory tract. Under comfortable conditions the expired air emerging from the nose still contains more heat and water than the inspired air. Breathing therefore results in a latent heat loss, E_{re}, due to the evaporation of water, and a dry respiration/heat loss, L, due to the increase in the temperature of the expired air.

$$K = R + C \quad (4.5)$$

where R is the heat loss by radiation from the outer surface of the nude or clothed body, as the case may be;
C is the heat loss by convection from the nude or clothed body; there may also be a small heat loss or gain due to conduction, for example, when standing barefoot on a metal surface.

Equations (4.4) and (4.5) are discussed in more detail by P.O. Fanger (Ref. 4.3, pp. 19–37), who contributed to the work on the comfort criteria currently used in the United States and later extended it [Section 4.7]. If the heat loss from the body is excessive, we feel uncomfortably cold; if it is insufficient, we feel uncomfortably hot. This book is not concerned with the more severe conditions of heat imbalance, which may result in illness or death.

Evidently, temperature is the main criterion of human comfort. It is difficult to be comfortable at a temperature higher than the central body temperature of 37°C (99°F) even if one wears very light clothing. There is a greater tolerance for a lower temperature because it can be compensated to some extent by heavy clothing and physical activity.

Humidity is next in importance, particularly at high temperatures, because the heat loss from the skin, and to a lesser extent by respiration, greatly depends on it. Sweat evaporates quickly in hot-dry weather, and the latent heat of evaporation cools the skin. Sweat does not evaporate so easily in high humidity.

Air movement causes heat transfer by convection from the skin and the clothing. In hot-humid weather it provides great relief, increasing the evaporation of sweat. However, when the air temperature rises above the body temperature of 37°C in a hot-arid climate, air movement increases the body temperature, and a hot-dry wind therefore creates discomfort. Air movement also increases the effect of a low temperature. This is called the *wind-chill index;* as a result a temperature around the freezing point when a high wind is blowing gives the same feeling of thermal discomfort as a much lower temperature in still air.

Radiation can also cause comfort or discomfort. Sunshine is pleasant on a cool day, but on a hot, clear day the temperature recorded by a thermometer exposed to the sun can be 20°C above the shade temperature that is quoted in meteorological data. Radiation from an open wood or coal fire or from an electric radiator is generally regarded as pleasant, but it heats only one side of the body. If the room has ill-fitting doors and windows that produce drafts in the room, the cold air currents chilling the other side of the body may cause discomfort that can be relieved only by turning one's back to the radiator to heat the other side of the body.

In addition to this hot radiation, the walls, floor, and ceiling or a room, and the human body all produce long-wave radiation [Section 4.2]. The human body gains or loses heat by radiation to or from the room surfaces.

Thus temperature, humidity, air movement, and radiation are the main external factors affecting thermal comfort. Their measurement was discussed in Sections 4.3 and 4.5.

Thermal comfort is greatly affected by clothing and by physical activity. Social and religious conventions impose limitations on the amount of clothing that can be shed in hot weather, and practical considerations impose limits on the protection that can be given to the face, the hands, and the feet in very cold weather. However, clothing can be of enormous help in creating comfortable conditions. Thus a man wearing a short-sleeved, open-necked shirt may be quite comfortable in a room that another man in a coat and tie finds stifling; women's formal clothes are generally better adapted to hot-humid weather. Conversely, a person wearing a sweater may be comfortable in a room that another person without one finds chilly.

The unit for measuring the effect of clothing on thermal comfort is the *clo*. One clo is defined as the insulating value of a man's lightweight business suit (long trousers and coat) and cotton underwear. The relative insulating values of other forms of clothing are listed in the *ASHRAE Handbook* (Ref. 4.1, pp. 8.6 and 8.7). They were determined by thermal measurements on a copper manikin, appropriately clothed, made at Kansas State University in 1973. Some values for men's clothing are given in Table 4.1

In interpreting data on thermal comfort it is important to note the type of clothing worn by participants and to bear in mind that "normal clothing" at the beginning of this century provided more insulation than "normal clothing" worn today.

Table 4.1

Insulating Effect of Men's Clothing

Type of Clothing	clo
Brief swimsuit	0.05
Shorts only	0.1
Shorts and short-sleeved shirt	0.2
Shorts, briefs, short-sleeved shirt, socks and shoes	0.3
Same, but long trousers	0.5
Same, plus singlet	0.6
Lightweight suit, cotton underwear	1.0
Three-piece winter suit, cotton underwear, wool socks	1.5
Same, plus overcoat	2.0–2.5
Heavy clothing designed for outdoor use beyond the Arctic and Antarctic Circles	4.0

Table 4.2

Approximate Metabolic Rate for Various Activities

Activity	Metabolic Rate	
	W/m^2	Met
Sleeping	41	0.7
Resting, lying down	47	0.8
Sitting and not working	58	1.0
Standing	70	1.2
Typing	80	1.4
Light manual work	120	2.0
Cleaning a house	180	3.0
Heavy work	270	4.5
Walking on level ground at 3 km/h	120	2.0
Walking on level ground at 6 km/h	200	3.5
Walking up a 15° slope at 3 km/h	270	4.5

The type of activity is also important. A person wearing a swimsuit lying on a beach may be very comfortable in weather that is most uncomfortable for a person doing heavy work.

The metabolic rate of the body, that is, the rate at which it produces energy, increases with the body's activity. It can be measured in watts per square meter or in *mets,* where 1 met is the metabolic rate of a person sitting, inactive, in a room with still air; it is approximately 58 W/m^2. Table 4.2 shows metabolic rates for some activities.

The metabolic rate varies with the surface area of a person's skin. This is usually determined by an empirical formula derived by D. and E.F. DuBois (Ref. 4.11). A man whose mass is 75 kg and who is 1.8 m tall has a DuBois surface area of 1.85 m^2 (Ref. 5.23, p. 6), so that 1 met corresponds for him to $58 \times 1.85 = 107$ W.

4.7 Criteria of Thermal Comfort

The effect of temperature on human comfort was already mentioned by Vitruvius in the first century B.C. Leonardo da Vinci recognized the influence of humidity in the fifteenth century. John Arbuthnot, a friend of the satirist Jonathan Swift, published in 1733 *An Essay Concerning the Effect of Air Movement on Human Bodies*. Thomas Tredgold, one of the pioneers in the use of iron structures, pointed out in his 1824 book *Principles of Warming and Ventilating Public Buildings, Dwelling Houses, Etc.* that people could be comfortable from the radiant heat of an open fire even when the air temperature was too low for comfort.

However, the various components of thermal comfort were not assembled into a single criterion until 1923, when C.P. Yaglou (Ref. 4.4) published the results of his experiments at the Pittsburgh research laboratory of the American Society of Heating and Ventilating Engineers (now renamed ASHRAE; see Ref. 4.1). A number of people were placed in a room where the temperature, humidity, and air movement could be controlled and measured. They were asked to nominate the conditions of temperature, humidity, and air movement that gave the same sensation of warmth while they were engaged in light physical activity and wore light clothing; the clothing worn in these experiments gave rise to the unit of 1 clo (Table 4.1). Those combinations of temperature, humidity, and air movement that gave equal sensations of comfort were then designated as having the same *effective temperature* (ET); the effective temperature is the dry-bulb temperature in that group of equal comfort sensations at which the relative humidity is 100% and the air movement is zero. In later observations the standard relative humidity was shifted to 50%.

The experiments determining the effective temperature have been progressively refined. The current standard is based on experiments carried out under the auspices of ASHRAE at Kansas State University and published in 1973 (Ref. 4.5). These experiments were conducted on college students wearing clothing equivalent to 0.6 clo and sitting in the test room without doing any work. The mean radiant temperature (MRT) was equal to the dry-bulb temperature, and the air velocity was low, less than 0.17 m/s. The main variables were the temperature and the humidity. This comfort zone can therefore be traced on the psychrometric chart (full line in Fig. 4.7.1). The recommendation of the ASHRAE Comfort Standard of 1974 (Ref. 4.6) is a little different (dashed line in Fig. 4.7.1); it is limited by the relative humidities of 60% and 20% and by the dry-bulb temperatures of 73°F and 77°F (22.8°C and 25°C). In 1975 as an energy conservation measure higher design temperatures were recommended for summer, and lower ones for winter (72°F, or 22°C); these are shown in Fig. 4.7.1 by dots surrounded by circles.

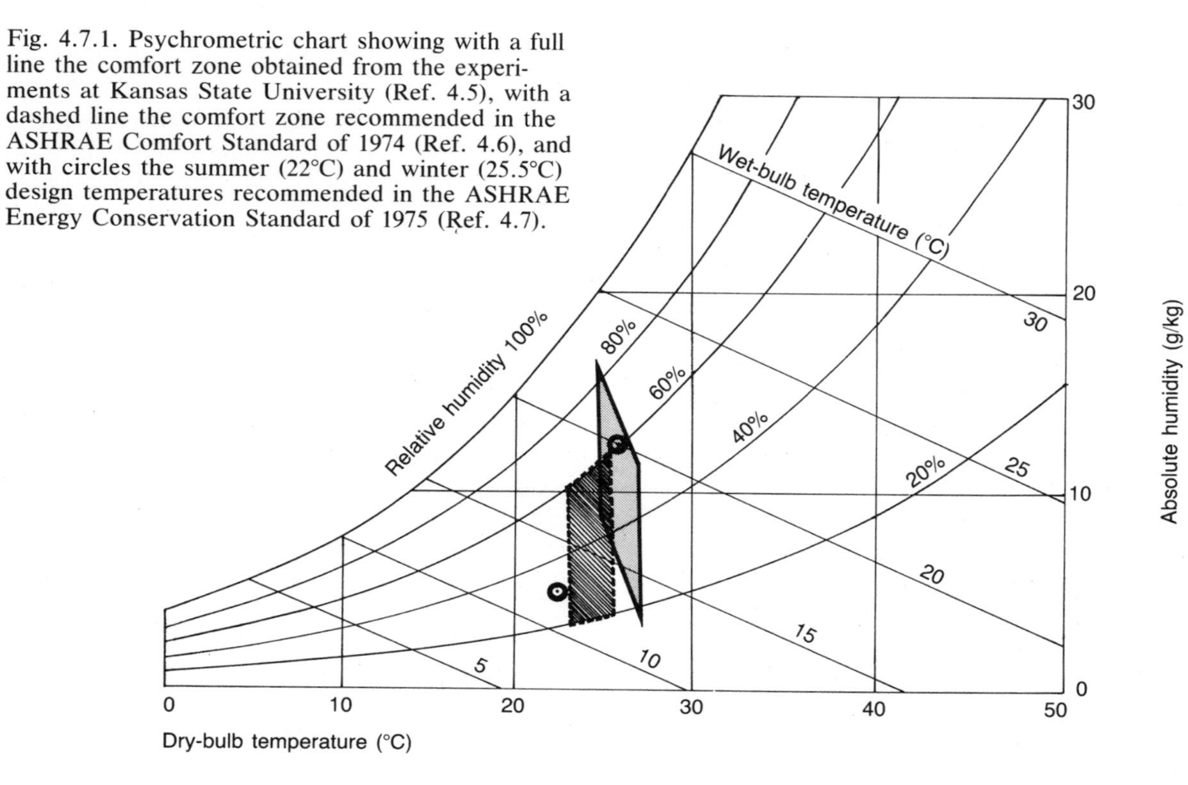

Fig. 4.7.1. Psychrometric chart showing with a full line the comfort zone obtained from the experiments at Kansas State University (Ref. 4.5), with a dashed line the comfort zone recommended in the ASHRAE Comfort Standard of 1974 (Ref. 4.6), and with circles the summer (22°C) and winter (25.5°C) design temperatures recommended in the ASHRAE Energy Conservation Standard of 1975 (Ref. 4.7).

The winter temperature is still considerably higher than that considered appropriate in England *before* the energy crisis of 1973. We mentioned in Section 1.4 the preferred temperature for the British House of Commons in 1840. When this was rebuilt after World War II, Dr. Thomas Bedford conducted an inquiry in the late 1940s and recommended a temperature of 67°F to 68°F (19.5°C to 20°C), which was found to be satisfactory. This was higher than the temperature recommended by Dr. Reid a century earlier, but it would have been too low for members of the U.S. House of Representatives. Differences in comfort votes are due to differences in the amount of clothing customarily worn, and this varies with time and place; but they also reflect different conceptions of comfort. A survey of people of both European and of Asian descent conducted in Singapore in 1952 (Ref. 4.8) indicated a preferred temperature of 27°C at a relative humidity of 80% (a high percentage) and an air velocity of 0.4 m/s. While the ASHRAE recommendations provide the best data available for the design of air-conditioned buildings in North America at present, they need to be modified for people who are used to higher summer or lower winter temperatures; this also conserves energy.

The concept of effective temperature and similar concepts elsewhere (equivalent temperature in Britain [Ref. 4.9] and resultant temperature in France [Ref. 4.10]) are based on experiments designed to determine the most comfortable conditions attainable, irrespective of the climatic conditions outdoors. As we pointed out in Section 1.4, it no longer seems realistic to assume that people will be able, or indeed will wish, to spend their entire lives in air-conditioned buildings and commute between them in air-conditioned cars. If allowance is made for a frequent interchange between the indoor and the outdoor environment, greater comfort can be achieved by wearing light clothes in summer, but much heavier clothes in winter, and adjusting the indoor design temperatures accordingly. The energy conserved thereby is an additional inducement.

It is open to question whether air conditioning is invariably the most desirable option, even when the client can afford the cost and the energy is freely available. The uniform conditions created by air conditioning lack the stimulus of the diurnal variations in a natural environment, which is particularly desirable in a residential building. In many parts of the world it is possible to create thermal comfort with only limited use of building services by designing the fabric of the building to make the best use of the climate. That is discussed in the next chapter. The design of heating and air conditioning plant is considered in Chapter 6.

References

4.1 *ASHRAE Handbook—1977, Fundamentals.* American Society of Heating, Refrigerating, and Air Conditioning Engineers, New York, 1977. 760 pp.

4.2 THOMAS BEDFORD: *Basic Principles of Ventilation and Heating.* H.K. Lewis, London, 1964. 438 pp.

4.3 P.O. FANGER: *Thermal Comfort.* Danish Technical Press, Copenhagen, 1970. 244 pp.

4.4 F.C. HOUGHTEN and C.P. YAGLOU: Determination of the comfort zone. *Trans. Amer. Soc. Heating and Ventilating Engineers,* Vol. 29 (1923), p. 361.

4.5 F.H. ROHLES JR.: The revised modal comfort envelope. *ASHRAE Transactions,* Vol. 79, Part II (1973), p. 52.

4.6 *Thermal Environment Conditions for Human Occupancy.* ASHRAE Standard 55–74. American Society of Heating, Refrigerating, and Air Conditioning Engineers, New York, 1974. 12 pp.

4.7 *Energy Conservation in New Building Design.* ASHRAE Standard 90–75. American Society of Heating, Refrigerating, and Air Conditioning Engineers, New York, 1975. 53 pp.

4.8 F.P. ELLIS: Thermal comfort in warm and humid atmospheres—observations on groups and individuals in Singapore. *J. Hygiene,* Vol. 51 (1953), pp. 386–404.

4.9 A.F. DUFTON: The equivalent temperature of a room and its measurement. *Building Research Technical Paper No. 13.* H.M. Stationery Office, London, 1932. 8 pp.

4.10 A. MISSENARD: Équivalence thermique des ambiences; équivalences de sejour. *Chaleur et Industrie,* Vol. 29 (1948), No. 276 (July), pp. 159–72, and No. 277 (August), pp. 189–98.

4.11 D. DuBOIS and E.F. DuBOIS: The measurement of the surface area of man. *Archs. Internal Medicine,* Vol. 15 (1915), pp. 868–81.

Suggestions for Further Reading

References 3.2, 3.4, 4.1 (Chapter 8), and 4.3.

4.12 N.S. BILLINGTON: *Thermal Properties of Buildings.* Cleaver Hume Press, London, 1952. 208 pp.

4.13 J.F. van STRAATEN: *Thermal Performance of Buildings.* Elsevier, London, 1967. 311 pp.

4.14 L.H. NEWBURGH (Ed.): *Physiology of Heat Regulation and the Science of Clothing.* Hafner, New York, 1968. 457 pp.

4.15 S.V. SZOKOLAY: *Environmental Science Handbook.* Construction Press, Lancaster, 1980. 532 pp.

4.16 R.K. MACPHERSON: What makes people accept a thermal environment as comfortable? *in* H.J. COWAN (Ed.): *Solar Energy Applications in the Design of Buildings.* Applied Science, London, 1980, pp. 13–32.

Chapter 5 Climate and Buildings

Climates are divided in this chapter into six categories. For small houses a suitable building type can be selected appropriate to the climatic zone, but a more accurate result is obtained by using actual climatic data. These must be modified to conform to the microclimate; this depends mainly on wind effects due to the topography of the site.

The thermal performance of buildings can be improved in most climates by good insulation, thermal inertia, appropriate color of the exterior surfaces, and correctly designed windows. In cool climates special care is needed to prevent condensation within the fabric of the building, and in hot-arid climates evaporative cooling is helpful. In hot-humid climates natural ventilation is of particular importance, and high thermal inertia and high insulation are a disadvantage, if air conditioning is not used.

Some applications of passive solar energy to the winter heating and summer cooling of buildings are described, and also the advantages and disadvantages of underground buildings. Finally, the significance of life cycle costing is explained.

5.1 Climatic Zones

For the purpose of architectural design, climates can be divided into six categories:

1. Very cold. People live there only to exploit minerals or other natural resources, or to staff installations for research, communications, or defense. It may be necessary to interconnect buildings with underground passages so that the occupants need not go outdoors unnecessarily. There are foundation problems due to permafrost, and structural problems due to high snow and wind loads.

2. Cold. Although brief periods of very hot weather may occur, summers are cool, and the design of the building should concentrate on providing warmth for winter. Some overheating may occur during the summer because of sunshine on walls facing west. The cost of heating can be greatly reduced by providing adequate insulation [Section 5.7], and passive solar devices may be economical [Sections 5.14 to 5.17].

3. Temperate. Some regions where the weather is a little too warm for comfort in midsummer and a little too cold for comfort in midwinter have truly temperate climates. In others classed as temperate, the weather is much too warm (and generally too humid) for part of the summer, and much too cold for part of the winter. However, neither period lasts longer than a few weeks. Winter heating or passive solar devices are essential, and good insulation should be provided [Section 5.7]. Sunshading to exclude summer sun and admit winter sun is helpful [Sections 3.4 and 3.5]. If the summer weather is both hot and humid, good ventilation or air conditioning is essential during the summer months.

4. Hot-humid. The coolest weather is not too cold for comfort, so that the design of the building can concentrate on providing relief during the hottest weather. Sunshading, light construction with low thermal inertia, and good ventilation [Sections 3.4, 3.5, 5.6, and 5.11] can often provide an acceptable environment without air conditioning. Although the relative humidity frequently exceeds 90%, the temperature rarely rises above 33°C. At the present time solar air conditioning is not economical [Section 6.7].

5. Hot-arid. Some heating may be necessary in winter, but summer cooling is more important. The interior environment can be greatly improved by sunshading that allows the sun to enter in winter [Sections 3.4 and 3.5], by insulation [Section 5.7], and by using building materials with high thermal inertia [Section 5.6]. Evaporative

Table 5.1

Amount of Heating Required in Various Cities, in Degree Days

City	Degree Days (Based on 18°C)	City	Degree Days (Based on 18°C)
United States		*European Continent*	
Miami	103	Rome	1317
Los Angeles	773	Paris	2892
Tucson	1000	Vienna	3148
Dallas	1315	Hamburg	3467
San Francisco	1743	Zurich	3502
Albuquerque	2230	Copenhagen	3836
Washington	2543	Stockholm	4546
New York	2943	Moscow	5315
Chicago	3490	*South Africa*	
Milwaukee	3933	Johannesburg	967
Minneapolis	4426	Bloemfontain	1222
Duluth	5402	Sutherland	2206
Canada		*Australia*	
Vancouver	3096	Darwin	0
Montreal	4666	Sydney	986
Winnipeg	6100	Melbourne	2440
Churchill	9527	Canberra	2760
Great Britain		*New Zealand*	
London	2947	Auckland	1171
Manchester	3049	Wellington	2050
Glasgow	3471	Invercargill	2983
Aberdeen	3745		

Degree days for all major cities in North America are given in the *ASHRAE Handbook—1976 Systems,* Chapter 43 (Ref. 5.1, Table 1, pp. 43.2–43.7).

cooling [Section 5.12] and passive solar energy can be used to advantage, and it may be possible thereby to avoid using other energy sources for heating and/or cooling [Section 5.16].

6. *Very-hot-arid*. In some parts of the tropics shade temperatures exceed 45°C. Few people live there unless their work obliges them to do so. It is impossible to create a comfortable interior environment without an air conditioning plant.

Even when air conditioning is used, winter heating is likely to be the major source of energy consumption for thermal comfort. The heat input required can be compared by calculating the difference between the mean daily temperature and a reference temperature, usually taken as 18°C (or 65°F), and adding the positive numbers for the entire year. This gives a measure of the amount of heating required during the year, in *degree days*. The number of degree days for various cities is listed in Table 5.1 and plotted in Fig. 5.1.1.

The number of degree days influences the pattern of energy consumption in industrialized countries. In regions with degree days (based on 18°C) above 3000, space heating uses more energy than any other single source, such as automobiles or industrial processes, and in consequence

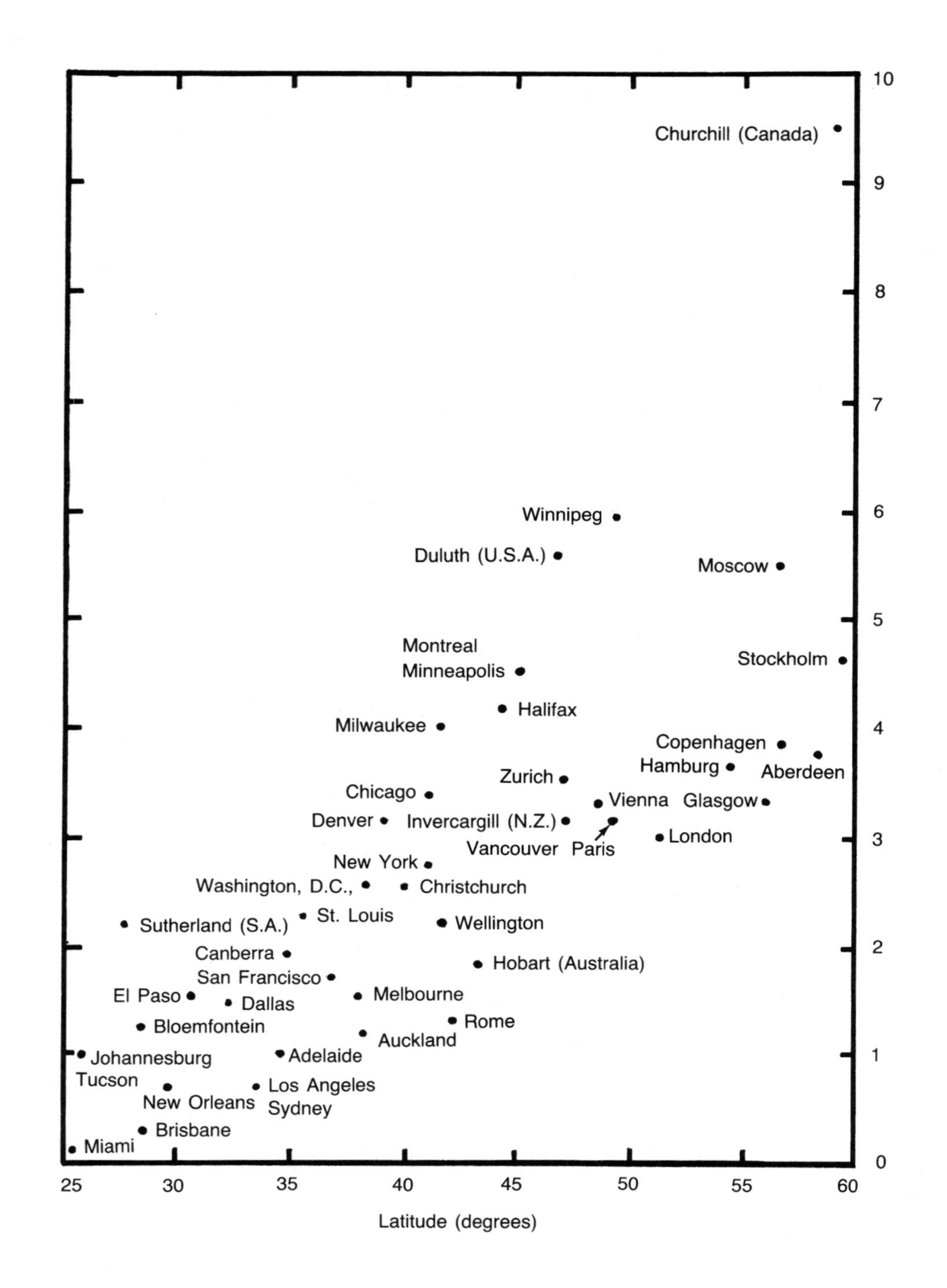

Fig. 5.1.1. Relationship between latitude and the amount of heating required, expressed in degree days based on 18°C, for the cities listed in Table 5.1, and a few others. All points fall within a strip whose center line is proportional to the latitude. The cities near the bottom of this strip, such as Brisbane, Los Angeles, Rome, and Glasgow, have a maritime climate. The cities near the top of the strip, such as Johannesburg, Denver, and Winnipeg, have a continental climate.

buildings consume at least 40% of the total energy used in the region. In areas with less than 1000 degree days, the energy used by buildings is less than a quarter of the total, and in the southern United States, Australia, and South Africa the supply of hot water consumes more energy than space heating.

The fabric of the buildings should be designed accordingly. Insulation can save a great deal of energy and money in climates that require a significant amount of space heating, but the expenditure is not worthwhile in warm climates [Section 5.7], where the lowest temperature at night is only a few degrees below the daytime maximum temperature.

5.2 Climatic Data Used in the Design of Buildings

For small buildings it is sufficient to select a design appropriate to the climatic zone. For more important buildings, actual weather data for the city or region are required.

Dry-bulb temperature, wet-bulb temperature, absolute and/or relative humidity, barometric pressure, wind velocity, and wind direction are available for a large number of weather stations [Sections 4.3 and 4.5]. However, solar radiation measurements are recorded by only a few of them. This information is of particular value in estimating heating and cooling loads, and it may be necessay to estimate the amount of direct and diffuse solar radiation from the recorded hours of sunshine or, more accurately, from observations of the amount of cloud cover and the type of clouds [Section 4.5].

Precise predictions of the heating and cooling loads for large buildings are to an increasing extent based on computer programs, and weather data on magnetic tape are now available from the National Climatic Center in Asheville, N.C., for associated weather stations (Ref. 4.2). The tapes contain hourly records for each day of one year, that is, $24 \times 365 = 8760$ sets of data for each type of observation.

Alternatively, weather data for one typical day of each month, or even one typical day for each period of the three months, can be used in the calculations, although the decision on what is "typical" involves some judgment and ideally requires a knowledge of the local weather.

5.3 Microclimate

The climate in a particular location may differ radically from the average climate of the region. The microclimate can be determined with certainty only by prolonged observations on the building site, which is rarely practicable. However, some conclusions can be drawn from the topography.

The difference between the microclimate of the site and the macroclimate of the region is mainly due to wind effects. Near large bodies of water, air movement results from the temperature difference between the surface of the water and the land. The land surface warms up more than the water when it is hot and vice versa. This produces a prevailing sea breeze in summer, particularly in the evening. In many islands and coastal regions in the tropics and subtropics this cooling breeze occurs regularly, and the cost of air conditioning can often be avoided if houses are oriented to catch it for cooling the bedroom [Section 5.11].

In the valleys of hilly regions prevailing winds are caused by the movement of air uphill during the day and downhill at night. This downward movement can occur even in small hollows in otherwise level ground; when the minimum nighttime temperature of the region, that is, its macroclimate, is near but above freezing, there may be frost in the hollow (Fig. 5.3.1) as a result of a pool of cold air flowing down the slope and collecting at the lowest point.

Human habitation and vegetation modify the local climate. In winter the exhaust from heated buildings increases the outside temperature, which is helpful. However, air conditioning makes the microclimate of streets less comfortable, particularly in a warm-humid climate, because the exhaust increases both the temperature and the humidity.

Buildings and automobiles produce air pollution. This reduces the solar radiation received at ground level, although stricter emission controls have

Fig. 5.3.1. A hollow in the ground collects cool air. When the macroclimate is near but above the freezing point, the temperature in the hollow may fall below 0°C, with consequent damage to vegetation or exposed water pipes.

Fig. 5.3.2. Wind tunnel study of the interaction of a proposed new building and the existing buildings that surround it.

reduced this effect in recent years. Pollution is unpleasant, but it can save energy by reducing daytime heat and radiant cooling at night. Generally the microclimate near the city center is more temperate in the winter but less agreeable in the summer.

Buildings deflect wind. The effect is difficult to calculate by theory because of the complex interaction of the building surfaces and the air currents; however, these can be predicted by model tests in a boundary-layer wind tunnel (Fig. 5.3.2). Such tests should always be made before major alterations are made to the urban landscape. Large buildings, openings that act as wind funnels, curved surfaces, and fountains particularly need to be investigated. High winds can ruin otherwise pleasant thoroughfares and outdoor spaces, for example by blowing off people's hats or by causing fountains to drench persons many meters distant. These effects can generally be predicted by a model study in a wind tunnel; they are often difficult to correct after the buildings have been erected.

Trees and other plants provide evaporative cooling in summer [see also Section 5.12], although their psychological effect is perhaps more im-

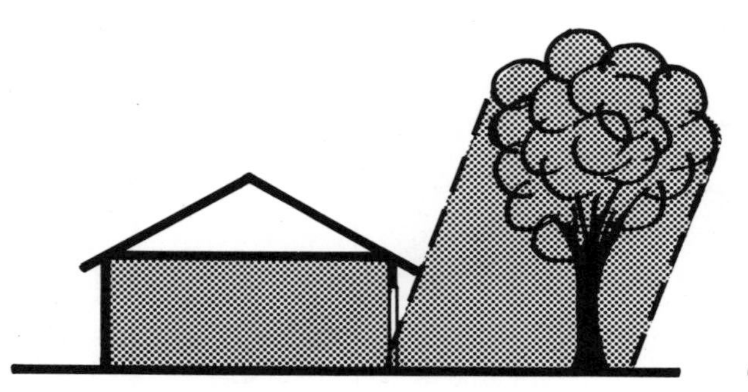

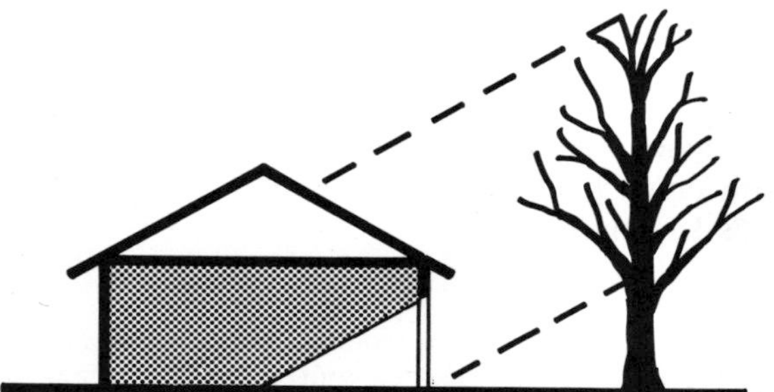

Fig. 5.3.3. Deciduous trees can provide shade in summer, while permitting sunlight penetration in winter.

portant than the reduction in temperature. The deciduous trees that are native to Europe and North America provide shading in summer but allow sunlight penetration in winter, when they have no leaves (Fig. 5.3.3). These trees substantially aid the design of sunshading, and they can be introduced for this purpose in countries where all native species are evergreen. In Australia this was done in the nineteenth century, with good results; but most architects and landscape architects today prefer evergreen Australian trees and argue that exotic trees do not belong in the Australian landscape.

Trees to be planted near buildings should be carefully selected, because the roots of certain trees seriously damage the foundations if they are planted too close to a building; other trees shed heavy branches, which can damage roofs and injure people.

Trees and grass generally increase the humidity, and they reduce the runoff after a heavy fall of rain.

5.4 Orientation of Buildings and of Rooms Within Buildings

We noted in Section 1.1 that since the time of Vitruvius architectural textbooks have suggested the optimal orientation for various rooms in residential buildings. Many of these rules are still valid. Bedrooms facing east receive the morning sun in both the Northern and Southern Hemispheres, and this is pleasant in most climatic zones. In all but the coolest climates the western sun is disagreeably hot in summer, and external sunshades are generally not effective because the sun's altitude is too low in the west (Fig. 5.4.1). This can be a particular problem for kitchens. We noted in Section 3.4 that rooms facing south in the Northern Hemisphere (or north in the Southern Hemisphere) can be designed by suitable sunshading to receive the winter sun but be protected from the summer sun. This is a good orientation for living rooms.

We noted in Section 3.4 and Fig. 3.4.1 that the most favorable orientation for a multistory building, whether a residential or a commercial building, is for the long walls to face north and south, with suitable shading on the sunny side; the short walls should face east and west, preferably without windows on the western facade. This reduces the heating and cooling loads to a minimum. In developing countries where cooling, and sometimes heating, are luxuries available only to a few, this orientation greatly improves the thermal environment, even though it does not save any energy if none is used (Fig. 5.4.2).

It is frequently not possible to employ the most favorable orientation, because of an existing street pattern that was designed without reference to climate. There may also be considerations more important than climate, for example, a spectacular view.

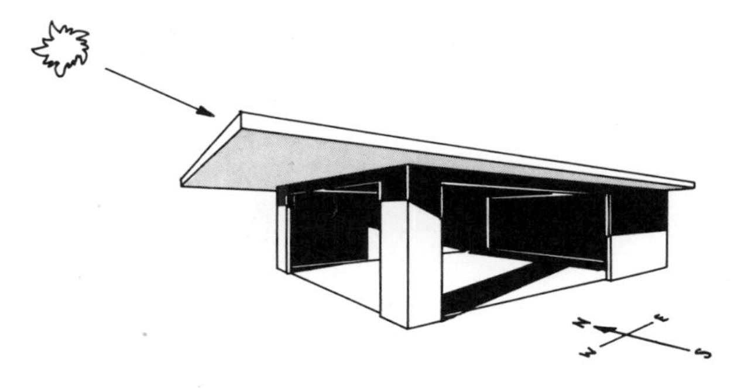

Fig. 5.4.1. Horizontal sunshades are generally not effective against the western sun, because its altitude is too low in the west. The solar radiation is approximately the same from the eastern morning and western evening sun, but in the early morning it is pleasant, whereas in the heat of a late summer afternoon it may be unwelcome.

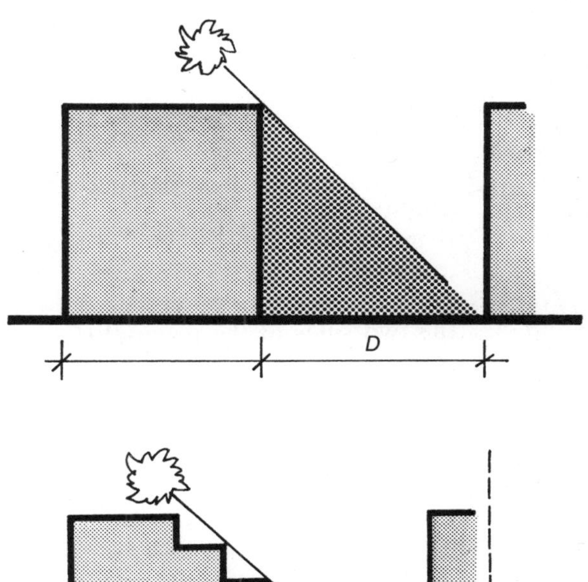

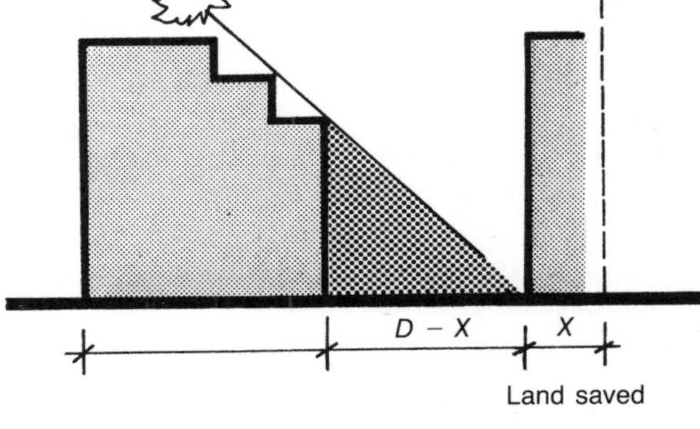

Fig. 5.4.2. High-density housing estate in Shanghai, China, with the longer facades oriented north and south. The spacing between the buildings is arranged so that the entire southern face of the building receives the winter sun, as the dwellings are not heated in winter. If the northern facade is set back, the buildings can be placed closer together, which saves land in an inner city area with a very high population density. (*From* Architectural Science Review, *Vol. 24 pp. 85–89*).

5.5 Thermal Transmittance of Walls, Floors, and Roofs

We noted in Section 4.2 that the quantity of heat conducted through a piece of material with two parallel boundaries, such as a concrete slab or a sheet of particle board, is

$$Q = \frac{k(T_1 - T_2)}{d} \tag{5.1}$$

where Q is the quantity of heat passing through a unit area (W/m^2);
k is the thermal conductivity of the material (W/mK);
$T_1 - T_2$ is the temperature difference between the two sides of the material (K);
d is the thickness of material (m).

We further noted in Section 4.2 that the quantity of heat transferred by convection from the surface of a material is

$$Q = h(T_1 - T_2) \tag{5.2}$$

where Q is the quantity of heat transferred by convection to or from a unit area (W/m^2);
h is the boundary layer heat transfer coefficient (W/m^2K);
$T_1 - T_2$ is the temperature difference (K).

For the purpose of determining the heat transfer through a wall, roof, or floor, it is convenient to divide it into the layers of the different materials of which it is composed, and the boundary layers on each side.

If Q is the quantity of heat passing through a unit area of the wall, roof, or floor (W/m^2), and the temperatures on the two faces are T_o and T_i, then the thermal transmittance is

$$U = \frac{Q}{T_o - T_i} \text{ watts per square meter per degree Kelvin } (W/m^2K) \tag{5.3}$$

Examining the individual layers (Fig. 5.5.1) when the temperature on either side is steady gives us

$$Q = h_o(T_o - T_1) \tag{5.4}$$
$$= \frac{k_1}{d_1}(T_1 - T_2) \tag{5.5}$$
$$= \frac{k_2}{d_2}(T_2 - T_3) \tag{5.6}$$
$$= \frac{k_3}{d_3}(T_3 - T_4) \tag{5.7}$$
$$= h_i(T_4 - T_i) \tag{5.8}$$

where h_o and h_i are the boundary layer heat transfer coefficients (in W/m^2K) outside and inside the wall, or above and below the roof or floor; k_1, k_2, and k_3 are the thermal conductivities of the materials in layers 1, 2, and 3 (in W/mK); and d_1, d_2, and d_3 are the thicknesses of layers 1, 2, and 3 (in meters). Combining these equations, we have

$$\frac{1}{U} = \frac{1}{h_o} + \frac{d_1}{k_1} + \frac{d_2}{k_2} + \frac{d_3}{k_3} + \frac{1}{h_i} \tag{5.9}$$

where U is the *thermal transmittance* of the composite wall, roof, or floor. This equation is sometimes expressed in terms of the *thermal resistance,* R, which is the reciprocal of the thermal transmittance:

$$R = \frac{1}{U} \tag{5.10}$$

The units of the thermal resistance are m^2K/W.

The thermal conductivities of common building materials are available in tables in various handbooks and textbooks (for example, Ref. 4.1, Chapter 22, and Refs. 5.2, 5.3, 5.4, and 5.5). Some are reproduced in Table 5.2.

The boundary layer heat transfer coefficient depends on the temperature difference between the surface concerned and the air in contact with it, and the velocity of the air, which in turn determines whether the air flow is laminar or turbulent. The theory is complex, but values of

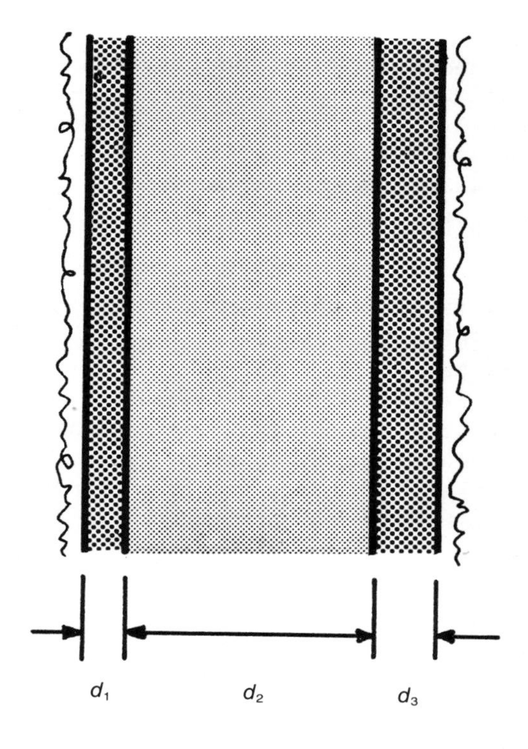

Fig. 5.5.1. Wall or roof consisting of three different materials, whose thicknesses are d_1, d_2, and d_3 respectively.

h for various inclinations of surface, various surface emissivities, and various wind velocities are listed in the *ASHRAE Handbook—Fundamentals* (Ref. 4.1, Table 1, pp. 22.11–22.12).

For nonreflective surfaces the value of h ranges from 6 to 9 W/m^{2}K in still air. It is much larger in a high wind; for example, for a wind velocity of 10 m/s the value of h for a nonreflective wall surface is approximately 50 W/m^{2}K. However, for these high values the effect of the term $1/h$ becomes insignificant in relation to the terms d/k, so that it has little effect on the value of U in Eq. (5.9).

The thermal transmittances (U-value) or thermal resistances (R-value) of standard methods of building construction, for example, timber frame walls with gypsum wallboard, or built-up roofing, are given in many textbooks and handbooks (for example, Ref. 4.1, Table 4A–4K, pp. 22.18–22.22). In addition, manufacturers of cladding and roofing systems normally supply the U-values or R-values of their products, determined by independent testing laboratories.

Example 5.1 *Determine the thermal resistance* (R-*value) and the thermal transmittance* (U-*value) of a concrete wall 150 mm thick.*

Solution *The thermal conductivity* (k-*value) of concrete is given in Table 5.2; k = 1.20. We also require to know the heat transfer coefficients for the inside and the outside of the wall, which are given in Table 1 of Chapter 22 of the* ASHRAE Handbook—Fundamentals *(Ref. 4.1, Table 1, p. 22.11). We will take the coefficient for the outside of the wall to be* $h_o = 20$ *W/m^{2}K and the coefficient for the inside of the wall to be* $h_i = 8$ *W/m^{2}K.*

From Eq. (5.9) we obtain the thermal resistance,

$$R = \frac{1}{U} = \frac{1}{20} + \frac{0.15}{1.2} + \frac{1}{8} = 0.30\ m^2K/W$$

and the thermal transmittance, $U = 3.3$ *W/m^{2}K.*

Example 5.2 *Determine the thermal resistance and the thermal transmittance of a 200-mm concrete wall.*

Solution *The thermal resistance is*

$$R = \frac{1}{U} = \frac{2}{20} + \frac{0.2}{1.2} + \frac{1}{8} = 0.34\ m^2K/W$$

and the thermal transmittance is $U = 2.9$ *W/m^{2}K.*

This is a very small improvement for 33% increase in the wall thickness.

See Section 11.4 for a comparison of thermal and sound insulation.

Example 5.3 *Determine the thermal resistance and the thermal transmittance of a wall consisting of 150 mm of concrete and 50 mm of mineral wool.*

Solution *The thermal conductivity of mineral wool, from Table 5.2, is 0.039 W/mK. The thermal resistance of the entire wall is*

$$R = \frac{1}{U} = \frac{1}{20} + \frac{0.15}{1.2} + \frac{0.05}{0.039} + \frac{1}{8} = 1.582\ m^2K/W$$

and the thermal transmittance is $U = 0.63$ *W/m^{2}K. This is a very substantial improvement;*

Table 5.2

Thermal Properties of Selected Building Materials

Material	Thermal Conductivity k (W/mK)	Density ρ (kg/m^3)	Specific Heat per Unit Mass c_p (J/kgK)	Heat Capacity per Unit Volume ρc_p (MJ/m^{3}K)	Diffusivity $\alpha = \frac{k}{\rho c_p}$ (mm^2/s)	$k \rho c_p$[a]
Copper	386	8900	390	3.48	110	1340
Steel	43	7800	470	3.67	12	158
Granite	2.80	2650	900	2.39	1.17	6.7
Limestone	1.50	2200	860	1.89	0.79	2.83
Concrete	1.20	2100	900	1.89	0.63	2.86
Brick	0.80	1800	840	1.51	0.52	1.21
Sand	0.40	1500	800	1.20	0.33	0.48
Gypsum plaster	0.10	1300	840	1.09	0.092	0.109
Water	0.60	1000	4180	4.18	0.144	2.51
Timber (pine)	0.14	650	2300	1.50	0.093	0.210
Fiber insulating board	0.048	380	1500	0.57	0.084	0.027
Glass or mineral wool	0.039	80	700	0.14	0.279	0.005

[a]This product has not been given a name. Its units are N^2/mm^{2}sK^2.
(From Ref. 5.2, pp. 136–38; Ref. 5.3, pp. 9 and 32; Ref. 5.4, pp. IV, 23–24; and Ref. 5.5, pp. 213–15)

a thin layer of lightweight insulation is far more effective than a thicker layer of the heavier material that is needed to build a wall.

In accordance with the static calculations used in this example, it does not matter whether the insulation is on the inside or the outside face of the wall, or whether it is in a 50-mm cavity between two 75 mm thicknesses of concrete:

$$R = \frac{1}{20} + \frac{0.075}{1.2} + \frac{0.05}{0.039} + \frac{0.075}{1.2} + \frac{1}{8}$$
$$= 1.582\ m^2K/W$$

However, if account is taken of the variable nature of heat flow, and of the problems caused by moisture [see Sections 5.6, 5.7, and 5.13], it is evident that the three methods are not identical. Variable heatflow within the fabric of the building is reduced, and condensation may be avoided by placing the insulation on the outer face of the wall. The lightweight insulation evidently requires a protective coating in that position.

An increase in insulation may or may not be desirable [Sections 5.6 and 5.7]. It depends on the climate of the place and the occupancy of the building.

Example 5.4 *Determine the thermal resistance and the thermal transmittance of a cavity wall consisting of two walls each 75 mm thick, separated by a 50-mm air space.*

Solution *The thermal resistance of air spaces is listed in Table 2 of Chapter 22 of the* ASHRAE Handbook—Fundamentals *(Ref. 4.1, Table 2, pp. 22.12–22.13). It varies appreciably in accordance with the mean temperature, the temperature difference, and the emittance of the surfaces. We will use* $R = 0.2\ m^2K/W$.

The thermal resistance is

$$R = \frac{1}{U} = \frac{1}{20} + \frac{0.075}{1.2} + 0.2 + \frac{0.075}{1.2} + \frac{1}{8}$$
$$= 0.50\ W/m^2K$$

and the thermal transmittance is $U = 2.0\ W/m^2K$.

This is an improvement on the thermal performance of the 200-mm thick concrete wall considered in Example 5.2, but greatly inferior to the performance of a wall of 150 mm of concrete and 50 mm of insulation.

The preceding examples illustrate the value of insulation for increasing the thermal resistance of existing buildings. Filling cavities with insulation evidently results in a substantial improvement of thermal performance. However, in many buildings cavities are needed to provide ventilation for the inhabitants and for the materials in the fabric of the building. Thus timber may rot if air spaces are filled with insulation so that condensed moisture cannot be removed by air movement [see Sections 5.7 and 5.13].

5.6 Variable Heat Transfer

In Section 5.5 it was assumed that the temperature differential between the two sides of a wall or floor remains constant. When the temperature varies, it becomes necessary to consider not merely thermal insulation but also the heat capacity per unit volume, also called *thermal inertia*. This is the product of the density of the material (ρ) in kilograms per cubic meter (kg/m³) and the specific heat per unit mass (c_p) in joules per kilogram per degree Kelvin (J/kgK). The heat capacity, or thermal inertia, is measured in megajoules per cubic meter per degree Kelvin (MJ/m³K).

The external temperature generally reaches its lowest point at dawn and its highest in the early afternoon (2 to 4 P.M.). The interior temperature is kept constant in buildings whose heating or air conditioning is automatically controlled, but it fluctuates in most other buildings. Buildings such as schools and offices that are used only for part of the time are often allowed to cool during the night. Residential units whose occupants all go to work may cool during the day in winter.

When a building is first heated in cold weather after a cooling period, a considerable part of the heat is needed to heat walls with a high thermal inertia. When heating is reduced or discontinued, the heat stored is released gradually. Thus thermal inertia reduces the fluctuations in the internal temperature, but it also slows the response of a cold room on heating.

The flow of heat under variable conditions is controlled by two material properties: its thermal conductivity, k, and its thermal inertia, or heat capacity per unit volume, ρc_p (Table 5.2). The ratio of these two quantities,

$$\alpha = \frac{k}{\rho c_p} \qquad (5.11)$$

is called the *thermal diffusivity* of the material (Table 5.2). The larger the diffusivity, the more rapidly a temperature change is propagated through the material.

The precise theory of variable heat flow is complicated, but an approximate solution can be obtained. In 1930, while working at the British Building Research Station, A.F. Dufton derived an equation for the time taken for the flow of heat through a wall, floor, or roof (Ref. 4.12, p. 53):

$$t = \frac{0.8d^2}{\alpha p^2} \qquad (5.12)$$

where t is time in seconds for the flow of heat through the wall, floor, or roof;

d is the thickness in millimeters of the wall, floor, or roof;

α is the thermal diffusivity in square millimeters per second (mm^2/s);

p is the ratio of the rate of heating and cooling that actually occurs to that under steady-state conditions.

The result is dependent on the correct choice of p, which is squared in the denominator, and the result is therefore only approximate if the value of p must be guessed.

Example 5.5 *Determine the time taken for heat to flow (a) through the wall of a galvanized steel shed, (b) through the uninsulated wall of a timber hut, (c) through a wall of concrete 200 mm thick, and (d) through a wall of limestone 350 mm thick.*

Solution *We will assume that the galvanized steel is 1 mm thick and that the timber planks are 25 mm thick. We will assume that the value of* p *is 2. From Table 5.2 the values of the thermal diffusivity for galvanized steel, timber, concrete, and limestone are 12, 0.093, 0.63, and 0.79* mm^2/s*, respectively.*

(a) For the galvanized steel walls, 1 mm thick, the time delay to the passage of heat is

$$t = \frac{0.8 \times 1^2}{12 \times 2^2} = 0.02 \text{ seconds}$$

(b) For the timber walls, 25 mm thick, the time delay is

$$t = \frac{0.8 \times 25^2}{0.093 \times 2^2} = 1344 \text{ seconds} = 22 \text{ minutes}$$

(c) For the concrete walls, 200 mm thick, the time delay is

$$t = \frac{0.8 \times 200^2}{0.63 \times 2^2} = 12\,698 \text{ seconds} = 3 \text{ hours } 32 \text{ minutes}$$

(d) For the limestone walls, 350 mm thick, the time delay is

$$t = \frac{0.8 \times 350^2}{0.79 \times 2^2} = 31\,013 \text{ seconds} = 8 \text{ hours } 37 \text{ minutes}$$

This time delay caused by a thick masonry or concrete wall is important for passive solar design [Sections 5.14 and 5.15].

The product of the conductivity and the heat capacity,

$$k\rho c_p$$

does not have a special name, but it is also of interest (Table 5.2).

A high thermal inertia is desirable to smooth out the response to the daily variations of temperature. A low thermal conductivity is also desirable to reduce the overall heat loss (or heat gain in summer). Thus the value of $\alpha = k/\rho c_p$ should be as low as possible.

However, if the heating is only intermittent, because a room or a building is not in use for part of the time, as discussed earlier in this section, the time taken to reach an acceptable temperature may be of greater importance than a low value of α. The diffusivity α has the dimension mm^2/s; given walls or roofs of equal thickness, the lowest diffusivity gives the shortest time of response. In lightweight construction with a low thermal inertia the necessary insulation can be achieved with less thickness, so that the response is faster; thus lightweight cladding panels are preferable to insulated load-bearing walls of brick or concrete.

The solution for homogeneous and simple composite walls has been derived by Billington (Ref. 4.12, pp. 52–83), who showed (Ref. 4.12, pp. 55) that the product $k\rho c_p$ should be as low as possible when rapid heating is required, that is, the wall should consist mainly of low-density insulating material.

The same considerations apply when the exterior temperature is above the desired interior temperature and the building is cooled by air conditioning.

However, there are two important differences between the heating and the cooling of buildings. Buildings are heated by the expenditure of energy even in most primitive societies, although the indoor temperatures are often lower than in industrially developed countries. Cooling by the expenditure of energy is still the privilege of the upper class in underdeveloped countries, and it is likely to remain so for some time. It may become less common in industrial countries as the price of energy increases and greater use is made of the fabric of the building in environmental design ("passive" solar energy; see Sections 5.14 to 5.17). Furthermore, the difference between the desired indoor temperature and the maximum exterior temperature is much less than for the minimum exterior temperature. In a temperate climate it is less than 10°C for a hot day but more than 20°C for a cold night. In an extreme climate it is about 15°C for a hot day and 50°C for a cold night.

We must distinguish between hot-arid climates, which have low humidities and high shade temperatures, sometimes above the temperature of the human body, and hot-humid climates, in which the shade temperature is usually less than the central temperature of the human body but the humidity can be very high.

The traditional method of construction in hot-arid countries employs thick walls, and sometimes thick roofs of stone, brick, adobe, or mud [Section 1.2]. These have a high thermal inertia, and in this respect an equal thickness of brick or concrete performs as well. The interior temperature during the hottest time of the day is reduced by the thermal inertia, although a higher nighttime temperature must be accepted. This is a disadvantage only in very hot climates. In that case a good, if somewhat more expensive, solution is the use of masonry materials only for the living quarters, while bedrooms are built from light panels with a low thermal inertia.

Reference has already been made to use of sunshades and to the utilization of thermal inertia of structural concrete floors by sunlight penetration in winter [Section 3.4].

In hot-humid climates the difference between the maximum daytime and minimum nighttime temperatures is much smaller, and a structure with a high thermal inertia does not improve the thermal environment during the hottest time of the day. Indeed, since the relatively high night temperature and very high humidity often make sleep difficult, a high thermal inertia is undesirable.

When air conditioning is not used, thermal comfort in hot-humid climates depends mainly on air movement [Section 5.11].

5.7 Insulation

The term *insulation* is normally used for lightweight materials, such as glass wool, expanded polystyrene, or urea formaldehyde foam, that greatly reduce heat transfer by conduction because of their very low thermal conductivity. For most insulating materials k is between 0.02 and 0.05 W/mK. All these materials contain pockets of trapped air, or occasionally of another gas. Still air is a very poor conductor, and thus the porous insulating materials are also very poor conductors. The inclusion of a relatively thin layer of insulation in a composite wall or roof greatly reduces its thermal transmittance (U), that is, it greatly increases its thermal resistance (R); see Example 5.3.

Because of their porous nature, most insulating materials are very light. The lightest and most effective materials have little or no strength. Some can be placed loose on a suspended ceiling under a roof, some can be injected in a cavity, and some can be included in a composite wall or roof, protected by a layer of stronger material, such as metal or plaster (Fig. 5.7.1).

Insulation can be added to most buildings after their completion, and this is one of the simplest and cheapest methods of improving the thermal efficiency of a building designed at a time when energy conservation was considered of slight importance; this is called *retrofitting*. However, it is cost-effective mainly in buildings that need to be heated in winter or cooled in a hot-arid summer.

The insulation is most effective if it is placed on the outside of the building (see Example 5.3), because the fabric of the building is then kept close to the nearly uniform indoor temperature instead of varying with the outdoor temperature. This is also likely to prevent the formation of condensation within the fabric of the building [Section 5.13]. However, it is much easier to place insulation in a cavity between two layers of the fabric, on a suspended ceiling below the roof, or on an inside wall or ceiling. Insulation on the outside of a building has to be protected from the weather, and this adds to the cost [Section 5.13].

The term *insulation* is also used for *reflective insulation,* which is essentially a mirrorlike material, such as bright aluminum foil, that reflects radiant heat. It is effective for reducing radiant heating during the day and radiant cooling during the night, if it is placed in air gaps where the heat transfer is mainly by radiation, such as roof spaces or cavities in walls. It is more effective against downward than upward radiation. Reflective insulation can be combined with resistive insulation, for example, by placing an aluminum foil on top of a fiberglass mat in a roof space.

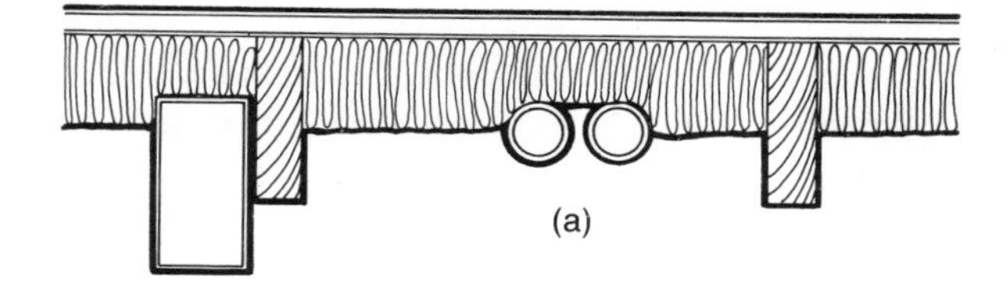

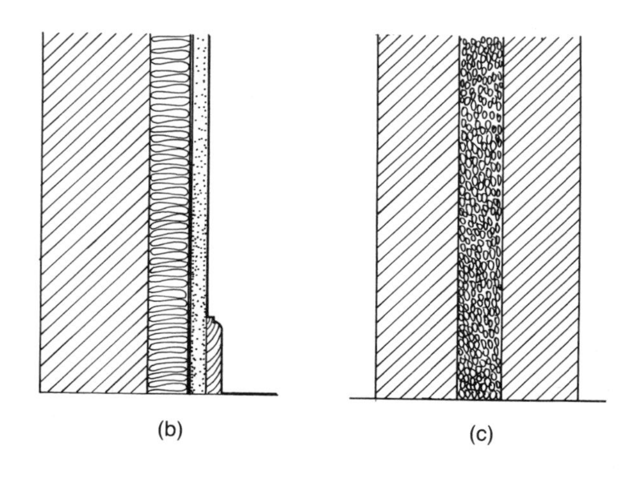

Fig. 5.7.1. Insulation.
(a) Blanket insulation pushed around ducts and electrical boxes. The vapor barrier is on the warm side.
(b) Blanket insulation installed between furring strips on a masonry wall. The vapor barrier is again on the warm side.
(c) Insulation blown into masonry cavity wall.
(*Drawing by Forrest Wilson.*)

However, this may cause condensation [Section 5.13]. Horizontal surfaces lose their reflective properties after a period of time unless it is possible to sweep the dust off reflective insulation.

Most of the existing legislation on the thermal performance of buildings deals with insulation. Some was enacted before the energy crisis of 1973, for example, in Great Britain, with the object of ensuring that people who bought or rented dwellings would be able to heat them to an acceptable temperature in winter without unnecessary expenditure on fuel. After 1973 several European countries, New Zealand, Canada, and several states of the United States laid down minimum standards of insulation that had to be met by all new buildings. Some governments also offered direct grants, interest-free loans, or tax incentives to encourage the installation of additional insulation in *existing* buildings.

The ASHRAE Standard 90-75 (Ref. 4.7) was developed as a "National Voluntary Consensus Standard" for energy conservation in the design of new buildings in the United States, and it has been widely accepted as a guide to good practice in other countries. It lays down maximum permissible thermal transmittances for the walls, the roofs, and the lowest floors of various types of building, in accordance with the number of heating degree days of the city [Section 5.1].

5.8 Reflectivity of Building Surfaces

The solar radiation received by the surface of a building is partly absorbed and partly reflected. The radiation emitted by the sun has wavelengths of less then 3 μm. This heats the surfaces on which it falls, but as they remain much cooler than the surface of the sun, the radiation emitted by them has a much longer wavelength [see Section 4.2]. For the same wavelength, the absorptivity and emissivity of a surface are equal; but their values vary with the wavelength. Materials differ greatly in their absorptivity of solar radiation and the emissivity of the long-wave radiation.

Thus concrete, black paint, and white paint all have the same emissivity, that is, they are cooled equally at night by radiation to the sky. However, the proportion of the solar radiation that they absorb during the day is quite different. A surface painted black becomes much hotter on a sunny day than a surface painted white; concrete is intermediate between the two. Evidently the reflectivity of solar radiation by concrete can be improved by using exposed white aggregate or white paint. However, it depends on the climate whether a high reflectivity is desirable. The absorption of solar radiation reduces the demand for heating, if heating is required.

In a cool climate some fuel economy could be achieved by painting houses black or a dark color, since most of the solar radiation would thus be absorbed, while the emissivity of the long-wave radiation is hardly affected by the color of the paint. It is a matter of opinion whether the fuel economy compensates for having to live with dark-colored houses in a climate where the winter daylight intensities are likely to be low.

The sol-air temperature was devised to obtain a measure of the effect of solar radiation. It is defined as the hypothetical external shade temperature that would have the same effect on the internal temperature of the building as the actual shade temperature *and* the solar radiation.

The building absorbs solar radiation, and it reradiates part of it as long-wave radiation. In most problems the reradiation can be ignored, as a first approximation, and the sol-air temperature then becomes (Ref. 5.7, p. 215):

$$T_{SA} = T_{SDB} + \frac{a\,I}{h_o} \qquad (5.13)$$

where T_{SA} is the sol-air temperature with the reradiation of the long-wave radiation ignored;
T_{SDB} is the dry-bulb temperature outside the building, in the shade;
a is the absorptivity of solar radiation of the surface of the building under consideration, listed in Table 5.3;
I is the intensity of the solar radiation;
h_o is the boundary layer heat transfer coefficient on the outside of the wall, previously considered in Section 5.5 and in Examples 5.1 to 5.4.

Example 5.6 *Determine the sol-air temperature for the wall of a concrete building when the shade dry-bulb temperature is 22°C.*

Solution *The intensity of solar radiation can be measured with a solarimeter at any given time and location. It can also be estimated from meteorological records, from the time of the year and of the day, and from the orientation of the part of the building under consideration. We will assume that* I *is 300 W/m*^{2}.

We considered the heat transfer coefficient of the wall of a building in Example 5.1, and we will use the same value, $h_o = 20$ *W/m*^{2}*K.*

For a gray concrete surface, from Table 5.3, the absorptivity is $a = 0.65$*, and from Eq. (5.13) the sol-air temperature is*

$$T_{SA} = 22 + \frac{0.65 \times 300}{20} = 22 + 9.75 = 31.8°C$$

If the concrete is painted white, $a = 0.30$*, and*

$$T_{SA} = 22 + 4.5 = 26.5°C$$

If the concrete is painted black, $a = 0.90$*, and*

$$T_{SA} = 22 + 13.5 = 35.5°C$$

5.9 Windows

Windows perform an essentially different function in single-family houses and in large, air-conditioned office buildings.

In small houses natural light generally provides sufficient illumination during daytime [Section 9.3] except on a dull day or for intricate work. Open windows greatly contribute to comfort in warm, humid weather [Section 5.11]. These are positive advantages, and few people would suggest that a house, even an earth-integrated building [Section 5.17], should be built without windows.

Counterbalancing these advantages, ordinary windows of the type used in small houses transmit a great deal of heat through conduction, convection, and radiation. A window with single glazing of ordinary glass has a thermal transmittance (U) of 4 to 7 W/m^2K, depending on its exposure to wind. This can be reduced to less than half by double glazing; the air space between the two sheets of glass is an excellent insulator. Thus double glazing is worthwhile in climates where the heating requirements exceed 3000 degree days (Table 5.1). Triple glazing is often worthwhile to reduce heat transfer further (Ref. 4.1, Chapter 22, Table 8).

All openable windows must allow some leakage of air between the window and its frame, and this causes a heat loss or gain by convection. The *ASHRAE Handbook—Fundamentals* (Ref. 4.1, Chapter 21, Tables 2 and 4) gives data for air infiltration through standard windows. In small houses, particularly those built many years ago with wooden windows that are worn or distorted, the heat loss through air leakage can be appreciable.

The heat transfer through windows by convection presents problems mainly in cold weather, when it causes undesirable heat losses. The heat transfer through windows by radiation is troublesome mainly in warm weather, when it can cause excessive heat gain from solar radiation. The most effective solution is external sunshading [Sections 3.4 and 3.5]. The next-best method is the use of venetian blinds in the air space between the two sheets of glass of a double-glazed window. Blinds inside the building are least effective, because the shading occurs after the radiation has penetrated the windows.

Solar heat gain can also be reduced by the use of heat-absorbing glass, although the heat so absorbed is reradiated partly to the exterior and partly to the interior. Glass is made of silica and certain metallic oxides. Heat-absorbing glasses are produced by adding a substance, such as ferrous oxide, to the glass that greatly reduces the transmittance of infrared radiation; some reduction in the transmittance of visible light must be accepted.

Heat-reflecting glass is even more effective in reducing solar heat gain. It is produced by coating the glass surface with a thin film of metal. The metal coating is placed inside the sealed space of a double-glazed windows to protect it from deterioration. A gold coating gives a characteristic gold color to the outside of the glass, but the transmitted light has a bluish-gray color (Ref. 5.9, pp. 45–48). Heat-reflecting glass, particularly if it covers the entire facade of a building, is a dominant architectural feature, because it acts like a mirror to its surroundings (Fig. 5.9.1). The heat reflected by the glass is received by other buildings and by people nearby, adding to their discomfort.

In multistory buildings, particularly in air-conditioned office buildings, the advantages of

Fig. 5.9.1. A building with heat-reflecting glass acts like a mirror to its surroundings. (*By courtesy of PMG Industries, Pittsburgh.*)

windows are more debatable. In the nineteenth century offices were frequently built with light wells to provide daylight for the occupants. However, it is more economical today to use artificial light for people too far from windows, and indeed for all the occupants of an office building [Section 10.1]. If the building is air-conditioned, it is customary to seal the windows to reduce air leakage, although it is open to question whether there may be advantages in having openable windows even in air-conditioned buildings, so that natural ventilation can be substituted for air conditioning when weather is warm rather than hot.

In most air-conditioned, multistory buildings erected during the 1960s and 1970s windows have not contributed to lighting or ventilation, and their only function is to provide a contact with the outside world. However, in many buildings the venetian blinds are permanently drawn because they are needed at some time to control solar radiation or glare [Section 9.6]. This negates the very purpose of the windows.

The thermal performance of windowless buildings is superior to that of buildings with windows. Studies of human performance in windowless buildings have shown no decline as compared with buildings that have windows (for example, Ref. 5.8). Nevertheless, windowless buildings have not become popular, and few have been built. Daylighting is discussed in Chapter 9.

5.10 Natural Ventilation

The amount of air passing through a room can be measured directly if the room is well sealed and the air is introduced through a duct. It is only necessary to determine the air velocity with an anemometer [Section 4.5]. However, in many buildings the air is introduced mainly through numerous small openings and through leakage of the building fabric, so that an indirect method must be used, such as the measurement of the concentration of a tracer substance.

We have already mentioned the air leakage that occurs around unsealed windows [Section 5.9]. A similar leakage occurs around all doors, even doors fitted with draft excluders. Open fireplaces cause a large amount of ventilation. The air is sucked up the chimney, particularly when the fire is burning, and it has to be replenished by leakage. The drafts for which the stately homes of England were famous were largely due to their open fireplaces.

Air also infiltrates through timber floors, wood-frame walls, and brick walls. Plaster and paint greatly reduce this infiltration. The amount of air infiltration can be calculated from formulas given in a number of books (for example, Ref. 4.1, pp. 21.4–21.11), or it can be expressed empirically in terms of the number of air changes per hour.

One air change per hour means that the amount of natural ventilation equals the volume of the room each hour; that is, 2 air changes per hour for a room measuring 5 m by 6 m and 3 m high is equal to a rate of ventilation of $2 \times 5 \times 6 \times 3 = 180$ cubic meters per hour, or $180 \times 1000/3600 = 50$ liters per second (l/s).

Fig. 5.10.1. Roof ventilators that provide ventilation without the use of power-operated fans. The rectangular vents also open automatically to provide smoke venting in the case of fire. (*Photograph by courtesy of Email Ltd. and Colt Ventilators.*)

Fig. 5.10.2. Exhaust fan.

Table 1 of Chapter 21 of the *ASHRAE Handbook—Fundamentals* (Ref. 4.1) gives the number of air changes that occur under average conditions in residences, exclusive of air provided for ventilation; these range from 0.5 for rooms with no windows or exterior doors to 2 for rooms with windows or exterior doors on three sides. Table 6 of Chapter 21 gives the ventilation requirements for occupants for a great variety of building types, expressed in cubic feet per minute and in liters per second.

Natural ventilation is required to remove the carbon dioxide generated by people, undesirable odors, and other irritants. The amount recommended by ASHRAE in Table 6 for residential buildings ranges from 4 l/s per person for bedrooms to 20 l/s per person in toilets and kitchens. Similar requirements apply to commercial and institutional buildings, but higher rates are needed for some industrial buildings.

Thus the air leakage is often sufficient for the required natural ventilation in residential buildings employing traditional construction; however, some building regulations, notably in Great Britain, require that each room have two vents or air bricks to ensure adequate ventilation. In rooms where a large number of people assemble, such as schools and restaurants, additional ventilators are required, but adequate results can often be obtained by natural ventilation (Fig. 5.10.1) or by exhaust fans (Fig. 5.10.2).

Air conditioning [Section 6.6] is essential when electrical equipment whose functioning depends on the maintenance of a uniform temperature is used; this applies to many computers. It is generally economical in large libraries because the air conditioning equipment filters the air and thus eliminates the cost of dusting the books regularly. The air filtration, which is unobtainable with natural ventilation, also makes air conditioning desirable for hospitals and many factories.

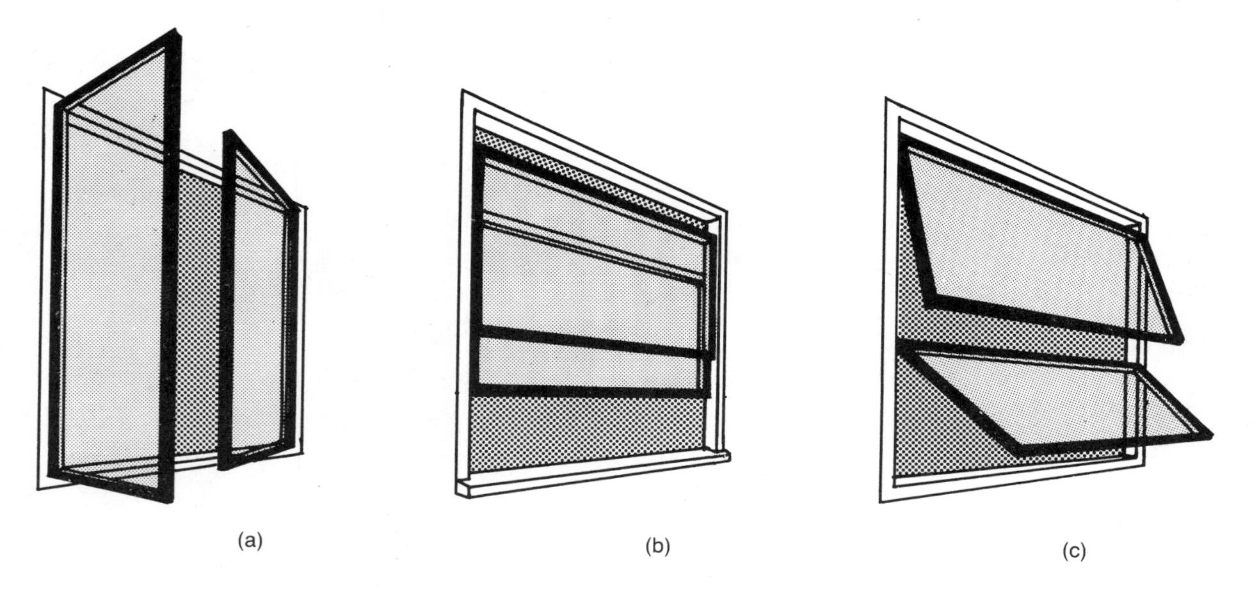

Fig. 5.11.1. Use of windows for ventilation in warm, humid weather.
(a) A casement window can be opened fully, and the swinging sash can be used to catch the prevailing breeze.
(b) A double-hung window can be opened for as much or as little as desired, but no more than half of the opening of the window.
(c) A hopper window can be used to direct a cooling breeze into the upper or lower part of the room, and it stops penetration of wind-driven rain to a certain extent.

5.11 Natural Ventilation for Thermal Comfort in Warm, Humid Weather

We have so far considered only the small amount of natural ventilation needed to replenish oxygen in the air and to remove odors and pollutants, and we noted that from 0.1 to 2 air changes per hour are required for this purpose. A much larger amount of natural ventilation, of the order of 100 air changes per hour or more, is needed in the warm, humid weather that occurs in many parts of the world for a short period in summer. The design of buildings is generally dominated by the winter heating requirements [Sections 5.6 and 5.7], but often a satisfactory thermal environment can be obtained in summer by opening the windows (Fig. 5.11.1), provided that they are suitably designed and that the house is appropriately oriented to catch the prevailing breeze. Although microclimate is not always predictable, the long-term residents, and builders and architects who have previously worked in the district, generally know whether there is a reliable cooling breeze and the direction from which it blows [Section 5.3]. It is essential to have an outlet as well as an inlet for natural ventilation at the high rate required. This is best achieved by a window on the opposite wall (Fig. 5.11.2), but internal doors arranged in a straight line with the windows are also effective.

The same considerations apply in the hot-humid tropics, but as winter heating is usually not a problem, the entire walls of a building can be designed with screens (Fig. 5.11.3) or with louvers to permit maximum ventilation (Fig. 5.11.4). Most parts of the humid tropics have large insect pop-

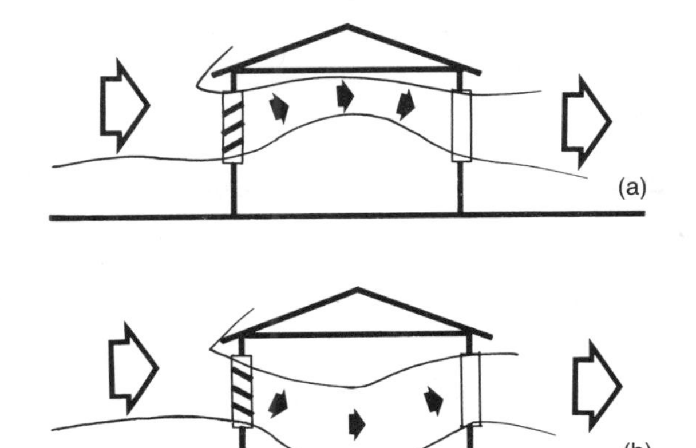

Fig. 5.11.2. Natural ventilation requires both an entry and an exit to be effective. This can be achieved by placing windows on opposite sides of the same room, or by suitably oriented internal doors. The direction of the internal air flow can be influenced by the angle of the windows through which the breeze enters. (*Wind tunnel test carried out in the Architectural Science Laboratory of the University of Sydney.*)

Fig. 5.11.3. Ventilating screen in Agra, India, cut from a solid slab of marble.

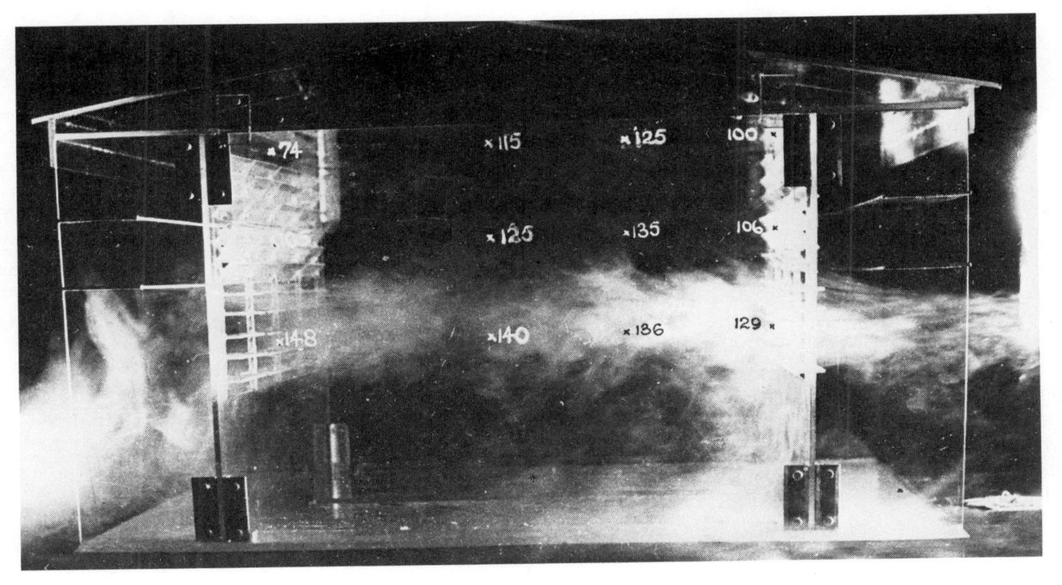

Fig. 5.11.4. Wind flow through building with louvered exterior walls. The angle of the louvers can be adjusted to direct the air flow to the part of the room most likely to be occupied.

ulations, which are most disturbing after dark, even when they do not bite. In addition, mosquitoes are a common problem, and in many countries they carry malaria. Openings therefore need insect screening, which reduces the air movement.

Fig. 5.11.5. Air scoops used in some parts of the Middle East to deflect air downwards through a sunshaded hole in the roof. (*By courtesy of Prof. R.M. Aynsley.*)

Fig. 5.11.6. Wind-tunnel model for a government building in the central Pacific, with quadrant-shaped ventilating hoods above the roof to catch the prevailing breeze and deflect it downwards. This made it possible to provide natural ventilation for a comparatively large building that would otherwise have been too deep for natural ventilation. (*Model test by Prof. R.M. Aynsley at the Department of Architectural Science, University of Sydney.*)

If the prevailing breeze is weak, or its access is blocked by surrounding buildings, it can be caught by means of an air scoop placed above the roof (Fig. 5.11.5), which deflects the air downwards through a sunshaded hole in the roof. Air scoops are traditional in some parts of the Middle East (Ref. 5.7, pp. 8–9), where they form a spectacular feature on the skyline. They have also been used in modern buildings (Fig. 5.11.6).

Natural ventilation is promoted by air pressure differences due to variations in air density with height. If the building is warmer than the outside, air moves up the chimney in an open fireplace or up an elevator shaft. This is known as the *stack effect*. It produces drafts in rooms with open fireplaces, and it may promote the spread of fire in a building with an elevator shaft. The stack effect cannot be utilized to improve ventilation in hot-humid climates, because the temperature difference between the outside and the inside is generally insufficient and because the air above the roof is usually warmer than the air indoors. However, overhanging eaves used for sunshading are useful for deflecting the wind to improve cross ventilation.

We noted in Section 4.7 and Fig. 4.7.1 that

the Kansas State University observations showed that for a lightly clad person in the absence of significant air movement a dry-bulb temperature of 26.5°C was comfortable at humidities up to 45%. At a humidity of 80% the comfortable temperature was reduced to 24.5°C.

Air movement greatly increases comfort at high humidities. An air velocity of less than 0.25 m/s is barely noticeable. A velocity of 0.25 m/s to 0.5 m/s is pleasant and not disturbing; it gives the impression that the temperature has been lowered by 1°C to 2°C (Ref. 5.6, p. 48). A velocity of 0.5 m/s to 1 m/s, corresponding to No. 1 on the Beaufort wind scale, gives the impression that the temperature has been lowered by 2°C to 3°C, but the air movement is noticeable, though not unpleasant. An air velocity of more than 1 m/s reduces the sensation of temperature even further but is likely to be disturbing indoors. On the other hand, a "gentle breeze" (No. 3 on the Beaufort scale, equal to 5 m/s) is pleasant in the open air and may give the impression of a temperature about 5°C lower. Since the wet-bulb shade temperature in the humid tropics is usually lower than the skin temperature [see Section 4.6], this is effective in giving at least a comparative sensation of comfort.

The simplest equipment for producing air movement artificially is a paper fan. This is widely used in China, where it is often the only method available; rapid movement of the fan gives a definite sensation of comfort in hot-humid conditions. In India during the British raj the punkah was in common use; this was a large wooden frame covered with cloth, slowly moved backward and forward with a cord operated by a man called a punkah-wallah.

It has been replaced by the slowly rotating ceiling fan (Fig. 5.11.7). Darwin, the capital of the Australian Northern Territory, has a very hot and humid climate, with temperatures ranging from about 33°C during the day, with a relative humidity of 60%, to about 25°C at night, with a relative humidity of 90%. Ceiling fans are now fitted to all rooms in new houses, and air conditioning is rare in homes. The fans are economical in the use of energy, and the rooms do not have the shut-up feeling resulting from full air conditioning.

Fig. 5.11.7. Slowly rotating ceiling fan for use in the humid tropics and subtropics. Because of its slow speed it has a low fan noise; but it requires slightly higher ceilings to ensure that the blades do not touch the heads of tall people. Ceiling lights should not be used in the rotating range of the fan, to avoid a flicker effect.

Portable fans on stands (Fig. 5.11.8) and desk-top fans are also effective methods of producing local air movement while a room is in use.

The swing frequently installed on the porches of houses in the southern United States in the past also produced air movement relative to the person sitting on it.

However, some people will prefer air conditioning for hot-humid weather, and for some tasks it is essential. Natural ventilation gives a feeling of comfort but does not reduce the humidity. This can be done only by cooling the air below the desired temperature until the dewpoint is reached, when unwanted water condenses; the air is then reheated to the desired temperature [Section 6.6]. This consumes an appreciable amount of electricity. Furthermore, electrical equipment deteriorates more quickly in the hot-humid tropics, and spare parts and trained maintenance staff are not always available in developing countries.

For multistory buildings air conditioning is essential for comfort in hot-humid conditions, because natural ventilation cannot penetrate the deeper floors used in multistory construction. It is also at variance with the requirement for sound insulation [Section 11.4].

Fig. 5.11.8. Portable fan on pedestal; it produces air velocities up to 4 m/s at 3 m from the fan, and 0.5 m/s at distances of up to 12 m.

When air conditioning is used, the fabric of the building should be designed accordingly. Air leakages, which are entirely acceptable when natural ventilation is used, should be reduced to a minimum, and the building should be insulated, particularly in the roof, to reduce conductive heat transfer [see Section 5.7].

5.12 Evaporative Cooling and Attic Fans

In hot-arid climates the nighttime temperature may be too low for comfort, and the daytime shade temperature may rise above the skin temperature of the human body [Section 5.6]. It is therefore desirable to have well-fitted doors and windows, suitably sunshaded, good insulation in the roof and the walls, and a fabric with a high thermal inertia to smooth out the response to the daily variations of temperature. The delay in this response can be calculated from the thermal diffusivity (Table 5.2 and Example 5.5).

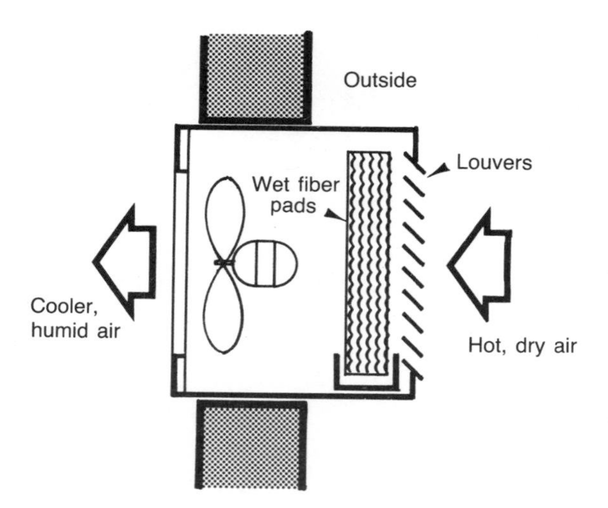

Fig. 5.12.1. Evaporative cooler, consisting of water-soaked fibrous pads placed on the suction side of a wall-mounted input fan. It is used in the hot-arid regions of the United States and Australia.

Fig. 5.12.2. Pavilion with veranda over a shallow ornamental lake in the Old Seraglio in Istanbul, used by the Turkish sultans until the mid-nineteenth century. In the middle of the lake there is a fountain, now out of order.

If this is not sufficient to produce comfortable conditions, a further improvement may be achieved by evaporative cooling. This has been used for centuries in some parts of the Middle East: Air is passed over an unglazed earthenware pot, from which water leaks slowly. In a modern version of this principle, evaporative coolers are used in conjunction with ventilating fans (Fig. 5.12.1).

The humidity in hot-arid zones is often too low, so that the evaporation of water increases comfort; it rarely raises the relative humidity above the comfort level. The latent heat required to evaporate the water is taken from the air, so that its temperature is reduced by several degrees.

Example 5.7 *The dry-bulb shade temperature at the hottest time of a summer day in a hot-*

arid climate is 44°C, and the relative humidity is 15%. How would the temperature and humidity be altered by evaporative cooling?

Solution *The extent of the cooling depends on the efficacy of the evaporative cooling device. Since no energy is supplied or absorbed by evaporative cooling, the heat content, or enthalpy, of the air remains constant. We therefore proceed along the line of equal enthalpy on the psychrometric chart (Fig. 4.4.5). The enthalpy of air at 44°C DBT and 15% RH is 74 kilojoules per kilogram of air; its moisture content is 8.5 grams per kilogram of air.*

An increase in the relative humidity from 15% to 20% at a constant enthalpy of 74 kJ/kg would require the evaporation of 1.3 g/kg to raise the moisture content to 9.8 g/kg. This would lower the dry-bulb temperature from 44°C to 41°C.

If the device is capable of evaporating more moisture, the temperature could be lowered further. At constant enthalpy, a relative humidity of 25% corresponds to a moisture content of 10.8 g/kg and a DBT of 38°C; a RH of 40% corresponds to a moisture content of 13.0 g/kg (requiring evaporation of 4.5 g/kg) and a DBT of 33°C.

In Islamic countries shallow ornamental lakes were frequently used to convey an impression of coolness (Fig. 5.12.2). They reduce the temperature by evaporative cooling, particularly if they contain a fountain that throws a spray of water into the air. However, the pleasure derived from the sight of water on a hot-arid day is perhaps more important than the actual slight cooling effect of the evaporation of water on the shade temperature of the adjacent spaces. The same applies to the small courtyard garden. The plants reduce the temperature a little by evaporative cooling, but the visual effect is more significant.

Comfort in hot-arid climates can also be improved by installing a low-speed *attic exhaust fan* in the roof space above the ceiling. It is switched on in the evening and left running for several hours during the night to draw air through the house into the roof space whence it escapes through vents. Both these methods have been used in the United States and in Australia.

5.13 Condensation and Rain

Condensation is a problem mainly in cool climates. As the psychrometric chart shows, the maximum absolute humidity of air decreases sharply with temperature, so that water vapor tends to turn into liquid water as air is cooled.

Example 5.8 *A house is heated to 22°C, and the relative humidity is maintained at 30%. The air comes into contact with the cold surface of a window. At what temperature does condensation occur?*

Solution *From the psychrometric chart (Fig. 4.4.5) the absolute humidity of the air is 5.0 grams of water vapor per kilogram of air. The air is cooled by contact with the cold surface of the window without changing its absolute humidity, but its relative humidity increases as it is cooled. When the relative humidity reaches 100%, the air becomes saturated with water vapor. In this instance the line of constant absolute humidity intersects the saturation line of 100% relative humidity at a dry-bulb temperature of 4°C. This is the dewpoint, at which moisture starts to condense. Therefore contact with a surface colder than 4°C produces condensation.*

Example 5.9 *Outside air at a temperature of 0°C and a relative humidity of 80% is heated to 24°C. If its absolute humidity remains unchanged, what is its relative humidity?*

Solution *Although a relative humidity of 80% may sound high, it corresponds at 0°C to an absolute humidity of only 3.0 g/kg. The psychrometric chart (Fig. 4.4.5) shows that when this air is heated to 24°C, its relative humidity is reduced to 16%, which is too low for comfort. To raise it to 25% requires a further 1.6 g/kg of water vapor.*

This can be added by evaporation from a dish of water, but the increase may occur automatically because of the activities of the occupants; for example, a person who remains in the house for a full 24 hours is likely to add 5 kg of water vapor to its air.

Condensation tends to occur in all heated buildings when the outside air is near or below the freezing point. The condensation that occurs on windows in cold weather is a well-known example.

Condensation is not necessarily harmful. If water droplets formed late at night are reevaporated by solar radiation during the day, they are unlikely to cause damage. However, if the water persists in liquid form, it is likely to produce deterioration in a wide variety of building materials through chemical or biological action, for example, corrosion in metals, peeling of paint, or dry rot in timber. As a result many building regulations make specific provisions for the prevention of condensation, as do the rules of some corporations that grant housing loans.

Condensation can be controlled by a vapor barrier (Fig. 5.13.1). This is an essential complement to insulation against cold weather [Section 5.7], and it must be placed on the "warm" side of the insulation, so that the air retains its low relative humidity until it strikes the barrier. The water vapor is thus retained inside the building. If the air were allowed to cool by passing beyond the insulation before it reaches the vapor barrier, its temperature might be reduced below the dewpoint. The vapor barrier would then cause condensation instead of preventing it.

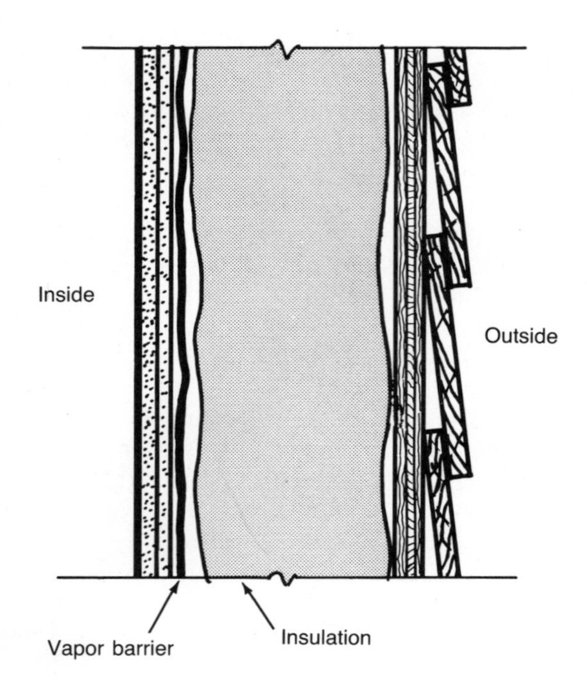

Fig. 5.13.1. Condensation is controlled by a vapor barrier. This must be placed on the "warm" side of the insulation in a wall or roof. The warm air from the inside of the building, at a relative humidity that is well below the dewpoint, is prevented by the vapor barrier from passing into the porous insulation and the other materials on the cooler side of the wall. If the vapor barrier were placed on the cooler side of the insulation, the temperature of the air would be reduced below its dewpoint (Fig. 4.4.5). Water would then condense on the vapor barrier. Thus the vapor barrier placed on the wrong side of the insulation causes condensation instead of preventing it.

A vapor barrier is an essential complement to insulation in cold weather [Section 5.7]. Buildings that are insulated only against summer heat and buildings that are not insulated do not generally need a vapor barrier in their walls and roof.

Consequently, a second vapor barrier should be avoided. Vapor barriers are not entirely impervious to vapor, and the air that passes through the vapor barrier and the insulation might still contain sufficient water vapor to reach its dewpoint when its temperature is reduced to that of the outside air. This would then be condensed at the second vapor barrier.

In practice it is not always possible to avoid the use of a second barrier, which may be needed to keep rainwater out. Condensation can then be prevented by venting (Fig. 5.13.2).

Vapor barriers can be made from films of polyethylene or other suitable plastic, aluminum foil, asphalt-saturated paper, or a combination of these.

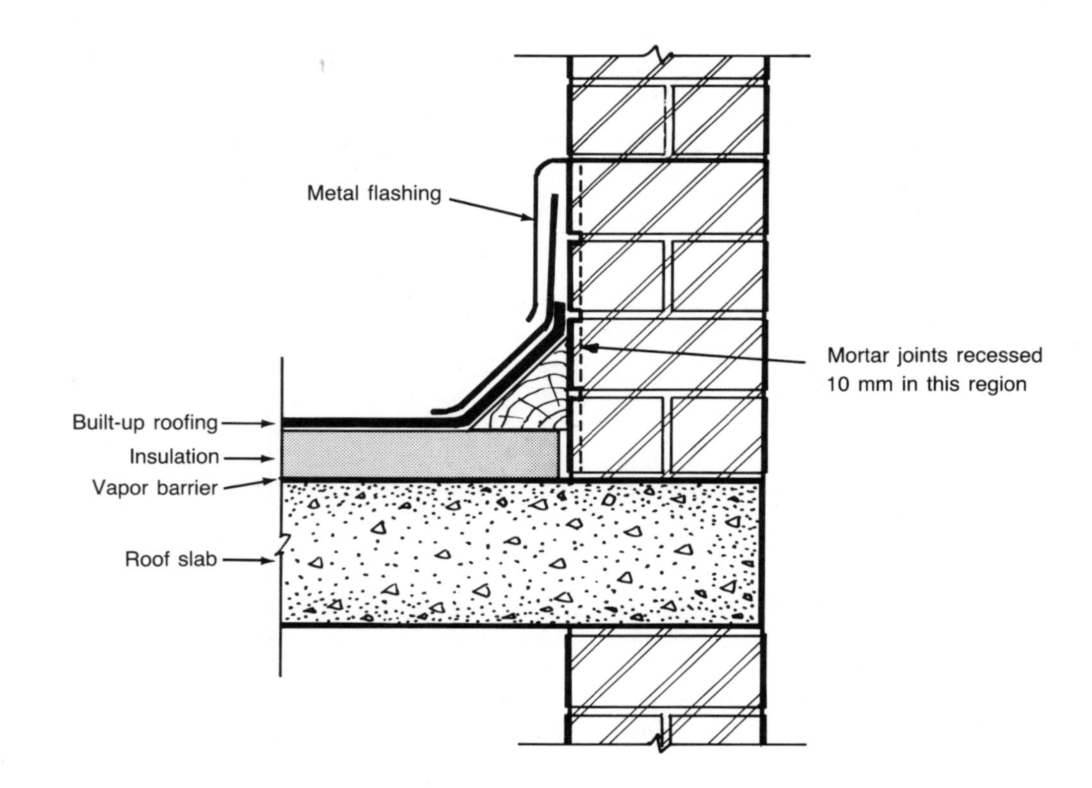

Fig. 5.13.2. Venting of insulation in built-up flat roofing. We noted (Fig. 5.13.1) that the vapor barrier should be on the warm side of the insulation. A second vapor barrier should be avoided because any vapor barrier on the cold side of the insulation is liable to cause condensation.

There is a conflict between this requirement and the need to provide a waterproof barrier to keep out the rain. This is resolved by venting the waterproof rain barrier to allow the water vapor to escape. This is more easily achieved in a wall and in a sloping roof than in a flat roof. In built-up roofs the venting can be accomplished by a suitable design of the flashings, which allow moist air rising through the insulation to escape while preventing the entry of rainwater.

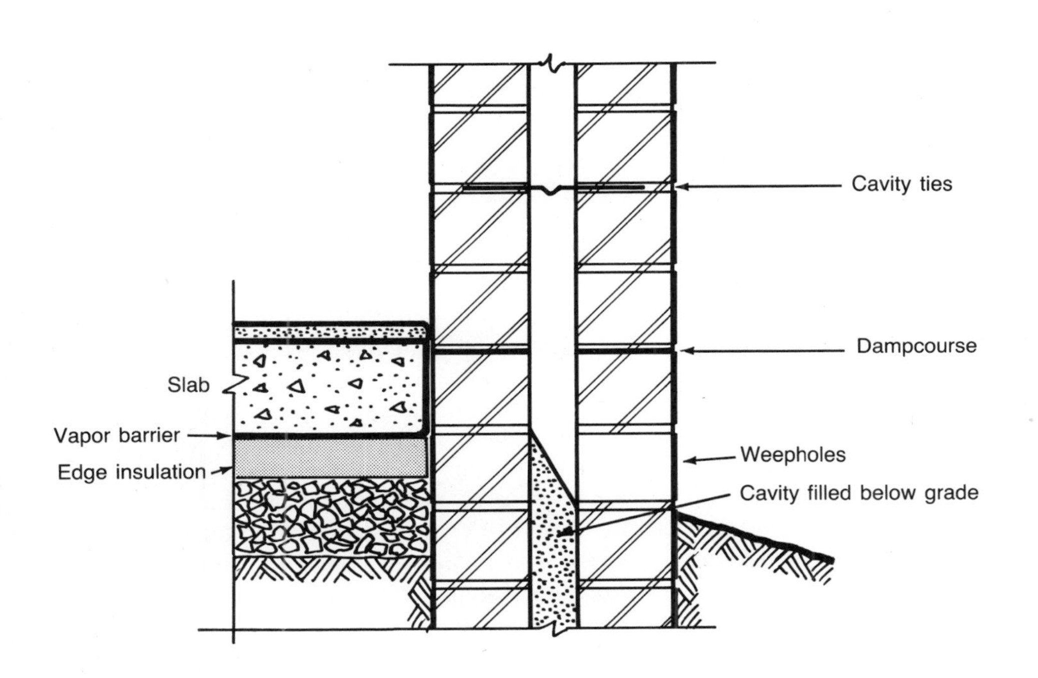

Fig. 5.13.3. Traditional British cavity wall. The air gap improves the insulation of the wall, and it catches both condensation resulting from heated warm air passing through the inner leaf of the wall and rain passing through the outer leaf. The water is drained through weepholes at the base of the outer leaf.

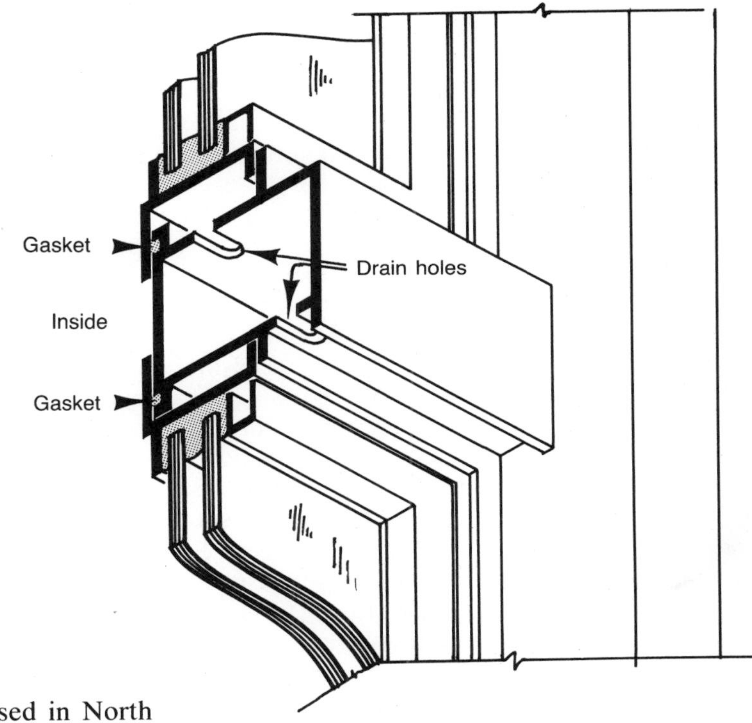

Fig. 5.13.4. Drained rebates in aluminum window frame to ensure that any rainwater that may penetrate can be drained and any condensation that may form can be removed. (*From Ref. 5.10.*)

Vapor barriers are not necessarily required to control condensation. The brick cavity wall (Fig. 5.13.3) that has been used in Great Britain for more than a century does so quite effectively in the British climate. Some water vapor may pass through the inner leaf of the brick wall in cold weather and condense in the cavity, but this is drained to the outside of the building by weepholes at the base of the outer leaf.

Cavity walls have rarely been used in North America or in central Europe, because the sustained lower winter temperatures may cause moisture in the cavity to freeze before it can drain and thus block the cavity.

We have noted that condensation in walls can be controlled either by a barrier or by allowing it to penetrate and then draining the water through weepholes. The same methods are available for disposing of rainwater, but the quantity of water to be handled is much larger.

Materials such as brick and natural stone absorb water. Some may penetrate into a cavity whence it is drained (Fig. 5.13.3). The remaining water absorbed by the brick or stone during the rain is evaporated after the rain has stopped. Other materials, such as glass and metals, are impervious to water, so that the amount of liquid running down the wall of a tall building during a heavy fall of rain is much greater.

Even when facades are sealed, it is desirable to provide holes so that any water that may penetrate and any condensation that may form can be drained (Fig. 5.13.4).

Rainwater running down a facade can cause serious discoloration on a number of building materials. Painted surfaces are particularly affected. In classical architecture and its historic revivals the water was thrown off the wall by a molding. There is disagreement about whether moldings were always intended to be decorative or whether they had a functional origin. Modern architects who disliked ornament rarely used moldings, and some of their buildings suffered from discoloration by rainwater.

Unless the facade consists only of materials that cannot be damaged by rain, a roof overhang or projection, decorated or plain, is needed at the top of a wall to throw the water clear of the wall (Fig. 5.13.5).

A sloping roof provides cheaper and more efficient rain protection than a flat roof. Thin, overlapping pieces of waterproof material, such as shingles, tiles, slates, or sheets of metal, are sufficient for a sloping roof. A flat surface requires built-up roofing that relies on layers of asphalt or pitch (bitumen) for waterproofing. These materials deteriorate and require periodic replacement. Although water will eventually evaporate from a perfectly flat roof as a result of solar radiation, it is preferable to give a slope of about 1 in 50 to the roof surface and drain the water into a gutter.

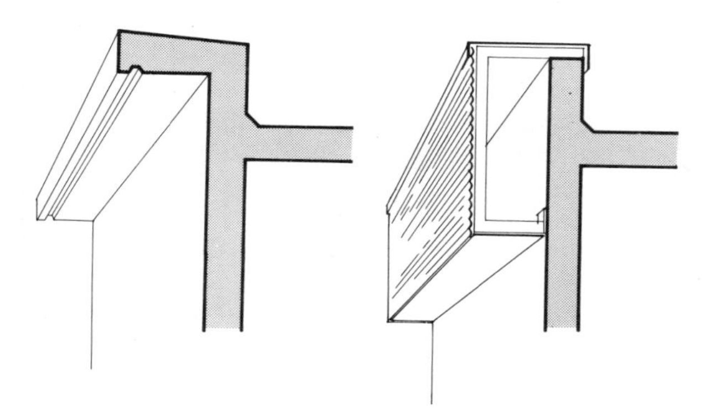

Fig. 5.13.5. A projection, whether decorated or plain, is needed at the top to throw the water clear of the wall.

5.14 Active and Passive Solar Energy

We noted in Section 3.1 that electricity can be produced from solar radiation by photovoltaic cells or by solar collectors that produce steam or hot water; some of these techniques are now economical in small communities remote from electricity supply lines, ports, and railroads, where the cost of conventionally produced electricity is consequently high. This does not affect the design of buildings, except to the extent that solar collectors may be mounted on the roof.

Solar collectors can also be employed to heat swimming pools or to heat water for washrooms, bathrooms, and kitchens [Section 2.4]. These applications are at present (1983) becoming economical in metropolitan areas where there is plenty of sunshine.

The use of heat derived from solar collectors for space heating [Section 6.4] is still more expensive than that produced from conventional energy sources, and this applies to an even greater extent to its use for air conditioning [Section 6.7].

All these applications employ solar collectors of various kinds, which may be mounted on the building or incorporated into its roof but are essentially separate from it.

In most books solar energy is classified as active if it requires the importation of energy not produced by the sun, for example, electricity to operate fans or pumps. Passive solar energy is energy produced without this aid.

In this book we use a sightly different distinction, bearing in mind that most buildings are connected to an electric power supply. Thus the small amount of electric power required to operate a fan for a few hours is not significant, and solar designs that are passive apart from the need for a fan are included in Sections 5.15 and 5.16.

On the other hand, we deal with systems that require the use of solar collectors in Sections 2.4, 6.4, and 6.7, even though some of these can operate without pumps. When a liquid-filled solar collector can be located below its storage tank, the hot liquid rises to the storage tank because it is lighter than the cold liquid, so that the system is "passive."

The applications of passive solar energy discussed in Sections 5.15 to 5.17 deal with designs in which the building itself is used to collect solar energy, without the use of solar collectors. This is not a new idea, because Xenophon already explained how buildings might be designed to make the best use of solar radiation [Section 1.1]. We considered more recent applications within the framework of traditional architecture in Sections 3.4, 5.1, 5.4, 5.7, 5.8, 5.11, and 5.12. However, the designation *passive solar house* applies more specifically to designs of unconventional appearance, developed during the last two decades to make the best possible use of solar energy without collectors, by using thermal inertia, thermal insulation, and ventilation to the best advantage.

These employ predominantly materials such as glass, concrete, stone, and brick that are durable and whose raw materials are available in abundance. By contrast, most solar collectors require metals that corrode and need to be replaced after several years.

Passive solar energy is at present generally cheaper for space heating and for cooling than solar energy obtained from collectors. However, it is not necessarily cheaper than the use of fossil fuels or of electricity. This aspect is discussed further in Section 5.18, which deals with life cycle costs.

Solar energy is inexhaustible and causes no pollution. Consequently some people feel that it should be substituted for energy derived from atomic power stations, and to a lesser extent from fossil fuels, wherever possible, even if it costs more. The building owner (but not the designer of the building) certainly has the privilege to make that choice.

However, solar energy has become a major source of inspiration to architectural designers since the mid-1970s, as structure had done a quarter of a century earlier. Some recent buildings designed to conserve energy have resulted in novel and interesting shapes and interior spaces, often at a substantial increase in cost. If passive solar energy inspires new forms, so much the better. Traditionally architecture has concerned itself not merely with firmness and commodity but also with delight. However, it is important not to confuse esthetic decisions with technical or economic ones.

In the following chapter a few applications of passive solar energy are described. Further examples can be found in Refs. 5.11, 5.12, and 5.19 and in many recent issues of the architectural journals. Ref. 5.13 provides a good survey of buildings heated or cooled by solar collectors. Daylighting and glare are discussed in Section 9.6.

5.15 Passive Solar Designs Utilizing Large Windows

The simplest concept is that of *direct gain* of heat or coolness. The main solar collector is a large window facing south (north in the Southern Hemisphere); the window should be oriented as nearly as possible in the direction of true south or north, which differs from magnetic south or north according to the local magnetic declination. This traditional method was discussed in Section 5.14, and in Sections 1.1, 3.4, and 5.4. In passive solar houses designed to eliminate the need for winter heating, additional windows facing south (north in the Southern Hemisphere) are required for rooms that do not face into the sun. This is done by means of roof lights or clearstory windows, or by staggering the rooms on a sloping site (Fig. 5.15.1). Furthermore, it is possible to make this design more flexible by the addition of a *rock bed*. This provides thermal storage into which excess heat or coolness can be pumped by means of a duct and a fan. Broken rocks and gravel have a high thermal inertia (Table 5.2), and the soil surrounding the rock bed provides the necessary insulation.

If the sun shines in summer on a concrete floor with hard paving [see Section 3.4] and on a concrete, brick, block, or adobe rear wall, the thermal storage of the floor and the wall are often sufficient to provide the required heat during the night. The rock bed store is used on cloudy days when the solar energy collected is insufficient. It provides the flexibility in thermal performance that the traditional houses derived from the classical tradition lack.

The rock bed can also be used to store coolness in summer, and this may be sufficient if summers are hot and arid with cool nights. For hot-humid summers houses should be designed to provide cross ventilation [Section 5.11].

The direct-gain method can provide a satisfactory thermal performance throughout the year, using only the energy required for operating a fan part of the time in climates that would otherwise require a substantial amount of heating. The rock bed and the thermal storage floor and rear wall impose relatively few restraints on the occupants of the building.

It is desirable that the rock bed be placed adjacent to the building, even though it is thermally more effective under the floor, where it is insulated by the building. Rock beds can breed bacteria and fungi, particularly when they are used for cooling as well as heating, as this tends to introduce moisture by condensation. They should therefore be inspected after several years and replaced or cleaned when necessary. It is generally not practical to filter the air in passive solar houses, because the velocity of the air flow is too low.

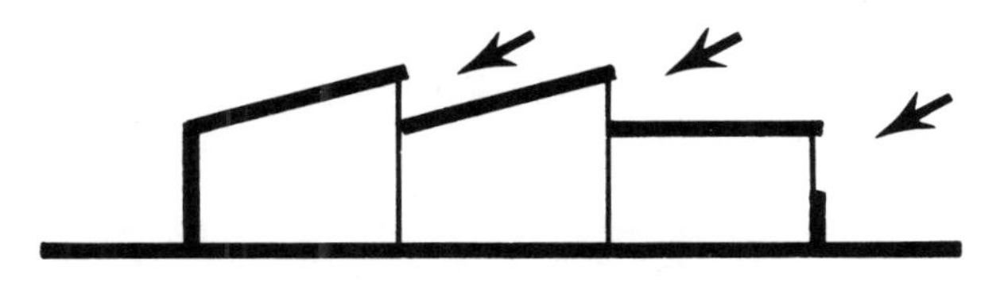

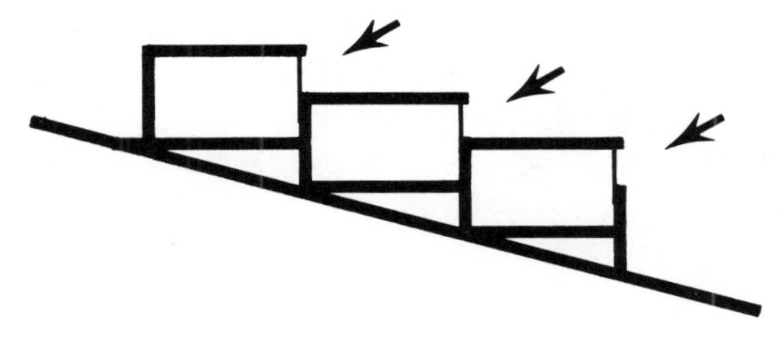

Fig. 5.15.1. Windows facing south for rooms that are not on the southern facade of the building (north in the Southern Hemisphere). On a sloping site large windows can be built if internal stairs are acceptable. On a flat site clearstory windows or rooflights can be utilized to collect solar radiation.

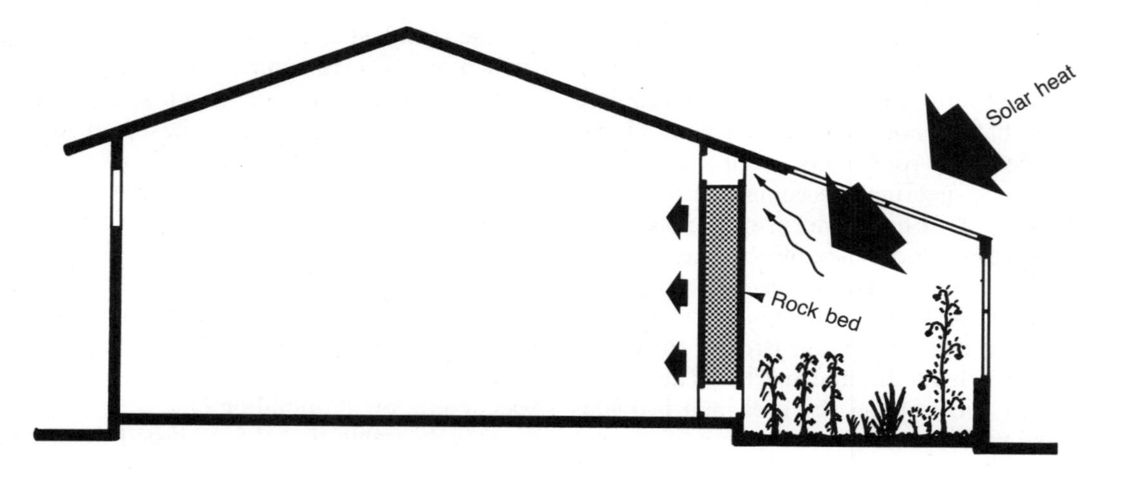

Fig. 5.15.2. The basic concept of the greenhouse, which may or may not be used for growing plants. It is a room with more glazing than is needed to heat that particular room. Hence it becomes too hot for comfort in the middle of the day. This excess heat is stored in a rock bed, which can be placed between the greenhouse and the rest of the house, or in the ground. A fan may be needed to circulate the air, particularly if an underground rock bed is used. Heat is transmitted from the greenhouse to the rest of the house by conduction, convection, and radiation, and the cold air is returned to the greenhouse.

The efficiency of thermal collection can be further increased by adding a sun porch or *greenhouse* to the building (Figs. 5.15.2 and 5.15.3). Since this room is intended to increase the solar collection, it must become too hot for comfort for several hours in the late morning and afternoon on a sunny winter day, and it can therefore be put only to limited use as a living room. In most passive solar houses it can be utilized as a greenhouse. The plants for this greenhouse need careful selection; pesticides and fertilizers can be used only to a limited extent, since the air from the greenhouse is circulated through the rest of the house.

An alternative approach is to use a *thermal storage wall* behind a large window facing south (north in the Southern Hemisphere). This can be constructed either of steel cylinders or drums containing water, or of concrete, adobe, block,

Fig. 5.15.3. Solar house in New Mexico with greenhouse on the ground floor and direct-gain windows on the upper floor. The house also has solar collectors for a hot water system and photovoltaic cells for generating electricity.

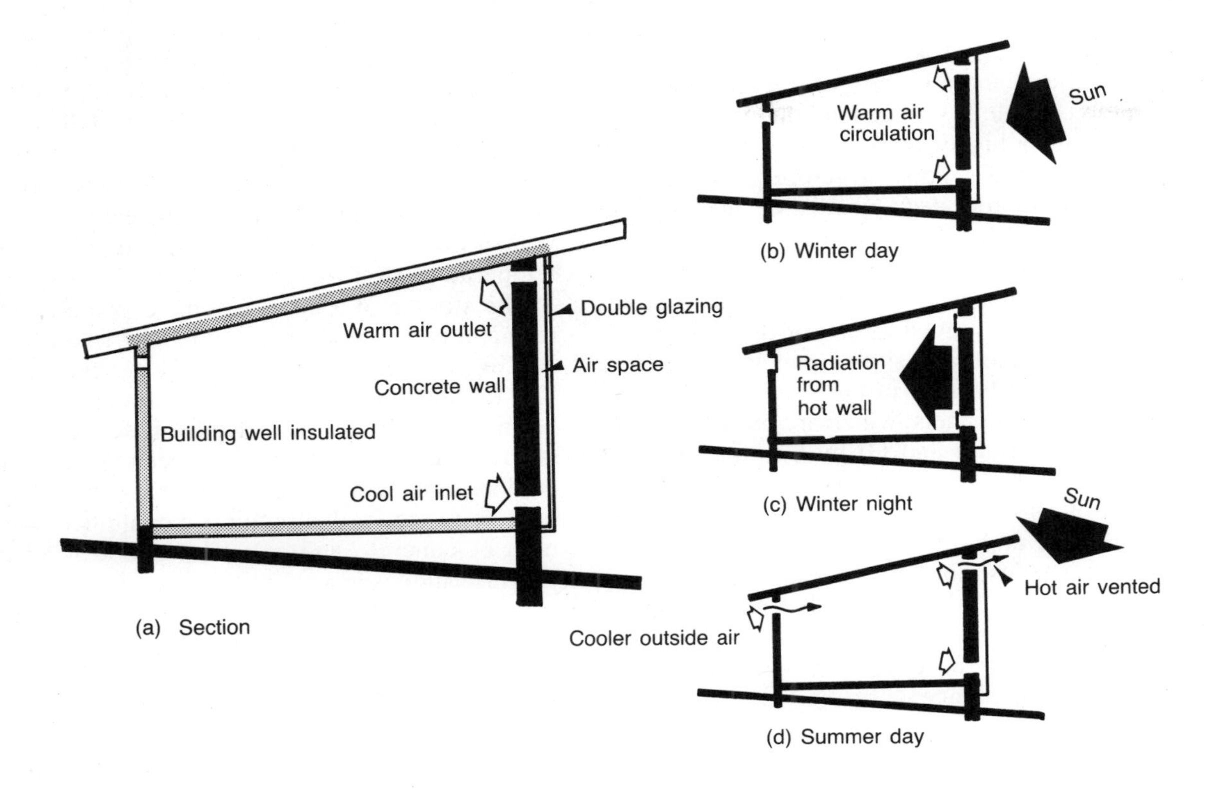

Fig. 5.15.4. House at Odeillo, southern France, designed by Dr. Felix Trombe and built in 1967. The house has a large, double-glazed window at the south face. However, there is no view from this window, because immediately behind it, with only a narrow passage for air circulation, there is a 300-mm concrete wall, painted black. This is now commonly called a Trombe wall (a). Heating and cooling are accomplished by natural ventilation (b, c, and d). During the summer the overhang of the roof shades the south wall completely, and there are vents opening outward that draw cool air from the north side of the house.

In winter during the day the south wall stores heat that is radiated at night. The total time of transmission of heat through the wall is about twelve hours. The thick concrete wall provides sufficient insulation at the south face. The vents through the wall are closed at night, and the glazing reduces heat loss by radiation from the wall. (*From Ref. 5.11, p. 636.*)

or bricks. This is often called a *Trombe wall,* after Dr. Felix Trombe, who first used it in the south of France in the 1960s (Fig. 5.15.4).

This can be made more effective by adding an insulating shutter that can be pulled up by hand (Fig. 5.15.5), or by filling the space between the two sheets of glazing of the window in front of the Trombe wall with insulation. This is done by blowing beads of expanded polystyrene 3 to 5 mm in diameter into the space with a fan, and sucking them out again with a vacuum to return them to their container, when the insulation is not required. The system needs regular cleaning to operate successfully. A sun sensor evaporating freon, the fluid used in refrigerators, can be used to activate the bead system automatically.

Thermal storage walls placed immediately behind large windows greatly improve the thermal performance of passive solar houses, but they do not add to their beauty. The large southern window has behind it a blank black wall or black water drums and provides from the interior no view of the landscape that it faces. The insulating shutter shown in Fig. 5.15.5 also does not improve the appearance of the building.

All the designs discussed in this section require large areas of transparent material that may be subjected to a wide variation of temperature, particularly if it is backed by a black wall acting as a thermal collector. It is therefore advisable to use safety glass, tempered glass, or a transparent plastic material, although the latter will

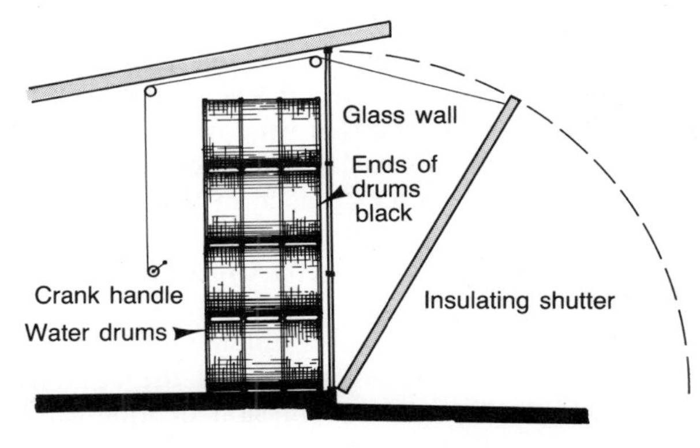

Fig. 5.15.5. Insulating shutter that can be used to improve the thermal efficiency of a Trombe wall built of water drums. The shutter is closed at night during winter to reduce radiation loss from the window, and during the day in summer to prevent radiation gain. (*Residence of S. Baer in Corrales, New Mexico, Ref. 5.12, pp. 50–53.*)

need to be replaced after several years because of deterioration.

Calculations for determining the sizes of the components used in passive solar houses are given in Refs. 5.12 and 5.14.

The designs described in this section can provide all the winter heating for climates that require less than 3000 degree days (based on 18°C; see Section 5.1), provided that the weather is predominantly sunny in winter (Ref. 5.14, pp. 142–258). The thermal inertia of the building and its rock bed determine the number of successive cloudy days for which heating can be provided. If more than one week of cloudy weather is possible, it is generally more economical to provide a standby conventional heating system [Chapter 6].

Although provision can also be made for summer cooling, large windows are particularly suitable for houses where the passive solar energy is required mainly for winter heating. This is likely to occur at latitudes greater than 30°, so that the greatest altitude of the sun in midwinter [Section 3.3] cannot exceed 90 − 30 − 23 = 37°. Thus solar radiation on windows facing south (north in the Southern Hemisphere) is more effective than on devices placed in the roof.

5.16 Passive Solar Designs Utilizing Roof Ponds or Roof Traps

The roof is particularly suitable for devices whose primary purpose is to cool the building during the night by the emission of long-wave radiation [Sections 4.1 and 5.8]. Although the building radiates in all directions, the roof, being most exposed to the sky, is the most effective long-wave radiator. However, it can be used for winter heating as well as summer cooling.

The *solar roof pond* (Fig. 5.16.1) consists of water bags containing a layer of water about 250 mm deep. The bags are generally made of black vinyl, which has an absorptivity of 0.9 for solar radiation and an emissivity of 0.9 for long-wave radiation (Table 5.3). The load due to these bags is about 25% higher than the live load normally specified for the floor structure of residential buildings. The roof deck must be detailed to drain any leakage from the water bags.

The water bags are covered with insulation when heat gain or loss is undesirable. This insulation can be moved in one of three ways: (1) By using horizontally sliding panels that can be moved by cords from inside; (2) by using horizontally hinged panels, which are pushed up with a bar from inside; and (3) by blowing beads of expanded polystyrene into a space between two sheets of glass or transparent plastic above the water bags, and sucking them out again with a vacuum, as described in Section 5.15.

The horizontally hinged panels evidently cannot be left in their upward position when a high wind is blowing, and the other two systems require periodic maintenance.

Water bags 250 mm deep can supply all the heating requirements in a sunny climate where the nighttime temperature does not fall below 2°C (Ref. 5.14, p. 150). In summer water bags can lower the indoor temperature to bring it within or close to the comfort range, reducing the indoor temperature up to 16°C below the outdoor shade temperature (Ref. 5.15).

The storage of a great volume of water in the roof can be avoided by a method devised by Dr. B. Givoni (Ref. 5.16), illustrated in Fig. 5.16.2. The thermal inertia of the flat concrete roof may be increased by a layer of gravel, so that there is a time lag of 6 to 8 hours between the extreme outdoor temperature and the release of the stored heat or coolness; this requires a combined thickness of concrete and gravel of about 300 mm. The insulation is in a separate second roof covered with corrugated sheets painted white that slopes to a window facing south (north in the Southern Hemisphere). Between the two roofs there is an insulated reflecting panel that is open during the

Table 5.3

Reflectivity and Absorptivity of Solar Radiation, and Emissivity of Long-Wave Radiation

Material	Reflectivity of Solar Radiation	Absorptivity of Solar Radiation	Emissivity of Long-Wave Radiation
Bright galvanized steel	0.75	0.25	0.25
Bright aluminum foil	0.95	0.05	0.05
Aluminum paint	0.50	0.50	0.50
White paint	0.70	0.30	0.90
Green paint	0.30	0.70	0.90
Black paint	0.10	0.90	0.90
White marble	0.55	0.45	0.95
Gray concrete	0.35	0.65	0.90
Brick	0.40	0.60	0.90

(From Ref. 3.9, p. 98, and Ref. 5.7, p. 108)

day in winter and closed during the night; in summer the process is reversed. The air in the space between the two roofs (hot in winter, cool in summer) can be drawn by a fan to a rock bed and released by reversing the direction of air flow. This provides a store of heat or coolness to tide the system over a few cloudy days. In summer the nighttime cooling can be further improved by sprinkling water over the gravel on top of the flat concrete roof and allowing it to evaporate [see Section 5.12]. However, there are no data on the effectiveness of the system, as no house has been erected that uses it, whereas a number of buildings with solar roof ponds have been occupied for several years.

Passive solar systems for cooling are effective only in hot-arid climates, where there is appreciable difference between the daytime and nighttime temperatures. In hot-humid climates, where the temperature difference between day and night is small, thermal comfort without air conditioning depends on ventilation [Section 5.11].

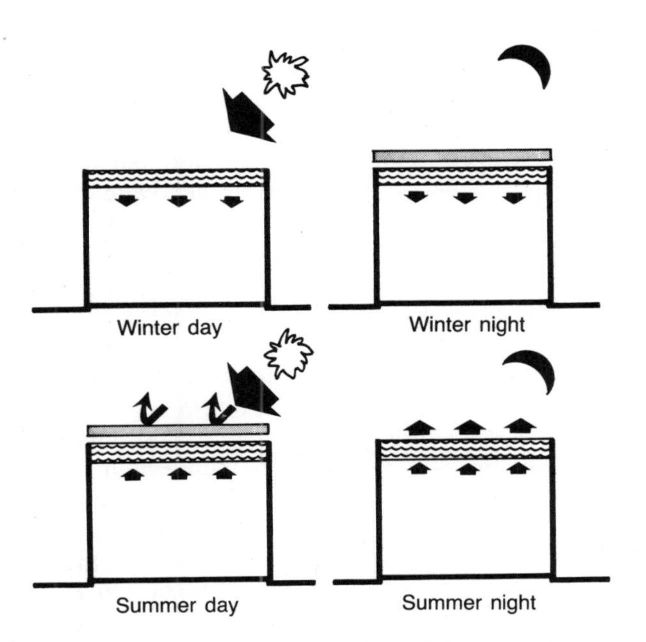

Fig. 5.16.1. Solar roof pond, formed by water bags about 250 mm deep. The bags are exposed to solar radiation during the day for winter heating and are covered with an insulating lid at night to prevent heat loss. In summer the process is reversed to prevent radiation gain during the day and to cool the building by radiation at night.

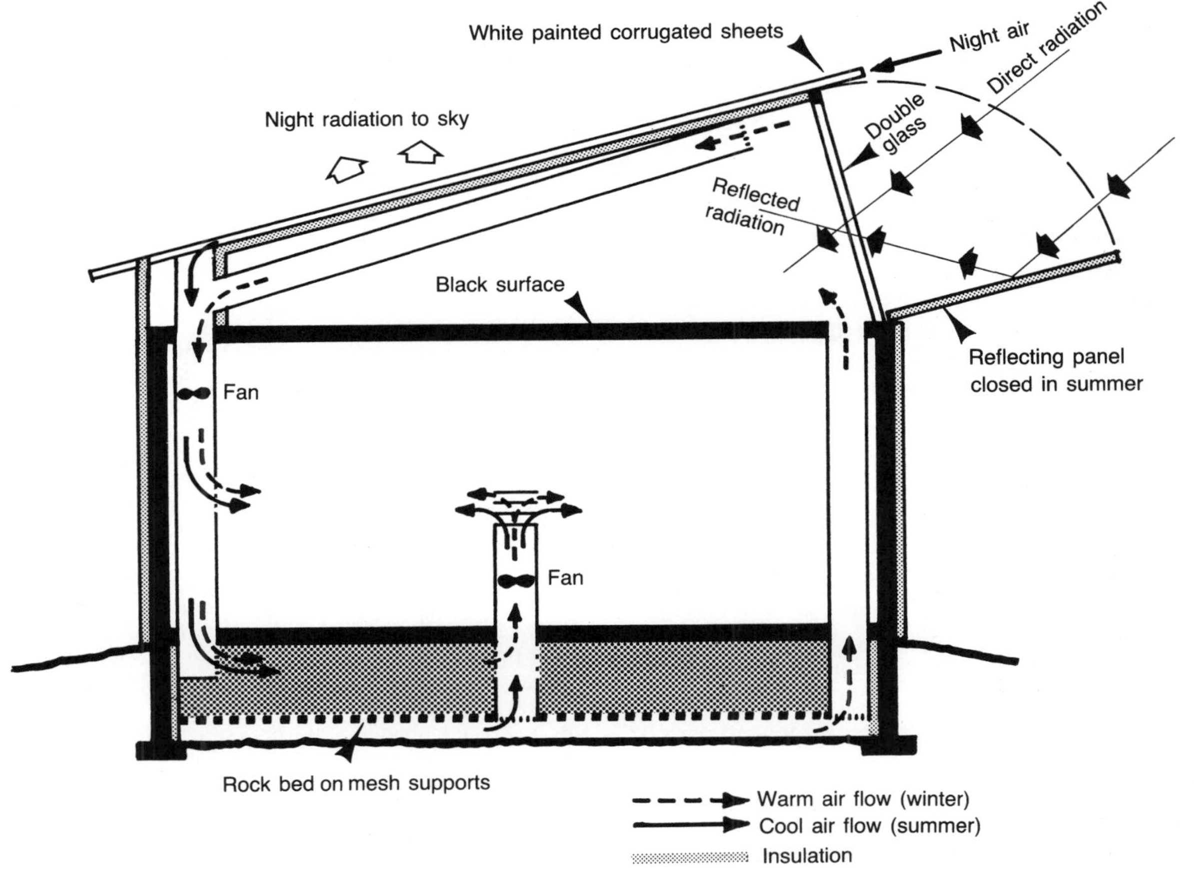

Fig. 5.16.2. Design for a radiation roof trap that does not require the storage of a great volume of water in the roof. In winter a hinged reflecting panel is opened during the day, so that solar radiation enters through a sheet of glass to heat the upper surface of the ceiling. The ceiling thus becomes a radiant heating panel. In addition, hot air can be drawn from the roof trap to a rock bed in the basement. This heat may be used at night and on cloudy days, and the rate of heat flow through the plenum ducts can be regulated with a fan.

The hinged panel is closed at night during winter to avoid heat loss and during the day in summer to prevent heat gain. During summer nights air can be drawn in under the white-painted corrugated roof sheets, where it is cooled by outgoing radiation on a clear night. The cool air can be fed directly into the building or into a rock bed. (*From Ref. 5.16.*)

5.17 Earth-Integrated Buildings

Earth is not a good insulating material. Its thermal conductivity (see Table 5.2) ranges from 1.5 to 0.3 W/mK, depending on the nature of the soil and its moisture content; its average value is roughly the same as that of concrete. The *k*-value of most insulating materials is of the order of 0.04 W/mK, better by a factor of 20. However, the layer of soil around an earth-integrated building is thick, so that its thermal resistance is excellent. Furthermore, the soil in contact with the building gives it a high thermal inertia.

At a depth of 1 meter the daily temperature range is reduced to less than 4°C, and spells of exceptionally hot or cold weather lasting only a few days are damped out. The difference between the average summer and winter temperatures, which in a cold continental climate may exceed 25°C, is reduced to about 12°C (Ref. 5.17).

The heating load for underground buildings is therefore much smaller than for above-ground buildings, and the heat losses through the fabric are low. Underground houses are normally heated by conventional means [see Chapter 6]; but if the site is favorable, passive solar heating can be used. The simplest method is by direct gain through large windows (Fig. 5.17.1). This can be done with only a small amount of excavation, if the building is on a sloping site facing within 20° of true south (north in the Southern Hemisphere). Alternatively, the house can be built on a flat site with its windows facing south, and earth heaped on top of it to create an artificial hill. The windows can be shaded to permit entry of the sun in winter and to exclude it in summer [Section 3.4].

If in addition to a cold winter prolonged periods of hot-humid weather may be expected in summer, it is advisable to have also a window on the opposite face to make cross ventilation possible [Section 5.11]. However, earth-integrated buildings confer no advantages in a climate that is predominantly hot-humid and shows little annual variation in temperature.

In a hot-arid climate an open courtyard or *atrium* in the center of the building (Fig. 5.17.2) is often more comfortable, and this form of construction is traditional in some parts of the Middle East.

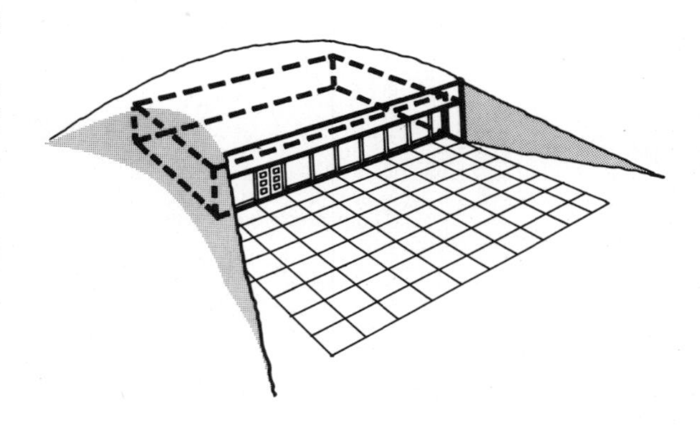

Fig. 5.17.1. Earth-integrated house built on a level site with its windows and entrance door facing south (in the Northern Hemisphere), and a paved courtyard in front of them. The roof overhang shades the windows from the summer sun but admits it in winter, to produce passive solar heating by direct gain. The building is covered with earth to form an artificial hill planted with grass.

Fig. 5.17.2. Earth-integrated building on a level site with its windows facing an open courtyard, or atrium, from which the building is entered. Provided that the atrium is wide enough, the windows on three sides of it can be used for passive heating in winter and shaded in summer, while the fourth side is always shaded. Internal courtyards can produce unexpected wind effects, and it is advisable to test a model of the design in a wind tunnel. The earth above the house can be shaped to deflect air currents and prevent undesirable turbulence inside the atrium.

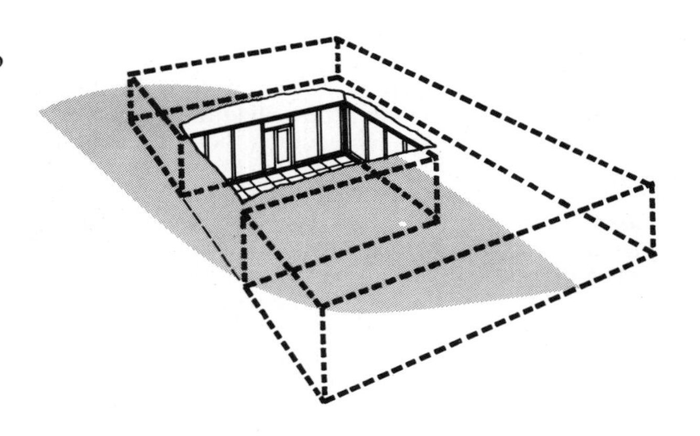

The interaction between the building and the soil and water requires careful consideration. Some clays expand on wetting and shrink on drying. If clay is present near the site of an earth-integrated building, an expert on soil mechanics should be consulted. The building must be erected above the level of the water table. If the water table is near the surface, the site is unsuitable for an earth-integrated building.

The excavation should be backfilled with a porous material (Fig. 5.17.3); if the excavated material is unsuitable, then it should be disposed of and a suitable backfill substituted for it. The drainage tiles should be placed *below* the floor level of the building. Drainage is easiest on a sloping site. On a level site it is generally necessary to pump the water.

Any cracks or holes in underground buildings must be filled during construction, and the walls waterproofed. Pitch (bitumen) or asphalt is traditionally used for this purpose, but in recent years membranes of polyethylene or butyl rubber have been used instead or in addition. Alternatively, a layer of bentonite, a fine-grained, expansive clay, can be employed; this keeps out moisture because it expands when it is wetted and thus seals its own cracks in the presence of moisture (Ref. 5.18).

Apart from the thermal advantages of an earth-integrated building, it enlarges the area available for a garden, which can be placed on its roof. In the center of a city or suburb open spaces can be created by placing commercial activities underground (Ref. 5.19). Moreover, earth-integrated buildings need no maintenance to most of the external surfaces and are relatively immune to vandalism.

Living and working underground evidently create some social and psychological problems. There should be sufficient windows to provide contact with the outside world, and it is advisable to have a level entrance to counter the feeling that one is entering a basement.

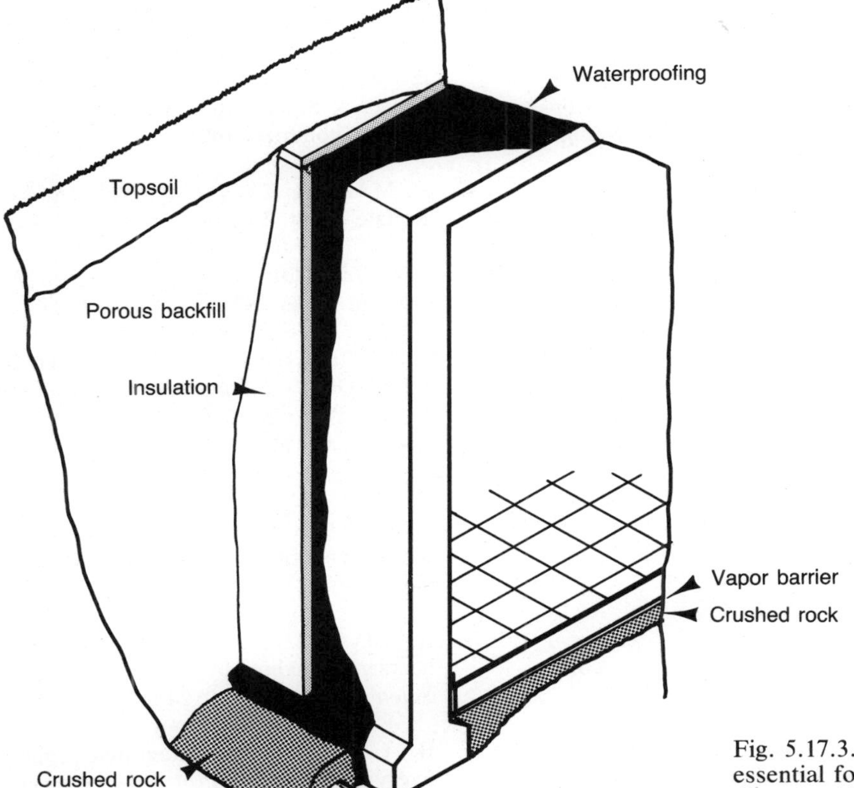

Fig. 5.17.3. Proper drainage and waterproofing are essential for earth-integrated buildings. The drainage tiles should be *below* the floor level, and there should be a porous backfill above them so that most of the water can be drained before it reaches the walls.

Earth-integrated buildings do not overshadow their neighbors, but the reverse can be a problem. Consideration should be given to the likely effect of a larger building that might subsequently be erected on a neighboring site.

5.18 Life Cycle Costing

Let us assume that the environmental system requires an initial capital outlay of C dollars and an annual charge for maintenance and energy (electricity, oil, and gas) of m dollars. This annual charge can be capitalized into an amount of M dollars by determining the number of dollars we would need to borrow from a bank, or withdraw from our investments, today to pay the annual bills later. Thus the total equivalent capital cost, or life cycle cost, is

$$E = C + M$$

The most economical design is one that gives the lowest value for E. Thus we may have to spend more money on the initial capital cost to save money on the life cycle cost.

Some energy-conservative measures discussed in this chapter cost little or nothing. It is only necessary to understand the problem correctly and make the right decision. However, others add significantly to the capital cost, and this may be justified if the life cycle cost is reduced. Better insulation adds a little to the cost of the building. Earth-integrated construction, solar collectors, and some aspects of passive solar design substantially increase the initial cost.

The method of calculating life cycle costs is explained in many textbooks on economics or building management (for example, Ref. 5.20). However, it involves a number of assumptions. For example, the cost of energy may continue to rise at the present rate, or faster, or slower; interest rates and the rate of inflation may rise or fall. An organization that promotes a particular product (for example, insulation or solar collectors) or a particular cause (for example, support for solar or for atomic energy) is naturally inclined to assume values that favor its argument, within the admissible range. Calculations of the value of increased capital cost in reducing running costs should not therefore be accepted unquestioningly.

Energy conservation is in the interest of humanity, particularly where the energy is derived from fossil fuels that cannot last forever. Governments can influence energy conservation by taxing energy sources or subsidizing energy-conservative measures, but it is unrealistic to expect any but a dedicated few to save energy if it costs more money. The life cycle cost is therefore important in determining the long-term economics of an unconventional design.

However, life cycle costing is likely to meet opposition from some clients. For a government or an institution that expects to be responsible for the operation of a building throughout its useful life, cycle costing is evidently economical. A young couple buying their first house may have great difficulty in borrowing the money, and they may prefer to pay for the extra energy needed to heat a cheaper house.

Whether a retrofit or a change in design is worthwhile can also be determined by calculating the number of years required to repay the capital investment from the energy savings achieved. This calculation is based on the same theory, and it involves the same uncertainty in the assumptions with regard to inflation, interest rates, and energy cost in the future. However, if the period of repayment is shorter than three years, the uncertainty is far less than if it were twenty years. Furthermore, the owner of a house or of a commercial building can form an opinion whether he is still likely to be in possession at the end of the payback period, as the new owner may not be willing to pay more for the building on account of an improvement that is not visible, for example, for additional insulation.

References

5.1 *ASHRAE Handbook—1976, Systems*. American Society of Heating, Refrigerating and Air Conditioning Engineers, New York, 1976. 760 pp.

5.2 B.H. JENNINGS: *Environmental Engineering*. International Textbook Company, Scranton, 1970. 765 pp.

5.3 O. FABER, J.R. KELL, and P.L. MARTIN: *Heating and Air Conditioning of Buildings*. Architectural Press, London, 1971. 562 pp.

5.4 J. PORGES: *Handbook of Heating, Ventilating and Air Conditioning*. Seventh Edition. Newnes-Butterworth, London, 1976. 15 sections.

5.5 P.W. O'CALLAGHAN: *Building for Energy Conservation*. Pergamon, Oxford, 1978. 232 pp.

5.6 W.J. McGUINESS, B. STEIN, and J.S. REYNOLDS: *Mechanical and Electrical Equipment for Buildings*. Sixth Edition. Wiley, New York, 1980. 1336 pp.

5.7 B. GIVONI: *Man, Climate and Architecture*. Second Edition. Applied Science Publishers, London, 1976. 483 pp.

5.8 *The Effect of Windowless Classrooms on Elementary Schoolchildren*. Architectural Research Laboratory, University of Michigan, Ann Arbor, 1965. 110 pp.

5.9 R. PERSSON: *Flat Glass Technology*. Butterworths, London, 1969. 167 pp.

5.10 G.K. GARDEN: *Rain Penetration and Its Control*. Canadian Building Digest No. 55, Division of Building Research, National Research Council of Canada, Ottawa, 1964, 4 pp.

5.11 KAIMAN LEE: *Encyclopedia of Energy-Efficient Building Design: 391 Practical Case Studies*. Environmental Design and Research Center, 940 Park Square Building, Boston MA 02116, 1977. 2 volumes, 1023 pp.

5.12 E. MAZIRA: *The Passive Solar Energy Book*. Rodale Press, Emmaus, Pa., 1979. 435 pp.

5.13 S.V. SZOKOLAY: *World Solar Architecture*. Halsted Press, New York, 1980. 278 pp.

5.14 W.C. DICKINSON and P.N. CHEREMISINOFF (Eds.): *Solar Energy Technology Handbook. Part B: Application, Systems and Economics*. Marcel Dekker, New York, 1980. 804 pp.

5.15 B. GIVONI: Passive cooling of buildings by natural energies. *Energy and Buildings,* Vol. 2 (1979), pp. 279–85.

5.16 B. GIVONI: Integrated-passive systems for heating of buildings by solar energy. *Architectural Science Review,* Vol. 24 (1981), pp. 29–41.

5.17 B. GIVONI: Earth-integrated buildings—an overview. *Architectural Science Review,* Vol. 24 (1981), pp. 42–53.

5.18 S. CAMPBELL: *The Underground House Book*. Garden Way Publishing, Charlotte (Vermont), 1980. 210 pp.

5.19 Underground architecture. *AIA Journal*, Vol. 67, No. 4 (April 1978), pp. 34–51.

5.20 P.A. STONE: *Building Design Evaluation—Costs-in-Use*. E. & F. Spon, London, 1967. 207 pp.

Suggestions for Further Reading

References 1.5, 4.12, 4.13, 5.7, 5.12, 5.13, and 5.18.

5.21 V. OLGYAY: *Design with Climate—Bioclimatic Approach to Architectural Regionalism*. Princeton University Press, Princeton, 1963. 190 pp.

5.22 M. EVANS: *Housing, Climate and Comfort*. Architectural Press, London, 1980. 186 pp.

5.23 D.A. McINTYRE: *Indoor Climate*. Applied Science Publishers, London, 1980. 443 pp.

5.24 T. ANGUS: *The Control of Indoor Climate*. Pergamon, Oxford, 1969. 110 pp.

5.25 R.M. AYNSLEY, W. MELBOURNE, and B.J. VICKERY: *Architectural Aerodynamics*. Applied Science Publishers, London, 1977. 254 pp.

5.26 T.S. ROGERS: *Thermal Design of Buildings*. Wiley, New York, 1964. 196 pp.

5.27 G. CONKLIN: *The Weather Conditioned House*. Reinhold, New York, 1958. 238 pp.

5.28 R.G. STEIN: *Architecture and Energy*. Anchor Press, New York, 1977. 322 pp.

Chapter 6 Heating, Mechanical Ventilating, and Air Conditioning

The limitations of natural lighting and ventilation and the need for artificial climate control are discussed. Space heating, of single rooms or by a central heating system, is first considered. Refrigeration machines are presented as a means of producing either heat or cold, and their application to air conditioning systems is described. Further details of air conditioning plant and the associated ductwork, and methods of zoning, are also discussed. The types of fuel available for heating and cooling, including solar energy, are listed. Sample calculations are given for determining heating and cooling loads and duct sizes.

6.1 The Need for Climate Control

The designer must decide, at an early stage in the design of a building, the extent of the artificial climate control to be used. Dependence on natural light and ventilation impose certain demands on the shape, size, and fenestration of the building. Artificial lighting and air conditioning, in theory, can be applied to any building, but in practice we wish to set reasonable limits on the cost of the plant and the energy it consumes. This imposes constraints on the building form that differ from those for a naturally lit and ventilated design (Fig. 6.1.1).

In the most favorable climates it is easy to design a small building that requires no heating, cooling, or mechanical ventilation (Fig. 6.1.2). In many climates that are mild but not ideal, a careful design incorporating passive solar techniques [see Sections 5.14 and 5.15] can be used to achieve thermal comfort with a minimum of added energy.

On the other hand, mechanical control is needed to make buildings comfortable in more severe climates. It is also needed in large buildings, where natural ventilation is not practical. Libraries, recording studios, and other special buildings require air conditioning to control the cleanliness, humidity, or sound isolation of the interior, quite apart from the comfort of the occupants (Fig. 6.1.3).

The major elements of climate control are heating, cooling, and ventilation. Heating and ventilation are sometimes supplied separately, but if cooling is required it is best to provide an air conditioning system with the facility to control all three factors, as well as the humidity of the air and its cleanliness.

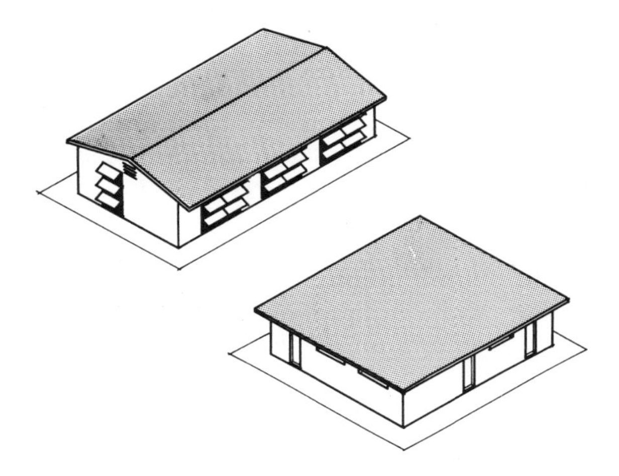

Fig. 6.1.1. A naturally ventilated building is designed differently from an air-conditioned building. In one the size of windows is increased and cross ventilation is encouraged by a narrow plan shape. In the other, windows are kept small and a shape is chosen to reduce total surface area exposed to the elements.

Fig. 6.1.3. Some buildings require extensive equipment because of their use, whatever the climate.

Fig. 6.1.2. In an ideal climate no heating or cooling is needed.

6.2 Local Heaters

Heat is produced by passing electricity through a resistance or by burning a fuel. A portable electric heater can be used from any suitable outlet, and it produces no harmful products. A fuel-burning heater needs a supply of fuel and produces combustion gases.

A portable heater, whether it uses electricity, gas in cylinders, or kerosene, places no special demands on the building designer. However, the output of portable heaters is limited. Fixed heating must be sized correctly for the heat load of the room and supplied with the appropriate fuel: a special electrical circuit, an oil pipe, a gas pipe, or a supply of solid fuel. Most oil and gas heaters also require electricity for ignition, and to operate a fan. Electrical heaters and some types of gas heater do not require a flue. The others require a flue or chimney to take away the products of combustion. Flues get hot and must be insulated from wood frame or other combustible material.

A stove or open fireplace burning wood, coal, or coke can be used to heat an individual room. The open fire uses fuel inefficiently, since most of the heat goes up the chimney (Fig. 6.2.1). However, it gives a cheerful appearance to a room, and if the householder can supply and cut his or her own wood, the cost is not a consideration. The work of bringing in the wood and cleaning out the ash must be taken into account. The enclosed slow-combustion stove is more efficient and therefore needs less fuel and produces less ash. Either type is a dominant feature in a room, and therefore its effect must be considered in the visual design of the interior as well as in the location of the chimney and provision for storing the fuel.

The main purpose of all space heating is to make the occupants comfortable [Section 4.7]. This can be achieved by radiant or convective heating. Radiant heating occurs when a hot surface such as the glowing element of an electric radiator emits heat in a particular direction. Convective heating occurs when the air in the room is heated and then circulated. Convective heating generally allows people to move about in the room and remain warm. However, it is less effective in large or high spaces, where the warm air tends to rise to the top; in these cases radiant heating may be more appropriate.

Heating may be continuous or intermittent. The thermal environment is partly influenced by the temperature of the surrounding surfaces, which contribute to the mean radiant temperature [Section 4.2], and of the furniture, which absorbs heat. Although it seems a waste of energy to heat an unoccupied space, it takes time to bring

Fig. 6.2.1. A fireplace can be a dominant feature in the design of a room.
(a) A traditional cast-iron fireplace. (*Photo by Warren Julian.*)
(b) A modern sheet-metal fireplace installed free-standing against a wall.

these up to the required temperature. If the occupancy pattern is known, it is advantageous to turn the heater on some time before the room is to be occupied, using a time-switch. Some background heating may be needed throughout the winter, to prevent condensation on cold surfaces or, in very cold climates, to prevent the formation of ice.

A room responds to intermittent heating more rapidly if the interior surfaces have a low thermal diffusivity [Section 5.6]; examples of such surfaces are carpet, and light plasterboard over insulating material. Solid brick and concrete respond slowly and are therefore better suited to continuous heating.

6.3 Central Heating Systems

To heat many rooms in a building, we can install a heater in each room, but it is often preferable to use a central heat source with a system of heat transfer to the rooms. There is no advantage in using a central system with electric resistance heating, but in the case of other fuels a single large furnace is more economical than many small ones.

The heat can be transferred by air, water, or steam, but the latter is often noisy and offers few advantages over hot water.

In a *warm-air plenum system* the air is circulated in ducts that serve supply registers in the floor. After passing through the rooms, the air is returned to the furnace through return air grilles near the ceiling and a return air plenum (Fig. 6.3.1). A fan is used to maintain the flow, and in most installations the air is filtered before passing over the hot surfaces of the furnace. The furnace may be controlled manually or by a thermostat located in the return air path.

The air ducts for the first floor of a small building can be accommodated in a basement or crawl space or buried under a concrete slab floor. Ducts

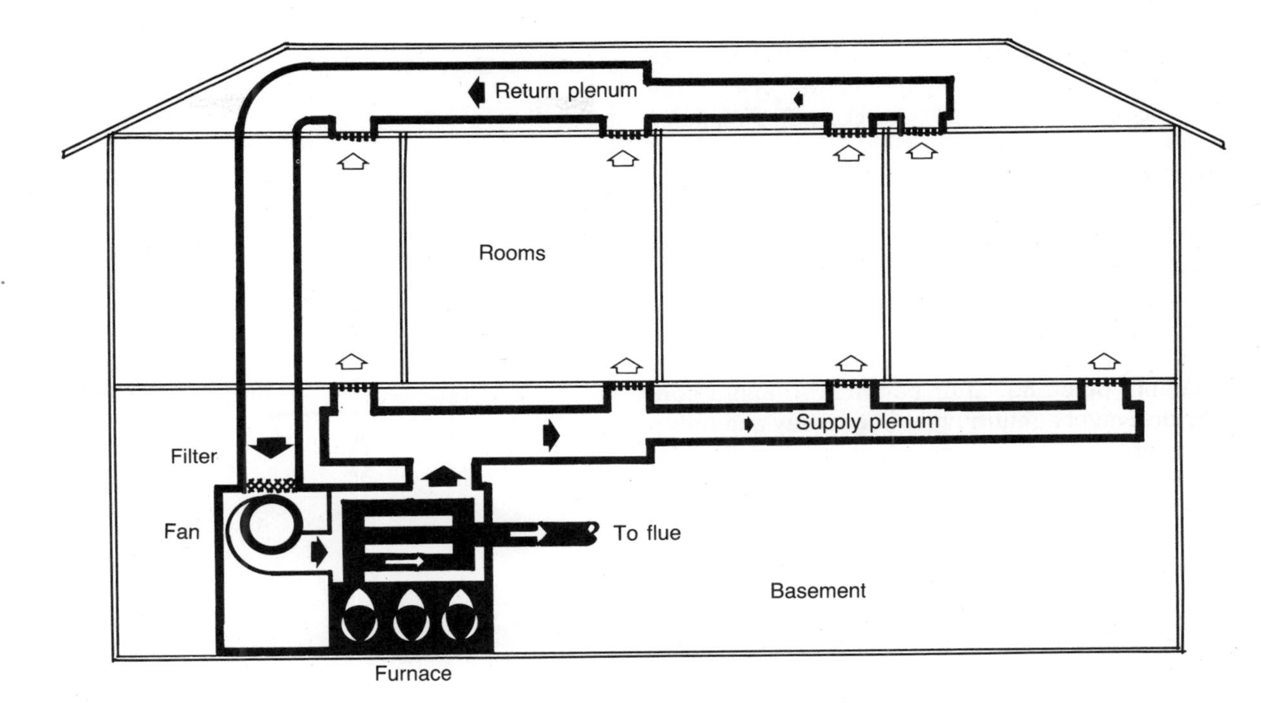

Fig. 6.3.1. In the warm-air plenum system, a fan circulates air around the furnace and distributes it to the rooms through ductwork. It is recirculated through a return grille back to the furnace.

to serve upper stories require special consideration if they are to be concealed.

In a *hot water central heating system,* the pipes are usually about 25 mm in diameter in a building of domestic size. Insulation increases this to about 50 mm, which is easily accommodated. Each room is heated with one or more convectors connected to the hot water circuit. Convectors may be free-standing or attached to the baseboard (Fig. 6.3.2). Where possible they are placed under the windows because these are a major source of heat loss. This also avoids interfering with the placement of furniture in the rooms, since large pieces of furniture are rarely placed in front of windows.

Fig. 6.3.2. A free-standing hot water convector.

The hot water flows in a closed loop. Some of it passes through the convectors in use, but it bypasses each one that is turned off. Circulation can be by thermosyphon (Fig. 2.4.6), but this is usually too sluggish for efficient operation, and so a small pump is preferable to maintain the flow. The piping may be arranged in a *single-pipe system* in which the water, after flowing through a convector, returns to the same loop (Fig. 6.3.3). Alternatively, a *two-pipe system* can be used, in which the water leaves the boiler in a supply pipe and, after passing through a convector, enters a return pipe. The supply and return pipes are located adjacent to each other for most of their length (Fig. 6.3.4).

Although most hot water central heating systems use convectors to provide heating to the rooms, it is possible to provide heating by embedding a coil of pipe in the floor, or in the ceiling, to produce a large, low-temperature radiant panel. However, the heating of ceiling panels by hot water is less common. Either ceilings or floors can also be heated with embedded electric cables, and this solution can be attractive if electric resistance heating is preferred to other energy sources. It is necessary to use insulation under the floor, or above the ceiling, to ensure that most of the heat is delivered in the direction it is intended.

Underfloor heating is pleasant, since it warms the feet, unlike most other heating systems. However, the response is slow, because the mass of the floor has to be heated. It is better suited to continuous than to intermittent heating.

Since underfloor heating performs better with bare floors than with thick, insulating carpets, it is sometimes used in selected rooms such as game rooms and bathrooms, while other forms of heating are used in the remaining rooms.

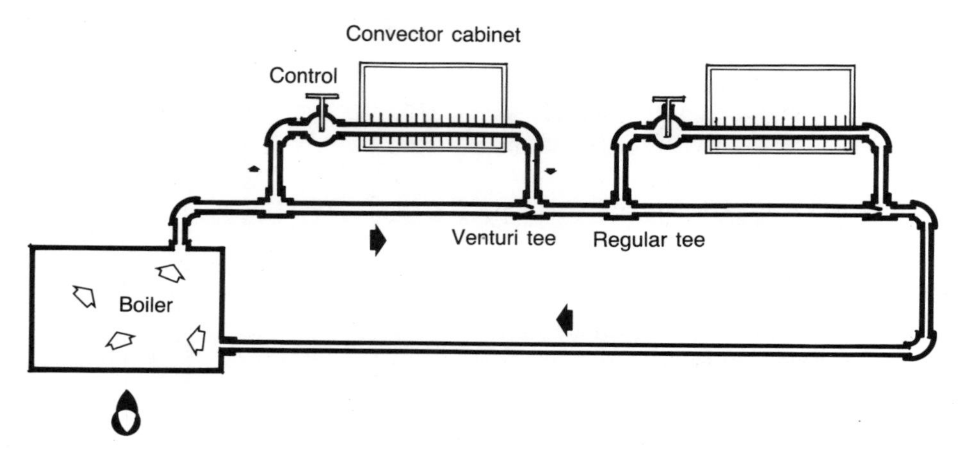

Fig. 6.3.3. The single-pipe system uses special tees that cause increased pressure on the upstream side, and a zone of reduced pressure on the downstream side, to assist flow through the convector.

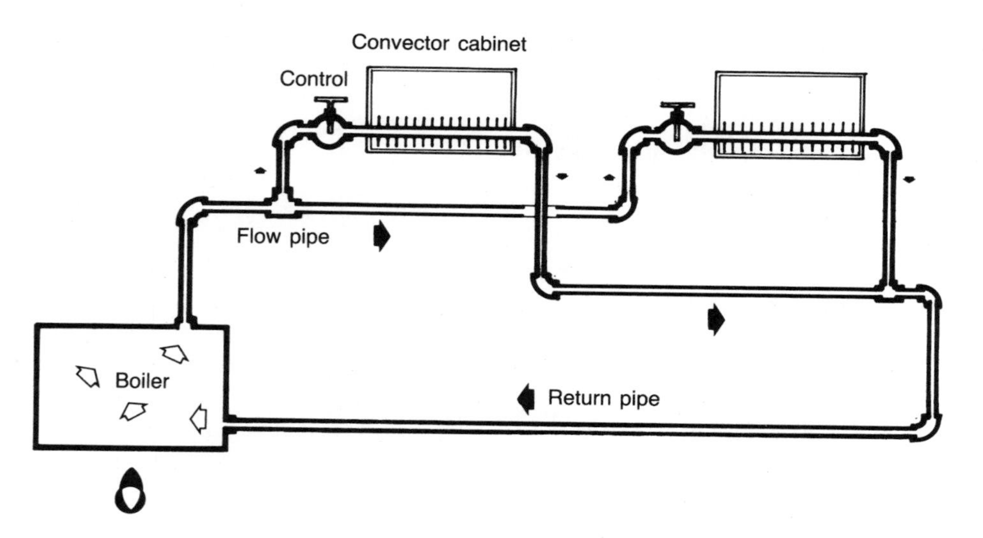

Fig. 6.3.4. In a two-pipe system the flow and return pipes are separate. There is a pressure difference across each convector.

6.4 Heat Sources for Central Heating Systems

The fuel for a furnace or boiler for a central heating system can be oil, gas, solid fuel, or electricity. A solid-fuel stove large enough to hold enough coal or wood for a day's operation is more convenient than a smaller one that needs frequent refueling. Electric resistance heating may be used, but a more economical use of electric energy is the *heat pump,* which is described in Section 6.6.

The source of energy may also be solar. We described solar collectors in Section 2.4. When used for domestic hot water, a collector area of about 5 m^2 per family is usual; when used for space heating, the area of the collectors may need to be half the floor area of the building. It is therefore important to choose an economical design, and it may be appropriate to use a collector that also serves as the roof covering. The durability of the collector materials, and the effect of a large collector on the appearance of the building, require careful consideration (Fig. 6.4.1).

If the working fluid of the collectors is air, the heat is stored in a rock bed; if it is water, the heat is stored in an insulated tank. For the same heat storage, water has about one fifth the mass and about one third the volume of a rock bed, and the circulating pipes are much smaller than the air ducts. However, air allows the use of cheaper materials, and the risk from leakage is less.

At present it is less economical to use solar energy for space heating than for domestic hot water. Each square meter of collector can collect only so much energy per day, but a hot water system uses the solar energy collected throughout the year, while the space heating system uses it only during the winter, with high peak demands. Climates that have a long heating season and clear winter skies are most likely to benefit from solar space heating. These occur in regions with a high altitude, moderate latitude, and inland rather than coastal location.

It is seldom economical to provide a solar heating system with sufficient storage to carry through all periods of bad weather. It is usual to seek a solar contribution of 60% to 75%, the remainder being supplied by conventional energy. The backup system is chosen to complement the solar installation, so that as little as possible of the plant is duplicated.

Fig. 6.4.1. If solar collectors are used to supply the space heating, they occupy a large part of the roof area. The illustration shows glass evacuated-tube collectors on an experimental house at Colorado State University.

6.5 Calculation of Heat Loads (Steady-State Conditions)

The amount of heat flowing into or out of a room can be calculated if we know the temperatures inside and outside and the thermal transmittance (or U-value) of the various elements of the construction [Section 5.5]. This calculation can be used to size the heater or to compare the effect of different forms of construction on the maximum heating load. It cannot be used to find the temperature that an unheated room will reach.

The assumption of steady-state conditions is realistic for winter, since the outside temperature may remain close to its minimum value for periods of one or several days. It is less realistic in summer, when the outdoor air temperature and solar radiation only reach their combined maximum values for a few hours. An example of calculating cooling loads is given in Section 6.11.

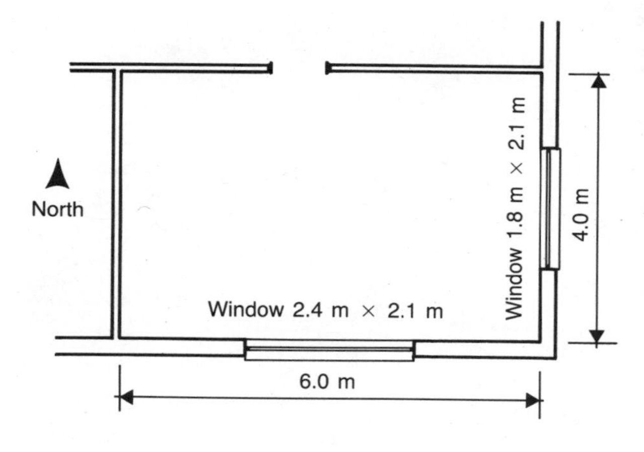

Fig. 6.5.1. The room for the heat loss calculation in Example 6.1.

Example 6.1. *Calculate the heating required to maintain the room shown in Fig. 6.5.1 at 22°C in winter near Washington, D.C. The construction details are:*

Exterior walls:	*200 mm brick, 25 mm expanded polystyrene, 12 mm gypsum plaster.*
Windows:	*metal-framed windows plus storm sash.*
Floor:	*100 mm concrete on grade, 50 mm edge insulation.*
Roof/ceiling:	*Built-up roofing on wood-framed flat deck, 150 mm rockwool insulation.*
Internal walls:	*gypsum plaster both sides of wood frame, no insulation.*

We could calculate the U-value *for each element from the properties of its materials, but it is usual to take them from a handbook such as the* ASHRAE Handbook—Fundamentals *(Ref. 4.1, Chapter 22). This also contains information about outdoor design temperatures and methods of estimating air infiltration. In the solution that follows, reference will be made to the page numbers in the* Handbook *where the values are found.*

Solution *Let us choose the weather conditions listed for Andrews Air Force Base, near Washington, D.C. The winter design temperature reached on all but 2½% of days is −10°C (+14°F) (p. 23.5). The* U-values *of construction are listed on pages 22.19 through 22.25. Areas are calculated from Fig. 6.5.1 and tabulated in Table 6.1. Each area is multiplied by the* U-value *of the construction and by the difference between 22°C and −10°C, which is 32 K.*

The heat loss through the floor can be reckoned by taking the length of the exterior perimeter, in this case 10 m. Page 24.5 of Ref. 4.1 suggests that, for outdoor temperatures down to −23°C, a loss of 38 W/m run should be allowed. This adds 380 W.

The room is infiltrated by cold air from outdoors. Page 21.5 of Ref. 4.1 gives an overall allowance of 1.5 air changes per hour for a room with two exposed walls; but this value is reduced to two thirds if the windows have storm sashes. Therefore we allow one air change per hour, that is, the volume of the room is exchanged with fresh air once an hour. The specific heat per unit volume of air is 1.2 kJ/m³K. The heat load represented by the volume of 57.6 m³ in 3600 seconds, with a temperature difference of 32 K, is

$$\frac{57.6 \times 1200 \times 32}{3600} = 614 \text{ W}$$

If the whole house is heated to the same temperature, no heat loss occurs at the internal walls. If we assume that the remainder of the house is only heated to 15°C, there is a temperature difference of 7 K across the internal walls. The total heat load is given in Table 6.2.

Therefore, a heater of 2.6 kW can just keep up with the heat losses under the design conditions.

Table 6.1

Element	Area (m^2)	*U*-Value (W/ m^2K)	δT (K)	Heat Loss (W)
South and east walls (excluding windows)	15.18	0.71	32	345
Windows	8.82	2.84	32	801
Ceiling	24.00	0.26	32	200
TOTAL				1346

We can also make some other judgments: If the room is occupied by four people, producing about 100 W of energy each, and there are two lights of 100 W and a television set of 200 W operating, this adds another 800 W of heat to the room. At these times, a heater of 1.8 kW would be adequate. On the other hand, if the room were left vacant with the heater turned off for a day, a 2.6-kW heater would be too small, because it has no reserve capacity to warm it up quickly. An excess capacity of 10% to 20% is usually allowed for this purpose. An electric resistance heater with three manual settings of 1, 2, and 3 kW, or one of 3 kW with thermostatic control, would be suitable.

Tables 6.1 and 6.2 show that the windows and the infiltration are by far the largest contributors to the heat loss. Further consideration could be given (1) to the use of double glazing, or (2) to improving the seals around the windows, or (3) to reducing their size; these changes may not be justified in Washington, D.C., but would be appropriate for a more severe climate. The total heat loss through the ceiling, which has 150 mm of insulation, is quite small. It may be more economical to save a little money by reducing the ceiling insulation and to spend it on one of the abovementioned items, where it would give more benefit.

Table 6.2

Element	Area	*U*-Value	δT (K)	Heat Loss (W)
Internal walls (excluding door)	22.32 m^2	1.64 W/m^{2}K	7	256
Internal door	1.68 m^2	3.12 W/m^{2}K	7	37
Infiltration	57.6 m^3/h	1200 J/m^{3}K	32	614
Floor edge	10 m	38 W/m		380
Exterior (from Table 6.1)				1346
TOTAL HEAT LOSS				2633

6.6 Heating and Cooling with Heat Pumps

The two refrigeration cycles of practical interest are the mechanical and the absorption cycles. Both rely on the latent heat of vaporization taken in by a liquid when it evaporates to become a gas. The cycle is operated as a *closed system,* that is, the gas is condensed back to liquid form and reused. The latent heat is recovered during condensation, but it is wasted in a normal refrigeration plant. A refrigeration machine is often called a *heat pump,* since it pumps heat energy from one part of the cycle (the evaporator) to another (the condenser).

The heat pump may be compared to a water pump. Water flows downhill of its own accord, but a pump is needed to make it travel in the opposite direction. Heat energy flows from a hotter to a cooler body of its own accord, but the heat pump makes it flow in the other direction. Heat is taken in at a low temperature at the cooling coils of a refrigerator and rejected at a higher temperature at the condenser coils at the back of the refrigerator, or in the cooling tower on top of an office building.

Therefore the heat pump can be used to take in heat from the cold outside air and release it at a higher temperature to heat the inside of a building. Provided the external heat source is not *much* colder than the required inside temperature, a heat pump can produce more useful heat than any alternative means of using the same energy input. In practice, outside air at temperatures down to 0°C can be used; if water is available in a lake, a stream, or a well, it may be used as a heat source in winter, as well as a sink for waste heat in summer.

In the *mechanical refrigeration cycle,* also called the *vapor compression cycle,* the heat pump is driven by mechanical energy, usually produced by an electric motor. The *working fluid* should change from gas to liquid at a convenient temperature and pressure and should have a high latent heat of vaporization. Water has a high latent heat, but to make it evaporate at 5°C, a pressure as low as 860 Pa (less than 1% of atmospheric pressure) is required. Ammonia was once widely used, but it is dangerous if a leak occurs. The refrigerants presently used are known by the generic name of *freon.* Freon is an organic fluid derived from methane (CH_4); one or more of the hydrogen atoms of methane are replaced by chlorine or fluorine.

Figure 6.6.1 shows a mechanical refrigeration cycle. Heat is taken in at the evaporator and given out at the condenser.

When used for cooling, the evaporator may be located in the airstream of the air conditioning plant. This arrangement, known as *direct expansion,* is common in small plants. In many larger plants it is convenient to pump chilled water to several air conditioning plants in different parts of the building. The evaporator is then immersed in the water of the *chilled water system.* It is not usual to circulate refrigerant further than within the one plant room.

The condenser of a small window-unit air conditioner is cooled by a fan, which passes outside air over it. The condensing temperature is therefore above outside air temperature, usually by 20 K or more. Since the refrigerator works more efficiently when the difference between the condensing and evaporating temperatures is as small as possible, it is better to use a water-cooled than an air-cooled condenser. The cooling water is allowed to cascade over a series of baffles, while a fan draws air through the water. A small amount of the water evaporates, and the remainder is cooled to the wet-bulb air temperature (Fig. 6.6.2).

If this cooling apparatus is located close to the plant room, it is possible to circulate the gaseous refrigerant through it, in a coil that is cooled by the water. This is called an evaporative condenser. In larger plants it is usual to locate the *cooling tower* on the roof and to circulate the cooling water from it to the plant room. The gaseous refrigerant is then cooled by giving up its heat to the cooling water in a heat exchanger.

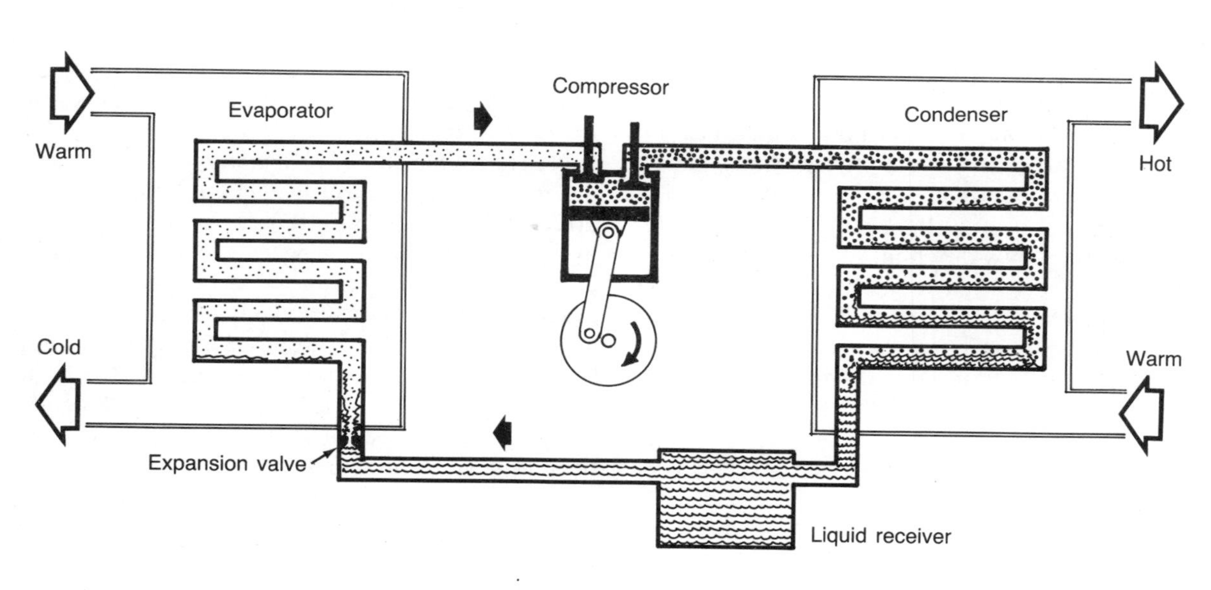

Fig. 6.6.1. The mechanical or vapor-compression refrigeration cycle. The refrigerant is kept under pressure in the liquid receiver. It is allowed to expand through a valve into the evaporator, which is at lower pressure. At the lower pressure it vaporizes and takes on latent heat. This cools the evaporator. A compressor then compresses the vapor, but this raises its temperature. It does not liquefy until it is cooled in the condenser. Heat produced in the condenser is available for use, if needed, or otherwise is discarded.

Fig. 6.6.2. A cooling tower uses the evaporation of water in the air to take away unwanted heat from the condenser of a refrigeration plant.

We do not measure the efficiency of a refrigeration machine in the same way as in other mechanical devices. The ratio of useful cooling to the energy used in the compressor is referred to as the *coefficient of performance* (COP). It can be greater than one, and for practical machines its value is usually between 1.5 and 3.0. This does not violate the principle of conservation of energy, since there is an input of energy both from the heat taken from the building by the evaporator and from the energy supplied to the compressor. This is balanced by the energy output, namely, the waste heat rejected by the condenser (Fig. 6.6.3).

Therefore

$$\frac{\text{Useful cooling}}{\text{Compressor energy}} = \text{COP}_{(\text{cooling})}$$

and

$$\text{Useful cooling} + \text{Compressor energy} = \text{Heat rejected}$$

If we use the heat pump for heating (i.e., in the *reverse-cycle* mode), the heat rejected becomes the useful output. Then

$$\frac{\text{Useful heating}}{\text{Compressor energy}} = \text{COP}_{(\text{heating})}$$

It can be shown from the above that

$$\text{COP}_{(\text{heating})} = 1 + \text{COP}_{(\text{cooling})}$$

COP values from 2.5 to 4.0 are possible in reverse-cycle heating, although a value as high as 4 is hard to achieve when the outside source temperature is low. It is an advantage to use a lake, a stream, or the ocean as the heat source.

A heat pump is therefore a more efficient way of using electrical energy for heating than a simple resistance heater. Furthermore, in most large buildings the interior of the building requires cooling, while the perimeter sometimes requires heating on all sides, or cooling on the sunny side and heating on the other sides. By careful design of the air conditioning plant and its controls, one heat pump can be used to supply both heating and cooling. There can be a great saving in energy consumption if the relationship between the heating and cooling requirements is favorable.

In the *absorption-refrigeration* cycle the refrigerant is evaporated as before, but the vapor is absorbed by another fluid, instead of being removed by a compressor. Ammonia and water are used as a refrigerant-absorber pair in some small refrigerators. As a refrigerant, ammonia can operate below 0°C and therefore produce ice. For air conditioning, however, the present practice is to use water as the refrigerant and lithium bromide as the absorber. The evaporator operates at a very low absolute pressure, so that the water boils at around 5°C.

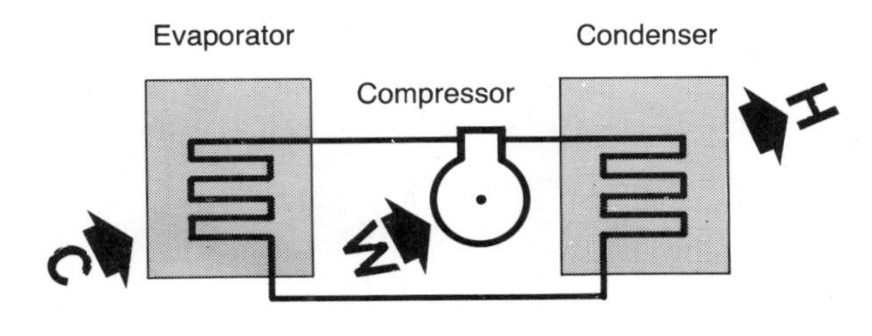

Fig. 6.6.3. The heat balance of a heat pump. C is the cooling available at the evaporator, W is the work done on the compressor, and H is the heat available from the condenser. The principle of conservation of energy requires that $C + W = H$. The coefficient of performance for cooling is $\frac{C}{W}$. The coefficient of performance for heating is $\frac{H}{W} = 1 + \frac{C}{W}$.

The four main parts of the system are shown in Fig. 6.6.4. The evaporator and absorber have already been mentioned. The lithium bromide becomes diluted by the water vapor it absorbs. To maintain its concentration, this solution is pumped to the generator, where the water is evaporated. The water vapor is then condensed and returned to the evaporator.

Heat is taken in at low temperature at the evaporator, and the driving heat energy is supplied at higher temperature at the generator. Heat is rejected both from the absorber and from the condenser. The overall coefficient to performance for cooling is typically about 0.6. The system could be used in the reverse-cycle mode, but in practice this is not done because the COP is not high enough to justify its use.

Absorption refrigeration can be used for air conditioning wherever there is a cheap source of heat at a temperature just above the boiling point of water, say, 130°C. The source may be waste heat from an industrial process, steam that has passed through a turbine but not condensed, or the jacket and exhaust heat from a diesel engine. Natural gas may be used if the price is low enough. The use of solar energy for absorption refrigeration is discussed in Section 6.7.

Since it is not practical to pass the evaporating refrigerant through coils in the airstream of an air conditioning plant, absorption refrigeration is used to produce chilled water, which in turn is used to cool the air.

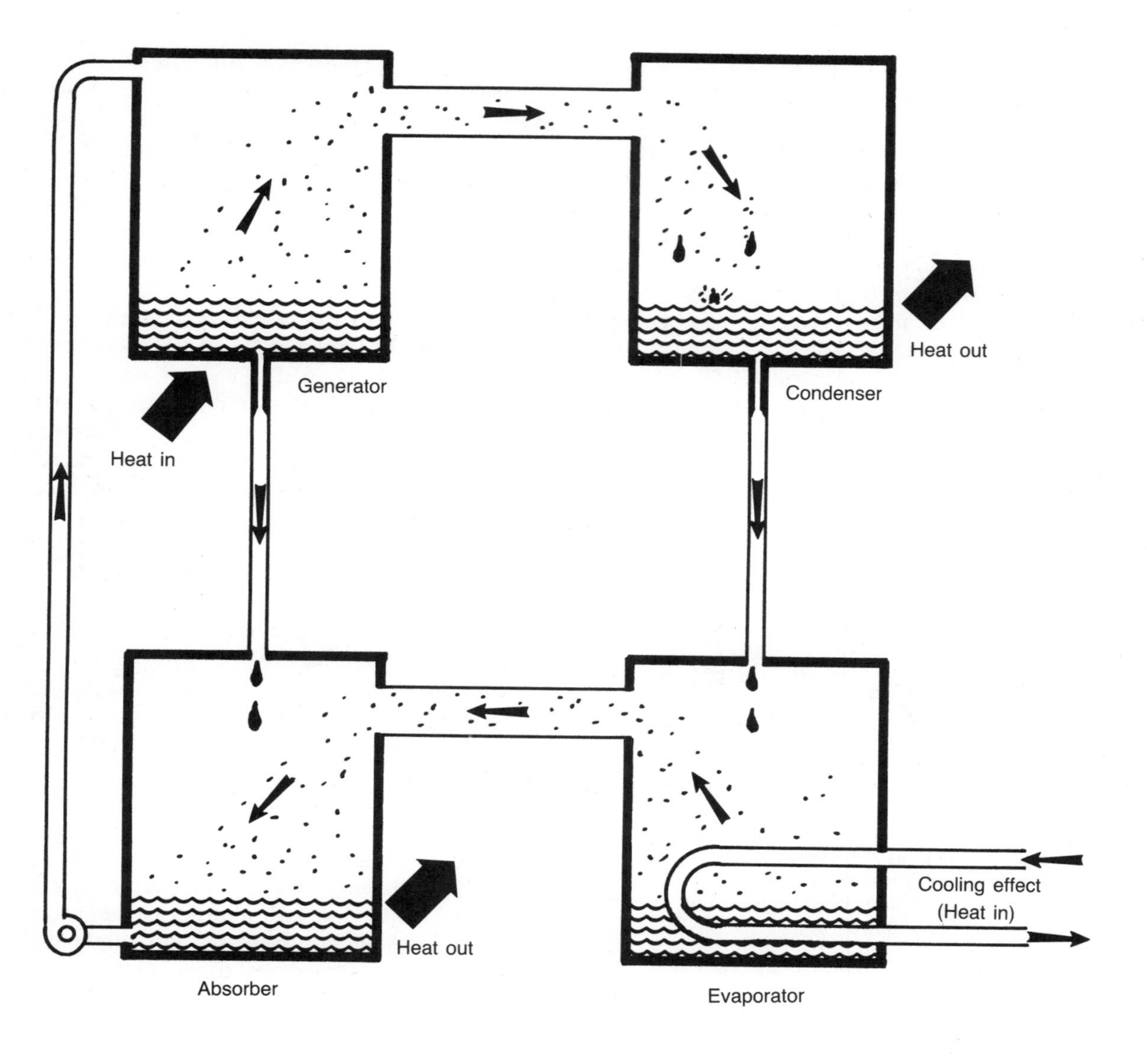

Fig. 6.6.4. The absorption refrigeration cycle. The refrigerant evaporates, under low pressure, in the evaporator. The latent heat taken in causes a cooling effect. The vapor is removed by absorption into a solution that has a strong attraction for that vapor. To prevent the absorber solution becoming too diluted, some of the vapor is driven off again in the generator. This is condensed and returned to the evaporator. The process is driven by the heat supplied to the generator, and waste heat is given off by the absorber and by the condenser.

6.7 Solar Energy for Cooling

Solar collectors used for heating water are described in Section 2.4, and solar energy for space heating in Section 6.4. Of all the applications of solar energy, space cooling is the one for which the supply most closely matches the demand. In theory an integrated system in which solar collectors produce hot water to power absorption-refrigeration machines in summer, and to provide space heating in winter, complements this supply-and-demand match with year-round usage.

In practice the amount of energy needed to run ancillaries, such as pumps and cooling tower fans, is a substantial part of the energy that would have been required to run a mechanical refrigeration plant to do the same job. We noted that the coefficient of performance of a well-designed mechanical refrigeration can be four or five times as great as that of the absorption machine. If we also consider the energy used for boosting the solar hot water when necessary, it is evident that the solar plant may use more electrical energy than a good nonsolar plant.

The lowest temperature at which the absorption machine can reasonably be operated, 85°C to 90°C, is at the upper end of the practical performance of flat-plate collectors. The returning water is at 75°C to 80°C, and unless the collector can raise its temperature above this, the solar contribution falls, not to a fraction of the input, but to zero. Fixed flat-plate collectors produce useful energy at these temperatures only in bright sunshine and when the angle of incidence is not too far from the normal; even then the efficiency of collection is not high.

Concentrating collectors (Fig. 6.7.1), which focus the sun's rays onto a linear tube, operate with improved efficiency and at higher temperatures. The initial cost is higher, and they have to track the sun throughout the day. The overall thermal efficiency is better than for flat-plate collectors, but there are conflicting opinions about the long-term cost efficiency, because of the cost of maintaining the tracking mechanism and keeping the reflecting surfaces clean.

Another type of collector particularly suited to solar refrigeration is the *evacuated tube* (Fig. 6.7.2). We noted in Section 2.4 that heat is lost from the surface of a collector by conduction, convection, and radiation. If the surface can be located in an evacuated space, the first two can be eliminated, and reradiation losses can be reduced by using a *selective surface,* one that has a high absorptivity in the solar spectrum but a low emissivity in the long-wave region [see Section 4.2].

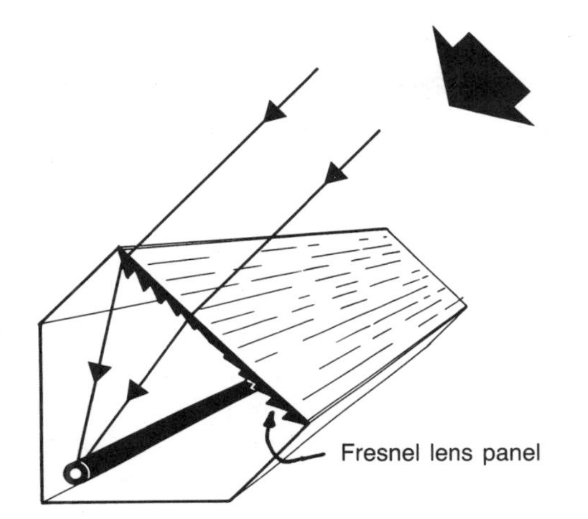

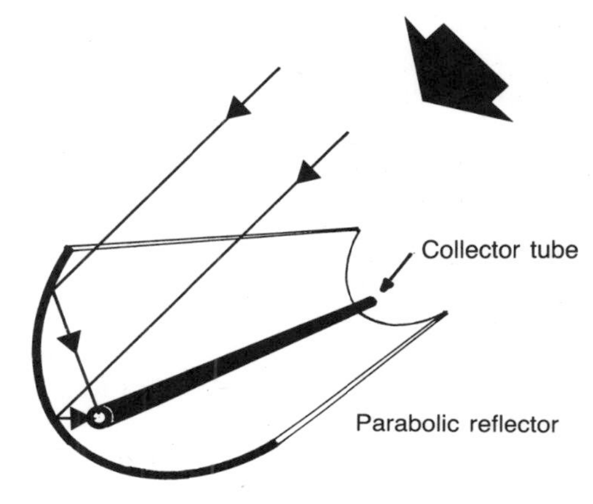

Fig. 6.7.1. Solar energy can be concentrated by a linear fresnel lens (left) or a parabolic trough mirror (right). These linear concentrators can increase the intensity by a factor of about 10. They must be steered to follow the sun's motion across the sky, but they require no adjustment for the seasons. In comparison, a spherical mirror concentrates the sun's rays almost to a point, increasing the intensity by a factor of 100 or more, but it must follow the sun both for time of day and for season of the year.

Fig. 6.7.2. The evacuated tube collector eliminates convection losses from the collecting surface by enclosing it in a vacuum. The efficiency of collection is increased over that of a normal flat plate, and higher temperatures can be obtained. (*Photo by courtesy of University of Sydney, School of Physics.*)

It is not possible to evacuate the space inside a flat-plate collector, because the air pressure would crush any glass cover unless it was as strong as the picture tube in a television set. However, it is relatively easy to make a glass tube, like a fluorescent lamp, that resists air pressure by virtue of its cylindrical shape.

Several manufacturers offer evacuated tubes, in diameters from 50 to 100 mm and lengths from 1 to 2 m. If they could be mass-produced in quantities comparable to fluorescent lamps, with which they have much in common, the cost could be reduced so as to make them competitive with other forms of collector. Because of their high thermal efficiency at temperatures well above 100°C, the economic view of solar cooling would then need to be revised.

Most solar refrigeration so far has been based on the absorption cycle. However, collectors operating at temperatures around 150°C could be used to run a steam engine to drive a mechanical refrigeration machine directly.

6.8 Air-Handling Plant and Ductwork for All-Air Systems

With an all-air system, the room temperature is entirely controlled by the temperature of the air supplied to it. With an air-water system, some of the room air is recirculated through a hot water or cold water coil within the room. With some hybrid systems, normal convection heaters are used in conjunction with a supply of conditioned air.

When the outdoor air is much warmer or much cooler than the required temperature, it is economical to recirculate most of the room air through the conditioner, adding only enough fresh air to meet the requirements of freshness. However, the plant should be capable of taking more fresh air from outdoors when it is close to the required temperature. The return air from kitchens, bathrooms, and other unpleasant areas is not recirculated.

The air-handling plant for a central station conditioner is located in one or more plant rooms. Figure 6.8.1 shows the main components. The return air and fresh air are admitted to the intake end in proportions controlled by dampers. The mixed air passes through a filter. If it is too cold and dry, it may need to be preheated. A water spray in conjunction with the cooling coils ensures that the air is close to 100% relative humidity at the temperature of the cooling coils. If drier air is needed, the cooling temperature is lowered so that more moisture condenses out of the air. If the resulting air at the correct absolute humidity is too cold, the reheat coil is used. Adjustment of humidity levels can be done with less energy input, by passing only some of the air through a very cold coil, the remainder bypassing the coil. When these two portions are mixed, the temperature should be correct, in which case the reheat coil need not be used.

After passing through all sections of the conditioner, the air enters a fan that increases its velocity and forces it along the supply ducts to all the conditioned rooms. The air enters the rooms from the supply ducts, through a number of supply registers. After mixing with the room air, some of it passes through return air grilles, into return air ducts, and back to the conditioner. A return air fan is located at the plant room end of the return air duct. Dampers are also required for the control of fire [see Section 7.2]. In the layout of the plant room, care must be taken to separate the spill-air outlet from the fresh-air inlet.

The return fan handles a little less air than the supply fan. This keeps the building slightly pressurized, so that unwanted leaks of air, rain, and dust are less likely to enter through doors and cracks. Air also flows *from* the occupied spaces

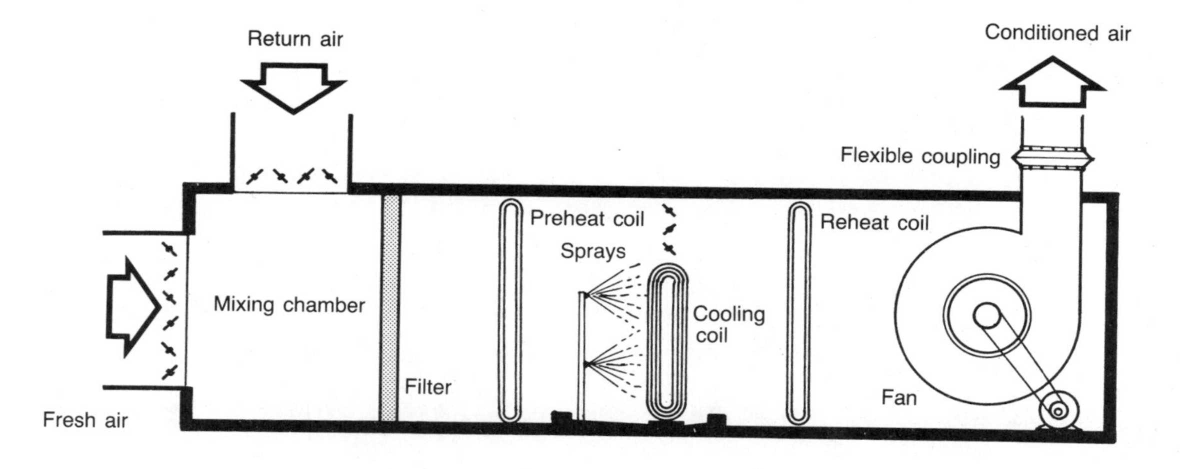

Fig. 6.8.1. In a central station conditioner, the air is filtered, and then according to requirements it can be preheated, humidified or dehumidified, cooled, reheated, and sent through the ductwork by a fan.

into the bathrooms (with their own separate exhaust ducts).

In planning a large building, the size of the air ducts is an important consideration. Their size is exceeded only by those of the elevator shafts and stairwells. The volume of air to be handled per second depends on the heat load [Section 6.11]. Once this is known, the sizes of the conditioner and the ducts are calculated according to the velocity of the air:

$$A = \frac{Q}{v}$$

where A is the cross-sectional area of the duct, in square meters (m^2);
Q is the rate of air flow, in cubic meters per second (m^3/s);
v is the velocity, in meters per second (m/s).

The velocity within the conditioner is kept low, to allow the air to exchange heat with the coils and to allow it to pass through the filters without excessive resistance. The velocity in the ducts can vary over a wide range.

Duct systems are classified as low-, medium-, or high-velocity. Most low-velocity systems use rectangular ducts. Circular ducts are preferred for medium-velocity systems, and they are used almost universally for high-velocity systems. Circular ducts are more efficient aerodynamically and are less prone to vibrations than those with flat sides.

In the ducts of medium- and high-velocity systems the speed of the air must be reduced, in the presence of sound-absorbing material, to minimize the noise produced in the room by the register.

The fans of a low-velocity system are driven by low-powered motors that do not consume much energy; but the ducts are large. High-velocity systems have the opposite effects. If the registers are close to people, the velocities through them must be kept low, to reduce noise and to reduce local drafts. On the other hand, higher velocities are needed through the registers in tall, large spaces to ensure that the supply air penetrates to the occupied part of the space. In a sports stadium or a transport terminal, where the level of background noise is high, quietness of the air conditioning system is not a major requirement.

A range of usual velocities is given in Table 6.3.

Example 6.2. *A conditioner serving part of an office building needs to handle 12 m^3/s of air. Determine preliminary cross-sectional sizes for the conditioner and the main riser duct.*

Solution *The overall cross section of the conditioner is determined by the air flow through the filters, unless they are arranged zigzag fashion to increase their face area. The area of the filters is*

$$A = \frac{Q}{v} = \frac{12}{1.5} = 8.0 \ m^2$$

A conditioner 2 m high and 4 m wide is well within the range of sizes encountered in large buildings. The provision of all the elements shown in Figure 6.8.1, together with walking space for maintenance between each element and a fan chamber at the end, results in a length of 10 to 12 m for a conditioner of this size.

The main duct area, for a velocity of 9 m/s, is given by

$$A = \frac{12}{9} = 1.33 \ m^2$$

Table 6.3

Location	Velocity (m/s) Low-Velocity System	Medium-Velocity System	High-Velocity System
Large main ducts	5–9	10–15	15–25
Branch ducts	4–6	9–12	13–20
Final branches	3–5	8–10	12–15
		(all systems)	
Filters in conditioner		1.5–1.8	
Heating and cooling coils		2.2–3.0	
Supply registers (quiet locations, or close to persons)		2.5–4.0	
Supply registers (noisy locations, or some distance from persons)		5–9	

Therefore a low-velocity duct from this conditioner requires a cross-sectional area of 1.33 m^2. A high-velocity duct, using 20 m/s, needs an area of 0.6 m^2, that is, a circular duct 875 mm in diameter.

6.9 Zoning in All-Air Systems

The conditioned air is supplied at a temperature above or below that required in the room, to allow for the heating or cooling load of the room itself. If the load varies from one part of the room to another, or from room to room, allowance must be made by varying either the temperature or the quantity of the supply air. This is controlled either at each supply register, or collectively within certain zones of the building.

When zoning is used, each facade with windows should constitute a separate zone, since the heat gain will change as the sun moves from one facade to another. The perimeter zone extends in for a distance of 3 to 5 m from the windows. The influence of the windows is considered to be limited to this area, and the remainder of the interior is taken as one central zone. In a multistory building a zone can extend through all the stories that have a similar occupancy. This arrangement can cause problems in perimeter zones of a building that is partly overshadowed by other buildings, since some of the windows may be sunlit and others shaded, on the same facade.

In a *constant-volume system,* in which the rate of air supply is fixed, its temperature must be regulated for each zone. This regulation can be done at the plant room or at locations closer to the point of use. Each zone is controlled by one thermostat located at a selected point in the zone. The thermostat compares the actual air temperature with a preset value and, if it differs, calls for increased heating or cooling in that zone. To prevent the conditioner from constantly cycling between heating and cooling, a deadband is built into the operation of the thermostat. The deadband refers to the range of temperature on each side of the set value, which must be exceeded before the thermostat signals for a change in the supply air.

If each zone is supplied with its own air from the plant room, the number of zones should be limited to about four. If the system is large enough, each zone can have its own conditioner. The return air can be collected in one or two risers and distributed among the conditioners. In a smaller system, a *multizone conditioner* can be used, as shown in Fig. 6.9.1. In this case both hot and cold air are produced and are mixed at the conditioner to give the conditions required for each zone. Because of the arrangement of the mixing dampers, the total volume of air passing through the fan is kept constant.

For energy-efficient operation, the hot deck of a multizone conditioner should be just hot enough to satisfy the zone calling for the greatest heating, and the cold deck just cold enough to satisfy the zone calling for the greatest cooling. The design and maintenance of a suitable control system is essential to the economical operation of the plant.

There is a limit to the number of zones that can be supplied from a multizone conditioner. If more localized zoning is required, a *reheat system* can be used (Fig. 6.9.2). All the supply air is supplied from the conditioner at the lowest temperature called for by any of the thermostats, and heaters are installed in the ducts serving each zone. These are usually electric resistance heaters, because they are compact and can be easily supplied with energy. It often occurs that one zone has a high solar heat load and therefore requires quite cool air, while the requirements

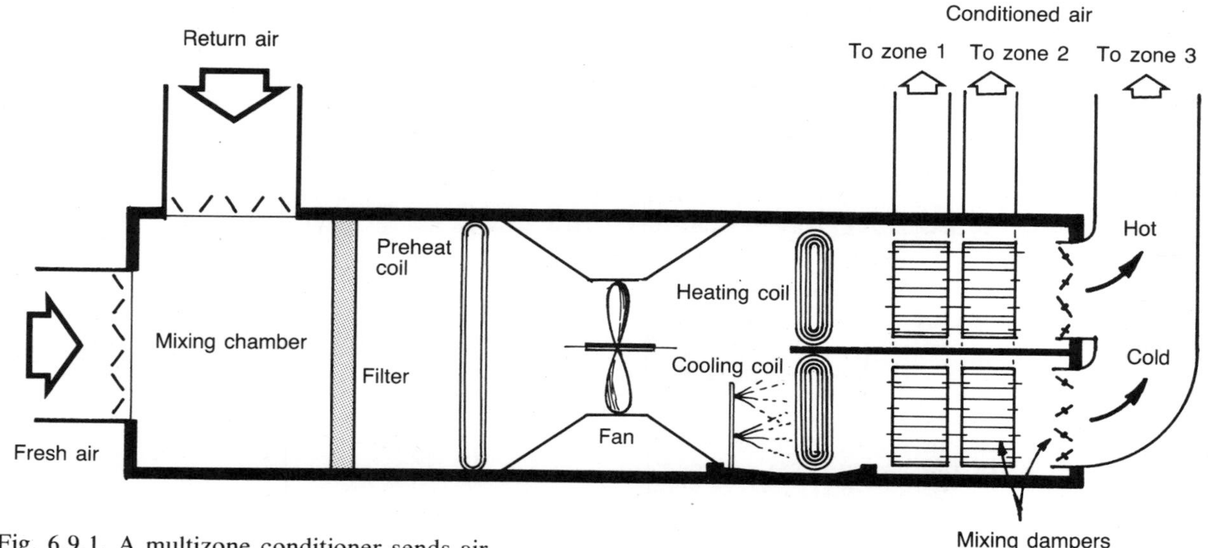

Fig. 6.9.1. A multizone conditioner sends air through three or four separate ducts at different temperatures, to suit the requirements of different zones.

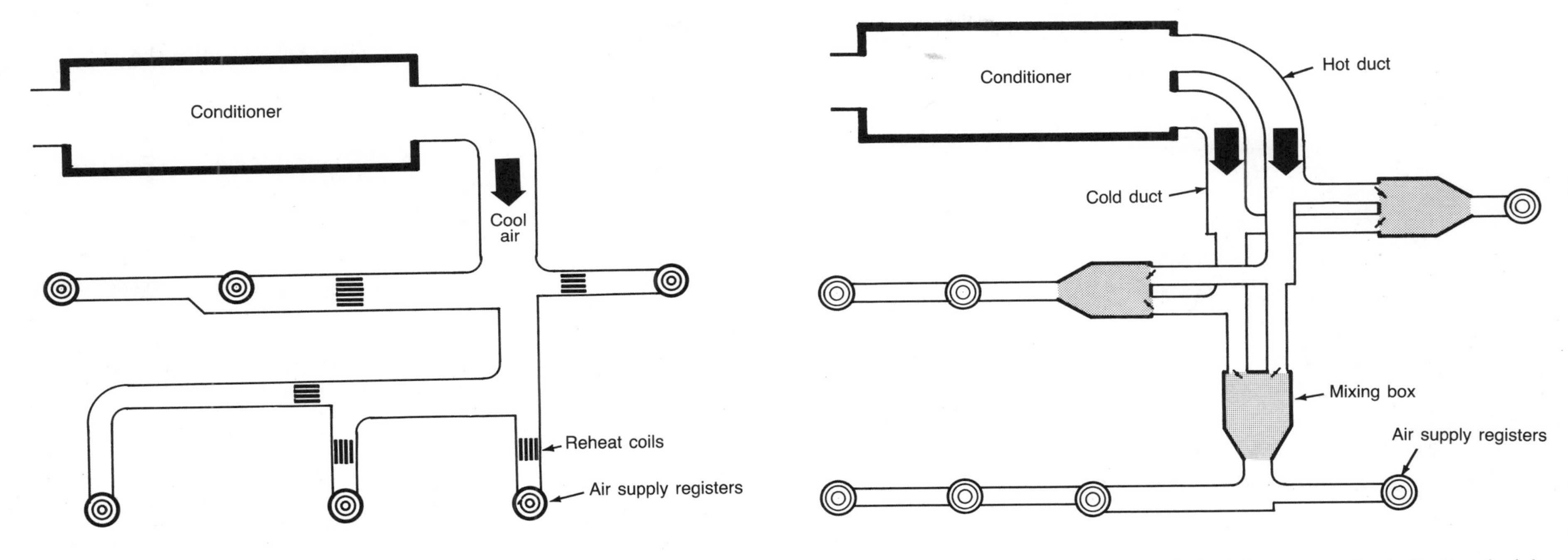

Fig. 6.9.2. A reheat system produces as many zones as necessary, by supplying cool air and reheating it at each zone. It can be wasteful of energy.

Fig. 6.9.3. A dual-duct system is similar in principle to a multizone conditioner, except that the hot and cold air are circulated throughout the building and mixed locally at each zone. A large number of zones are possible.

for the remainder of the building are much less severe. In this situation the reheat system must cool all the air to the lowest temperature needed and then reheat most of it. Since this represents a considerable waste of energy, reheat systems are not preferred at present. In previous times, when energy was much cheaper, the reheat system was quite common because of its simplicity.

The *dual-duct system* is an extension of the multizone conditioner in which the hot and cold decks become hot and cold ducts serving the whole of the building (Fig. 6.9.3). Each mixing box is controlled by one thermostat and can serve one or several registers. The dampers are arranged so that the total volume of air to each outlet is constant, whatever the mix of hot and cold. The amount of ductwork is nearly double that needed for a reheat system, and to keep the size of the ducts within limits, it is usual to use a medium- or high-velocity system. As with the multizone conditioner, it is important not to make the hot air too hot or the cold air too cold; but with the multitude of thermostats in a dual-duct system, it is more difficult to arrange a control system that can sense the extremes. Computer monitoring of the controls is usual if maximum plant efficiency is to be obtained.

In all the systems described so far, the objective has been to vary the temperature of the air between zones while keeping the volumetric rate constant. In the *variable air-volume* (VAV) system the opposite approach is taken. Until recently, VAV was not considered a suitable solution, because the throw of air from the supply registers varies as the volume changes, and also because the throttling of some outlets affects the performance of the fan and increases the pressure on the remaining outlets. It is also limited to cases where all outlets require cooling in varying amounts; it cannot supply heating to some and cooling to others.

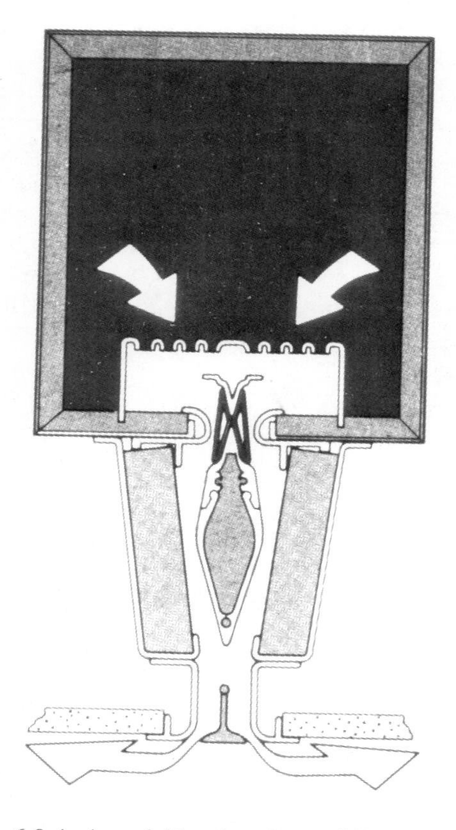

Fig. 6.9.4. A variable air-volume diffuser. To suit different conditions throughout the building, more or less air is supplied at each diffuser. The VAV system can accommodate a wide range of heating requirements or of cooling requirements, but not heating and cooling at the same time. With careful design of both the building and the system, VAV can be quite economical in the use of energy. (*Photo by courtesy of Carrier.*)

However, VAV offers considerable economies in energy consumption, because it avoids the wasteful practice of cooling air and then reheating some of it. This has led to improvements in the design of the registers and fans, and the system is now widely used (Fig. 6.9.4).

6.10 Air-Water Systems

In most parts of most buildings, the rate of air supply necessary to control the temperature is much greater than the rate of fresh air needed for ventilation. In the systems described so far, a large proportion of the air is returned through the return air ducts and recirculated through the conditioner. The effect of *air-water systems* is to reduce the amount of air handled through the central conditioner to the required fresh air component, called the *primary air,* while the recirculation, called the *secondary air,* takes place within the rooms of the building (Fig. 6.10.1).

As the room air recirculates, it passes through a filter and a coil containing either hot or cold water, which provides the temperature control. The temperature can therefore be controlled separately at each individual unit.

This system is frequently used in the perimeter zones of buildings, while the center zone may be served by an all-air system. The capability of individual control is useful when a facade is partly sunlit (as mentioned in the previous section). Furthermore, when there are separate private offices around the perimeter of a building, oc-

Fig. 6.10.1. Circulation of air within a room, (a) in an all-air system, and (b) in an air-water system.

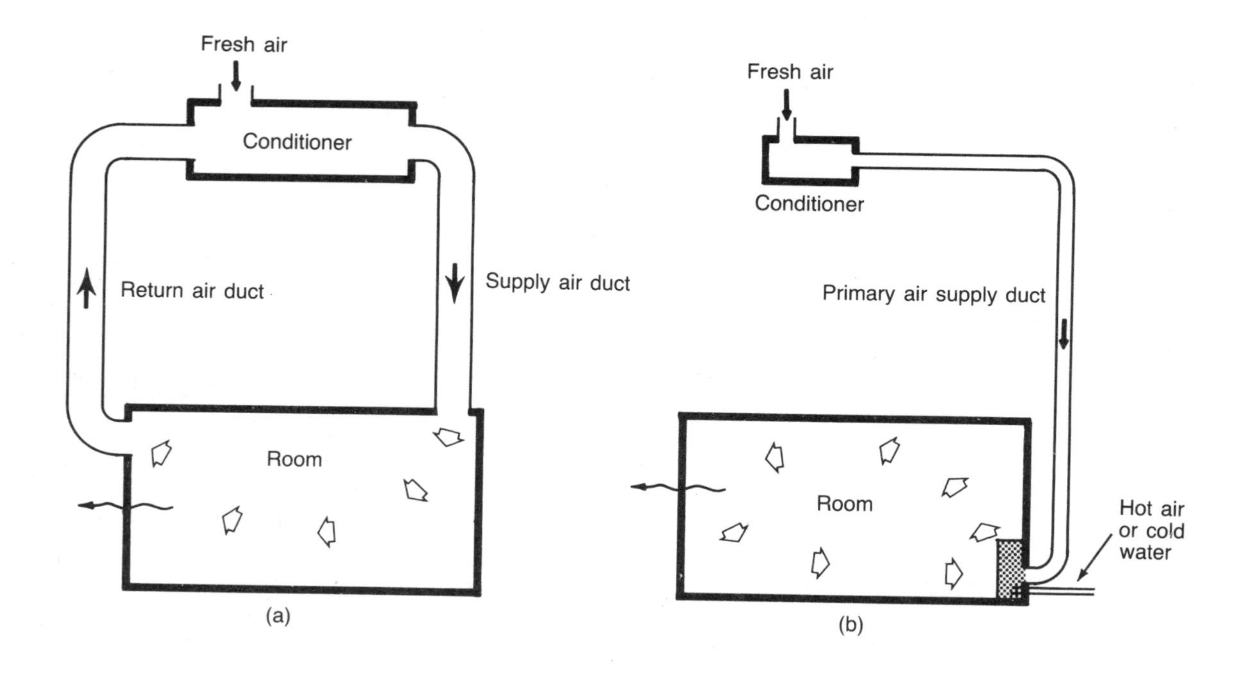

cupants can adjust the temperature of their own offices. It is also widely used in hotel rooms to allow guests to control the temperatures of their own rooms, and to serve specialty shops in a shopping center, where the heat load and hours of operation can vary from one shop to another.

There are two ways of providing the energy to circulate the room air. In the *induction unit* (Fig. 6.10.2.a) the primary air is supplied at high velocity. It is allowed into the cabinet through nozzles, and the resulting jets of air entrain some of the surrounding air, so that the total amount of circulation is many times the volume of the primary air. This produces some background noise, which is usually acceptable in a busy office but can be too loud for domestic use or a hotel room.

In the *fan-coil unit* (Fig. 6.10.2.b) the velocity of the primary air is not important. In smaller buildings it can be taken in through the outside wall, without passing through a conditioner. The energy for circulating the room air comes from a small fan. This is quieter than the induction system. It also has the advantage that the fan can be turned off completely if the room is unoccupied or if the occupant wishes to open the window. Because the fans are located in each unit, this does not interfere with the pressure balance in the rest of the system. With the fully ducted systems, opening any window can seriously upset pressures throughout the system.

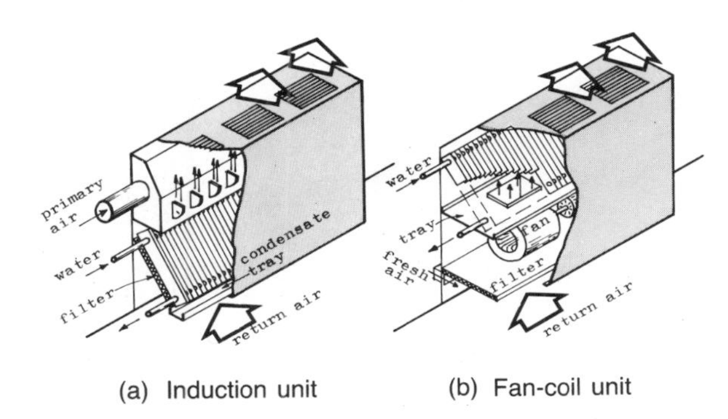

(a) Induction unit (b) Fan-coil unit

Fig. 6.10.2. (a) Air-water induction unit. The circulation of air within the room is achieved by the primary air, at high velocity, passing through jets and entraining room air with it. Heating or cooling is done by water passing through the coil.
(b) Fan-coil unit. The circulation of air within the room is produced by a fan, while heating or cooling is done by water passing through the coil.

An air-water system requires the circulation of hot and/or chilled water around the building, to serve each of the units. In a two-pipe system, *either* hot *or* chilled water is circulated, at the discretion of the plant operator or automatically, in response to the outside temperature. This allows the thermostat at each unit to modulate the temperature of the secondary air in one direction only. A three-pipe system circulates hot and chilled water but uses a common return pipe, and therefore wastes a little energy in comparison with a four-pipe system, where both hot and chilled water have their own return pipes.

Air-water systems require filtering and temperature control to be done in many places throughout a building. Maintenance is therefore needed at regular intervals, to clean filters and check the operation of the controls. This is labor-intensive, and it requires the maintenance technician to work in each private office, possibly disrupting work or reducing office security. On the other hand, the routine maintenance on a central station conditioner is done in the plant room and involves a few large items rather than many small ones.

6.11 Cooling Load Calculations

The cooling load for a sunlit building cannot be calculated by steady-state methods, since the solar heat load varies more rapidly than the building can respond to it. We must take into account the thermal inertia of the structure, and the cumulative effect of the heat load on the building for several hours previously.

Most of the methods suitable for hand calculation make simplifications that give acceptable results. These methods are used to find maximum cooling loads and thus to choose a suitable plant size. To obtain energy consumption as well as plant size, we must consider not only the maximum load but also all typical, part-load days. We thus require hourly climatic data for several typical days each month, or at least each season, as well as information about the full-load and part-load efficiencies of the refrigeration plant, cooling tower, heat exchangers and fans that make up the air conditioning system. Computer programs are usually used for these calculations (Refs. 6.1, 6.2); ASHRAE has also published a "Bin method" (Ref 6.3) that can be carried out by hand if desired.

One method suitable for hand calculations is given in ASHRAE's *Handbook of Fundamentals, 1977 Edition* (Ref. 4.1). It uses *cooling load temperature differentials* and *cooling load factors,* tabulated to allow for the time lag in typical construction systems. The method is used in Example 6.3.

Example 6.3 *The west face of an office building is 30 m long and 3 m high per floor. The facade is 80% solid and 20% glass. Calculate the cooling load of the western zone for the hours of noon, 2 P.M., 4 P.M., and 6 P.M. in July, in New York City.*

Solution *Page and table numbers are those in Ref. 4.1 (unless otherwise noted), and data have been converted to SI units. The perimeter zone gains heat through the solid wall by conduction, and through the glass by radiation and conduction. There are also heat gains from the lights, the occupants, and any machinery in the office.*

Fresh air taken in at outside temperature causes an additional cooling load in the plant room, but not in the conditioned space.

The assumptions made and values obtained from Ref. 4.1, Chapter 25, are given below, and the calculations are given in Table 6.4.

Solid Wall

Metal curtain wall with air space and 75 mm insulation. In Table 6 of Chapter 25, the U-value is 0.52 W/m²K, which falls into Group G for the purpose of looking up other tables.

Table 7 of Chapter 25 gives cooling load temperature difference (CLTD) values for the times of noon, 2 P.M., 4 P.M., and 6 P.M. as 11 K, 23 K, 37 K, and 37 K. These represent the difference between the outside, sunlit face of the wall and the inside design temperature of 24°C. They assume an outside daily average air temperature of 29.5°C. The New York July average is about 28.5°C (Table 1 of Chapter 23), and therefore these values are reduced to 10 K, 22 K, 36 K, and 36 K.

Glass Areas

Double glazing, heat absorbing plus clear sheet. From Fig. 6 of Chapter 26, the shading factor for an angle of incidence of 20° is 0.75. No external shading or internal drapes are assumed. At the time of peak loading (late afternoon in summer) the sun is close to normal incidence on a western wall. This leads us to choose an angle of incidence of 20° and also to assume that external shading would be of little use.

Table 14 of Chapter 26 gives the U-value as 3.74W/m²K for an outside wind speed of 1.5 m/s. For conduction, Table 9 of Chapter 25 gives CLTD's for the four times being considered as 5 K, 7 K, 8 K, and 7 K. As in the case of the solid walls, these values are reduced, because of the mean outdoor temperature, to 4 K, 6 K, 7 K, and 6 K.

For solar heat load, Table 10 of Chapter 25 gives the maximum solar heat gain factor (SHGF) for a western wall, at latitude 40°N in July, as 681 W/m², while Table 11 gives the four cooling load factors (CLF) as 0.14, 0.29, 0.50, and 0.55. The cooling load is calculated as SHGF × CLF.

Internal Gains

We have calculated the external heat load using only the dimensions of the external wall; but the internal gain depends on the depth of the zone, measured from the wall toward the interior of the building. It is usually taken as 3 to 5 m

Table 6.4

West Facade

Element of Heat Gain	Solar Time[a] 12 noon	2 P.M.	4 P.M.	6 P.M.
Conduction through solid wall Q = Area × U-value × CLTD = 72.0 × 0.52 × CLTD	× 10 = 374	× 22 = 824	× 36 = 1348	× 36 = 1348
Conduction through glass Q = Area × U-value × CLTD = 18.0 × 3.74 × CLTD	× 4 = 269	× 6 = 404	× 7 = 471	× 6 = 404
Solar heat through glass Q = Area × SHGF × SF × CLF = 18.0 × 681 × 0.75 × CLF	× 0.14 = 1287	× 0.29 = 2666	× 0.50 = 4597	× 0.55 = 5056
Gain from lights Q = Installed watts × CLF = 1800 × CLF	× 0.78 = 1404	× 0.81 = 1458	× 0.83 = 1494	× 0.85 = 1530
Gain from people (sensible heat) Q = No. of people × average output × CLF = 15 × 75 × CLF	× 0.74 = 832	× 0.80 = 900	× 0.85 = 956	× 0.89 = 1001
Total sensible heat gain of zone (watts)	4166	6252	8866	9339

[a]The mean time indicated by a watch differs from solar time [Section 3.3 of this book].

[Section 6.9 of this book]. The exact dimension is not critical; the central zone carries any load not taken up by the perimeter zone, and the total heating or cooling capacity supplied by the plant is therefore correct. In this example let us choose a zone width of 3 m. The floor is 30 × 3 = 90 m^2.

Lights

Assume a lighting level of 600 lux (i.e., 600 lumens per square meter; see Section 8.2 of this book), supplied by fluorescent lamps having a luminous efficacy of 60 lm/W and an overall coefficient of utilization and maintenance factor of 50% [see Section 10.5 of this book]. The electrical energy supplied per square meter is therefore 600/(60 × 0.5) = 20 W. When the lights are first turned on, most of the heat produced is absorbed by the structure of the building; but later in the day some of this heat is returned to the air in the space as a cooling load. This is reflected in the cooling load factors. From Table 15B of Chapter 25, assuming that the lights are switched on for 10 hours a day, beginning at 8 A.M., we get CLF's for the times of noon, 2 P.M., 4 P.M., and 6 P.M. as 0.78, 0.81, 0.83, and 0.85.

People

Assume 1 person per 6 m^2, which gives 15 people in 90 m^2. In Chapter 25, Table 16 gives an average heat output for people in offices as 75 W sensible plus 75 W latent heat. If people arrive in the offices at 8 A.M., Table 17 gives the CLF's for noon, 2 P.M., 4 P.M., and 6 P.M. as 0.74, 0.80, 0.85, and 0.89.

Calculations

It is convenient to carry these out in tabular form (Table 6.4).

Discussion

As we would expect, the solar heat load on the western facade increases during the summer afternoon. The direct radiant load through the windows accounts for more than half the total load. The very high figure for 6 P.M. could be ignored, since it represents 7 P.M. summer time and thus occurs well after normal office hours. However, this would create uncomfortable conditions for any persons working overtime.

It is usual to avoid western windows, if possible, because the low angle of the sun makes shading difficult, and because the solar load occurs after the building and its surroundings have heated up during the day. For comparison, Table 6.5 gives the values for the same calculations but for a facade facing south. The solar heat gains in Table 6.5 could be reduced further by simple sunshading [see Section 3.5 of this book]. The maximum solar heat loads for a southern facade

Table 6.5

South Facade

Element of Heat Gain	Solar Time							
	12 noon		2 P.M.		4 P.M.		6 P.M.	
Conduction through solid wall Q = Area × U-value × CLTD = 72.0 × 0.52 × CLTD	× 21 =	786	× 25 =	936	× 20 =	749	× 13 =	487
Conduction through glass Q = Area × U-value × CLTD = 18.0 × 3.74 × CLTD	× 4 =	269	× 6 =	404	× 7 =	471	× 6 =	404
Solar heat through glass Q = Area × SHGF × SF × CLF = 18.0 × 344 × 0.70 × CLF	× 0.52 =	2254	× 0.58 =	2514	× 0.47 =	2037	× 0.36 =	1560
Gain from lights (Same as Table 6.4)		1404		1458		1494		1530
Gain from people (sensible heat) (Same as Table 6.4)		832		900		956		1001
Total sensible heat gain of zone (watts)		5545		6212		5707		4982

occur in fall or winter, when the sun shines more directly on the windows. However, at these times the other loads are less because air temperatures are lower.

Sometimes a western facade cannot be avoided, because it provides the best (or the only) view from a site. In that case, internal drapes or venetian blinds between the glass should be used, and the glass area kept to a minimum [see Section 5.9 of this book]. External fixed sun shades are of limited use on the west [Section 3.5 of this book]; movable sunshades require a great deal of maintenance to keep them movable.

References

6.1 S.Y.S. CHEN: Existing load and energy programs. *Heating, Piping and Air Conditioning,* Vol. 47 (Dec. 1975), pp. 35–39.

6.2 K. LEE: *Energy-Oriented Computer Programs for the Design and Monitoring of Buildings* (2 Volumes). Environmental Design and Research Center, Boston, 1979.

6.3 T. KUSUDA: *A Proposed Simplified Energy Calculation Procedure for Use in Energy Conservation Standards Activities*. ASHRAE RP-205, American Society of Heating, Refrigerating, and Air Conditioning Engineers, Washington, D.C., 1981.

Suggestions for Further Reading

Reference 5.6, particularly *Indoor Climate Control*, pp. 89–348.

6.4 J.L. THRELKELD: *Thermal Environmental Engineering*. Prentice-Hall, Englewood Cliffs, N.J., 1970. 495 pp.

Two More Mathematical Textbooks

6.5 D.J. CROOME and B.M. ROBERTS: *Airconditioning and Ventilation of Buildings*. Volume 1. Pergamon, Oxford, 1981. 575 pp.

6.6 K. KIMURA: *Scientific Basis for Air Conditioning*. Applied Science, London, 1977. 269 pp.

Chapter 7 The Control of Fire

Some of the major fires of the past are described, including several that occurred this century. Correct construction limits the spread of fire, but in addition large buildings require automatic fire alarms and sprinklers. Tall buildings pose a special problem, because the ladders of the fire department do not generally extend above 30 m. All but the smallest buildings require provision for evacuating people from the seat of the fire. For tall buildings refuge areas protected from smoke must be provided.

7.1 Historical Note on Fires and Fire Regulations

Fires do not claim as many lives at the present time as automobiles or accidents in the home caused by falls and poisons (Ref. 7.1, p. 293); however, they are far more destructive than structural failures. Like other accidents, fires are caused mainly by people. Most are reported between 10 A.M. and 11 P.M., when people are normally awake and active (Ref. 7.2, p. 3). The majority are small, and most break out in private homes; collectively they cause more damage and loss of life than the more widely reported conflagrations in public places.

Many more fires are reported in North America than in Europe. The incidence per capita is approximately four times greater in the United States than in England (Ref. 7.2, p. 2.); the energy consumed per capita is also about four times as much (Ref. 7.2, p. 5). Thus the continuing high incidence of fires in modern times is at least partly due to new sources of energy that pose additional hazards.

The damage and loss of life from major fires is still a cause of concern. The burning of entire cities, frequently recorded by ancient and medieval historians, is no longer a danger, but the material damage and the number of deaths was in the past limited by the small size of the cities; Ancient Rome was an exception. The total destruction of Lugdunum (now Lyon in France) in A.D. 69 caused less damage than the partial burning of Rome in A.D. 64, because Rome was a much bigger city. The burning of most of London in 798, and again in 982, and again in 1212, involved fewer buildings than the fire that burned part of the city in 1666 because London had grown much larger.

Loss of life has generally been light in conflagrations that had a small beginning and spread only gradually; the Great Fire of London in 1666 caused only six deaths. Many casualties have occurred when crowded theaters with insufficient exits were destroyed; about 750 people were killed in a fire at the Burg Theater in Vienna in 1881. Heavy loss of life has also resulted from fires started by natural disasters or by military action. About 700 people died in San Francisco on April 18, 1906, but most of them perished in the fire that followed the earthquake. The Tokyo earthquake of September 1, 1923, caused a firestorm; estimates of the number of people killed range from 74 000 to 143 000. Deaths from the firestorm following the 1945 air raid with conventional bombs on Dresden were estimated at 35 000, and from the atomic bomb dropped on Hiroshima, in the same year, at 75 000 (Ref. 7.3).

No reliable figures exist for the property damage caused by ancient and medieval fires, but some modern fires have probably been more destructive. A fire in the covered bazaars in Istanbul on November 27, 1954, caused damage estimated at $178 000 000, and a fire in a shopping center in Ilford, a London suburb, on March 16, 1959, resulted in a property loss of about $30 000 000 (Ref. 7.3).

The fires that have influenced the design of buildings and the organization of firefighting services have not necessarily been those that produced the greatest destruction or loss of life. The Great Fire of Rome in A.D. 64 marked the beginning of regulations for fire-resistant construction. Following the Great Fire of London in 1666, the first fire brigades were organized by insurance companies, first in London, then in other European cities. However, they would only fight fires in buildings insured by the company.

In 1824 Edinburgh became the first British city to amalgamate its independent fire brigades (of which there were twelve at the time) into a single service; but before the new organization could establish itself, a fire broke out on November 15 of that year in a printer's shop in High Street. It spread to the tall buildings of the medieval city, some of which were twelve stories high. From there windblown firebrands spread the fire to Parliament Square and the Tron Kirk. The water supply failed, and contradictory orders hampered firefighters. On November 17 a torrential rain saved the city. This disaster led to rules for the conduct in the event of fire by firefighters, police, magistrates, and property owners (Ref. 7.4).

On December 16, 1835, a fire in New York destroyed 530 buildings in the business part of the city. Firefighting had been greatly hampered by a shortage of water, and this led to the construction of the Old Croton Aqueduct (Ref. 7.5), the first modern water supply comparable in magnitude to those of Ancient Rome.

In 1861 Cotton's Wharf and adjacent parts of London caught fire. The result was the passing of the Metropolitan Fire Brigades Act of 1865.

Two great fires started on October 8, 1871, in the American Middle West. The more destructive of the two, at Peshtigo, Wisconsin, wiped out seventeen small towns and caused 1052 deaths. A fire the week before had burned the telegraph lines, and news of it reached the outside world only when a boat arrived at Green Bay two days later. By that time the newspapers were occupied with the Great Fire of Chicago, which started on the same day but killed only about 300 poeple.

A smaller fire had broken out in a woodworking factory in Chicago on October 7; it was brought under control on October 8 at 3:30 A.M. after four city blocks had been completely burned. The fire that started at about 8:30 P.M. on the same day seemed at first a minor one by comparison. The subsequent inquiry established that the fire had broken out in a barn behind a laborer's shingled cottage on a small piece of land measuring 30 m by 7.5 m (100 ft by 25 ft), surrounded by other timber cottages. Most accounts state that a cow kicked over a kerosene lamp, which set fire to some straw (Ref.7.6). The fire was at first fought by the O'Leary family and their neighbors, and no alarm was given until 9:40 P.M. The firefighters were still tired from the earlier fire, and fire engines were slow to arrive. Although Chicago had several powerful steam-operated pumps, they were unable to extinguish the blaze, which was fanned by a high wind estimated by the U.S. Weather Signal Office at 27 m/s (60 mph). Windblown firebrands spread the fire to a timber store and a gas works. By 1 A.M. on October 9 it was out of control, and even buildings considered fireproof started to burn. At 3 A.M. a flaming piece of timber, said to have been about 3.6 m (12 ft) long, pierced the roof of the waterworks, which supplied the mains of the entire city. It was built of stone, and the shingle roof had recently been replaced by one of slate to render the building fireproof; yet the firebrand set the roof timbers alight, and the building was destroyed. The only building that survived was the crenellated water tower. On October 9 more fire engines arrived from Milwaukee, the wind changed direction, and at 11 P.M. rain started to fall. On October 10 the fire died down. The Chicago fire had far-reaching consequences for architectural design because it was responsible for the fireproof reconstruction of the city buildings that led, within a few years, to the invention of the skeleton frame (Ref. 1.7, Section 2.6).

The need for better protection of iron and steel frames was demonstrated first by a fire in Baltimore in 1904, when eighty city blocks were destroyed, and again by the fire that followed the San Francisco earthquake of 1906 (Ref. 7.7).

Reference has already been made to the fire in Vienna's Burg Theater, in which about 750 people were killed and which resulted in the introduction of regulations for fire endurance and for egress from places of public entertainment. They became more stringent, however, when on December 30, 1903, a fire at the Iroquois Theater in Chicago caused 602 deaths because most of the exits were locked.

These regulations have not always been sufficiently enforced, particularly in small restaurants and night clubs. On November 14, 1942, a fire in a nightclub in Boston caused six deaths and led to an order that all similar premises be inspected. The Cocoanut Grove Club was inspected on November 20. The inspector stated that the conditions were satisfactory and the exits adequate. At a subsequent inquiry he claimed that he had tried to ignite the imitation palm trees and the half-cocoanut light fittings and found them not flammable. The club was licensed for 500 people, but on the night of November 28 it held about 1000 (Ref. 7.8, pp. 26–30). One guest unscrewed a light bulb to dim the lighting. A waiter attempted to put it back, and because he could not see the fitting in the darkness he lit a match. A tree caught fire and the flames spread rapidly over the cloth-covered walls. One door to the street had been locked to prevent guests from leaving without paying their bills. A revolving door, which was unlocked, became jammed with people inside it, and the pressure of others behind blocked it. Most of the 493 people who died in the fire were killed by noxious fumes in less than a quarter of an hour.

In spite of the publicity this disaster received, similar fires, fortunately with less loss of life, have occurred since. In November 1970 at the Cinq-Sept dance hall in the small French town of St. Laurent-du-Pont a chair caught fire during a Sunday night dance, for reasons unknown. This ignited a foam-plastic cushion, and the fire spread to the walls, which had been sprayed with foam plastic to imitate a grotto. The dance floor had no windows. The entrance had a turnstile and the main exit was controlled by a pedal operated by a cashier; four other exits were locked. There was no emergency lighting, no telephone, and the hydrant was not connected to the water supply. Most of the 146 deaths occurred in the first five minutes.

Plastics were also a major cause of the fire in the Summerland leisure complex on the Isle of Man, off the west coast of England, which had a capacity of 5000. It was totally destroyed on August 2, 1973; most of the 3000 people inside escaped, but 50 were killed (Ref. 7.9). The fire was started by three schoolboys who claimed that they discarded a lighted cigarette, but the Commission of Enquiry thought that they might have been playing with matches. There was no suggestion of deliberate incendiarism. A fiberglass kiosk was ignited, and this set fire to the coating on the steel sheet that formed the wall behind. Because of the high thermal conductivity of the metal, fuel vaporized quickly from the coating, and the fire spread in the cavity between the sheet metal and the fiberglass lining. Within ten minutes substantial quantities of flaming vapors were ejected into the amusement arcade, and the fire then spread rapidly through the entire complex, which contained large quantities of acrylic plastic.

7.2 How Fires Spread

All the fires described in the previous section, except those started by military action or natural disasters, commenced as small blazes. A fire can remain small for many minutes, even hours, and while it remains small it is easily extinguishable. During that time hot gases are formed and rise to the ceiling. When these have accumulated in sufficient quantity, a *flashover* occurs. It is promoted by combustible wall surfaces, but it can occur even with entirely fire-resistant walls. Within a few seconds the room becomes filled with flames, and thereafter the fire could only be extinguished by a firefighter wearing special clothing and breathing apparatus. Eventually all the combustible materials in that room are burned, and the fire therefore stops for lack of fuel.

It is important that large buildings be designed so that a fire that gets out of control in one room does not spread to other rooms. For this purpose large buildings are divided into compartments with fire-resistant walls; the openings in these walls are closed with fire-resistant doors that close automatically in case of a fire and thereafter close automatically whenever someone opens them.

Terrace houses should be separated by fire walls extending above the roof.

The heat produced by fires cracks window glass. Wired glass remains in position for a limited period of time, but ordinary glass and toughened glass are shattered, so that the window presents no barrier to radiant heat. Most fire regulations require a minimum depth for a fire-resistant spandrel beam or for a fire-resistant horizontal projection. It must be large enough to keep the flame away from the window above (Fig. 7.2.1). Very short horizontal projections set up vortices that project the plume of the flame back onto the windows in the story above.

Windows in fire-resistant doors must be of wired glass, and they should be kept as small as possible (Fig. 7.6.1).

In multistory buildings the elevator shaft must be isolated in the event of a fire to prevent it from becoming a chimney stack that draws the fire up to the other floors [see Section 5.11]. Thus the shaft must be built with fire-resistant walls, such as reinforced concrete, and the doors of the elevator shaft must be fire-resistant and self-closing.

Ducts for heating, ventilation, and air conditioning can also convey fire from one compartment to another. Dampers that are released by fusible links must be placed within the ducts wherever a duct passes through a wall separating one fire compartment from another, and they must also be placed at each outlet from a duct (Fig. 7.2.2), unless the air-handling system has been specifically designed to operate as a smoke-spill system in an emergency (Ref. 7.11, p. 316).

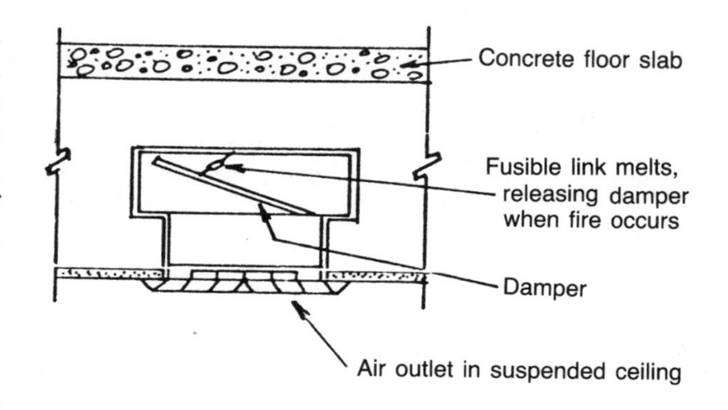

Fig. 7.2.2. A fire damper is required at each outlet of a duct used for heating, ventilating, or air conditioning. This is released by a fusible link that melts when the temperature rises to the predetermined level, and closes the duct. Fire dampers are also needed whenever a duct passes through a wall separating one fire compartment from the other.

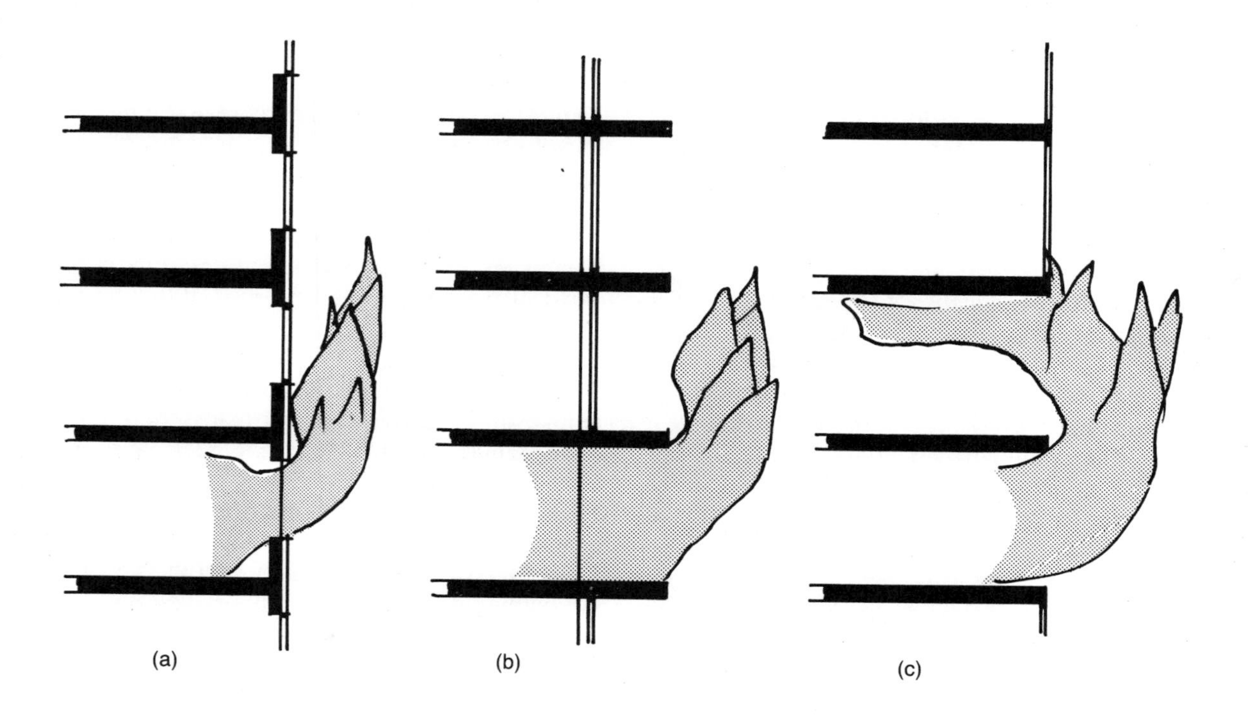

Fig. 7.2.1. In a multistory building the spread of fire to the story above can be delayed, and perhaps prevented, either (a) by having a spandrel of fireproof material or (b) by a horizontal projection of fireproof material, such as a balcony or sunshade. If there is neither a fireproof spandrel nor a horizontal projection of adequate size, (c) the flames emerging from the broken windows of the lower story will shatter the windows of the story above, so that the fire spreads to that story.

When pipes and ducts pass through fire-resistant walls or floors, there is generally a space between the pipe or the duct and the wall or floor. This must be stopped with a fire-resistant material, such as mineral wool (Fig. 7.2.3).

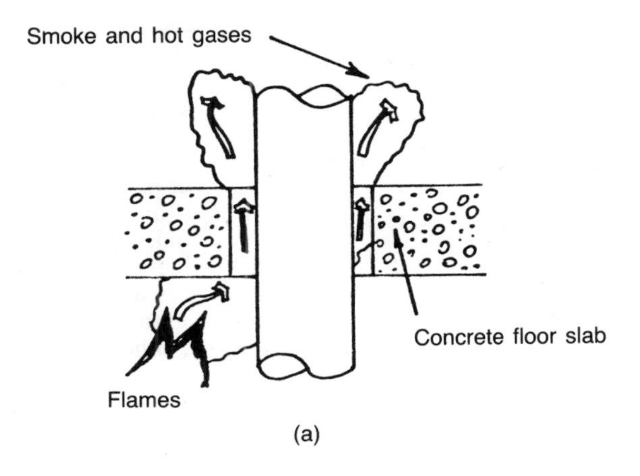

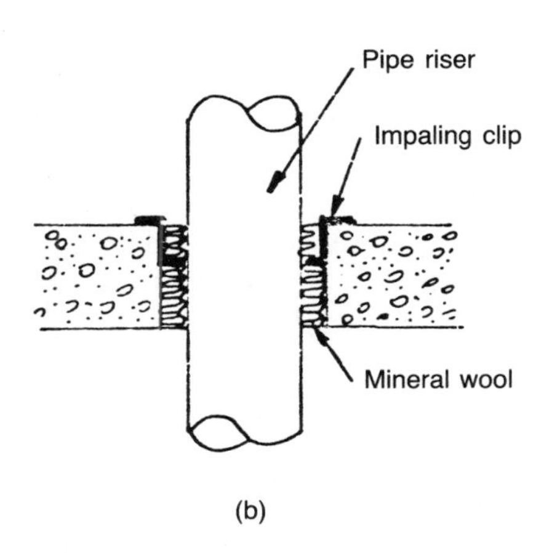

Fig. 7.2.3. The space around a pipe or duct permits the spread of fire (a) unless it is stopped with a fire-resistant material, such as mineral wool (b).

Fire can also spread by conduction through a floor. Thus a floor which has ample fire resistance, such as a reinforced concrete slab 100 mm (4 in.) thick, will eventually conduct sufficient heat to cause ignition of the floor covering above, such as a carpet. The thickness required depends on the time needed to burn all the combustible materials in the room below, assuming that the fire is not previously extinguished by sprinklers or by the fire department.

7.3 Fire Loads, Fire Tests, and Fire-Resistant Construction

The design of fire-resistant floors and walls depends on the combustible content of a fire compartment. This is called its *fire load*. It is not a load in the conventional sense, but the heat generated if all the contents were burned. It can be measured as the mass of combustible material per unit floor area in kilograms per square meter (pounds per square foot in American units), or in terms of the heat produced in megajoules per square meter (British thermal units per square foot in American units).

After flashover occurs [see Section 7.2], the temperature rises rapidly. It reaches 500°C (900°F) in a few minutes, and 1100°C (2000°F) after about four hours if there is sufficient combustible material. Thus a high fire load requires a structure and fire compartment walls with a high fire resistance. Buildings are classified in accordance with the activities that take place in them and the combustible materials that they contain. These are specified in the local building code, and they vary in different parts of the world. Generally 1½ hours of fire resistance are required for apartment buildings, 2 hours for office buildings, 3 hours for department stores, and 4 hours or more for warehouses.

To ensure that the construction used in a building has the requisite fire resistance, a sample is tested in a special furnace designed to imitate the conditions that are likely to occur in a building after flashover. The temperature in the furnace rises in accordance with a standard fire curve (Fig. 7.3.1) that was derived from tests carried out at the National Bureau of Standards in Washington, D.C., in the 1920s (Ref. 7.10, p. 5). The fire resistance rating of the construction is determined by the test (Fig. 7.3.2). For example,

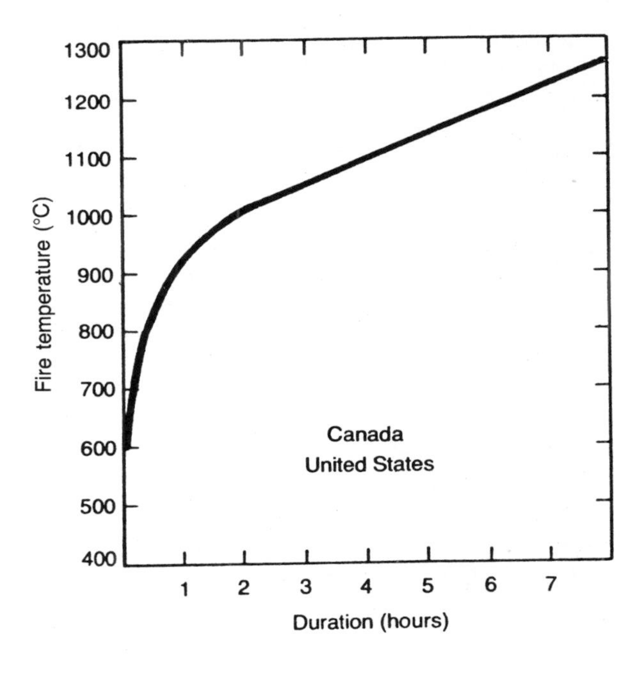

Fig. 7.3.1. Standard temperature-time curve used in the United States and Canada for testing building components. It was determined from measurements during a fire produced in a model room containing the furniture and other contents normally expected in an office building. The standard temperature-time curves used for fire tests in other countries are similar but not identical.

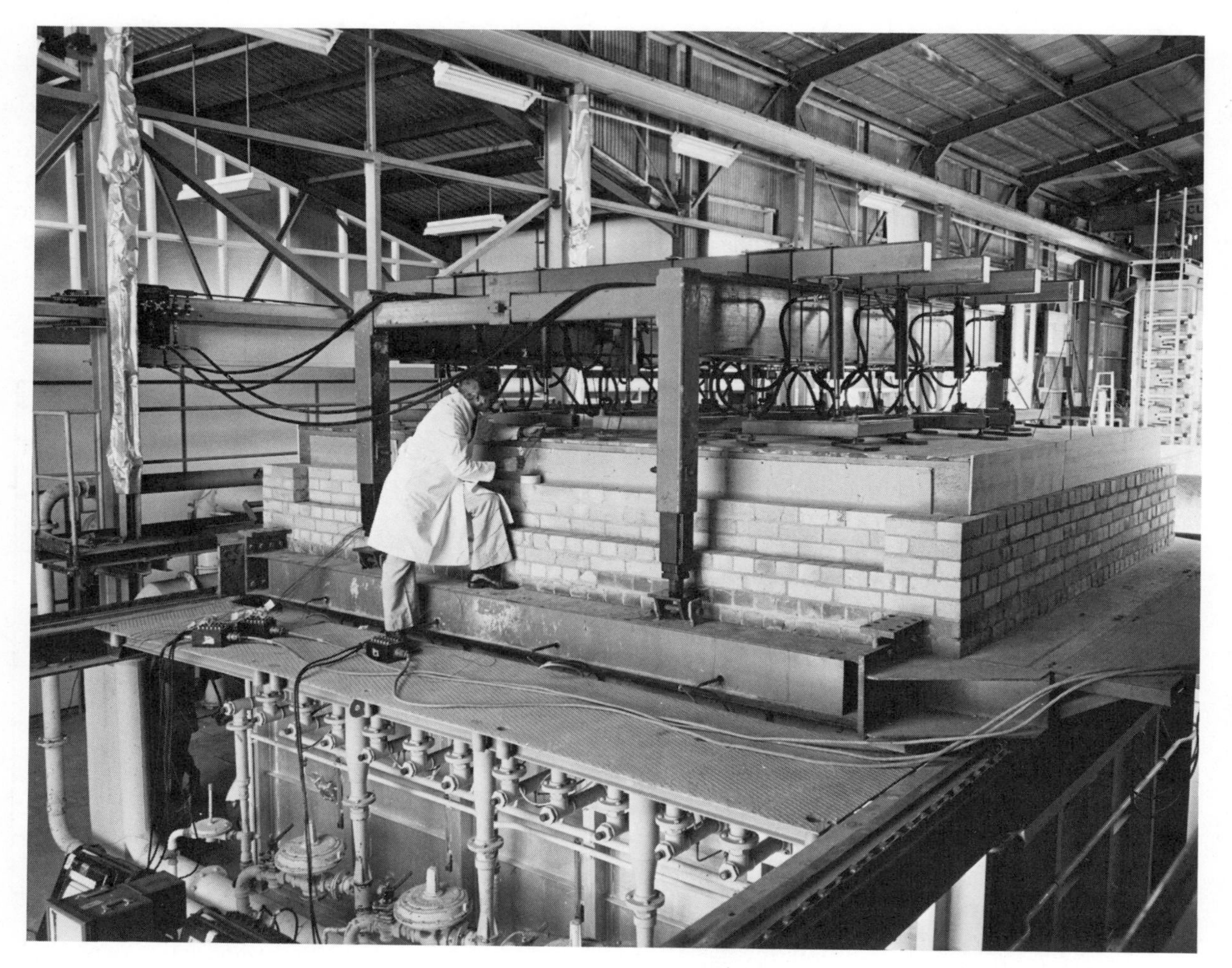

a reinforced concrete slab 100 mm thick with a standard 20-mm cover has a fire resistance rating of $1\frac{1}{2}$ hours (Fig. 7.3.2). The fire testing of materials and assemblies of materials is expensive, but there is no reliable theory for ascertaining their fire resistance.

Special consideration needs to be given to certain materials, such as plastics, that produce toxic gases during combustion [see Section 7.1].

Fig. 7.3.2. Fire testing furnace used by the Experimental Building Station, Sydney, Australia. The fire rating of a building component is the longest period during which the component in the furnace does not suffer significant damage or deformation, the component does not allow the passage of flames, and the temperature on the side not exposed to the fire remains below the temperature that would cause combustible materials to ignite. Structural members are required to support their normal service load during a fire test.

7.4 Stopping Fires Before They Become Too Large

Fire alarms are required in all large buildings, and the additional safety and reductions in fire insurance premiums may make their installation worthwhile even in buildings on the domestic scale. Fire alarms may be operated by a rise in temperature, by the appearance of a flame, by the appearance of flammable vapors, or by the appearance of smoke.

Heat detectors may utilize fusible links; bimetallic strips that deflect when the temperature increases; or thermocouples or thermistors [see Section 4.3]. Flame detectors measure reflected radiation. Detectors of flammable vapors operate on a catalytic principle.

Some smoke detectors utilize a beam directed on a photoelectric cell; the appearance of smoke reduces the amount of light. The most accurate smoke detectors contain a small quantity of a radioactive substance that ionizes the air; combustion products reduce the current flowing through the ionized air, and this sets off the alarm (Ref. 7.11, pp. 309–11).

In large buildings the alarm is generally sounded in the control room and is also reported directly to the fire department.

In many cities sprinkler installations are required for all multistory buildings and for all buildings that extend over a large area, even if they are entirely at ground level. Sprinkler heads are placed in pipes at intervals of 3 m to 5 m (10 ft to 16 ft), and they spray the floor with water once the sprinkler head is opened (Fig. 7.4.1).

The sprinkler head is normally operated automatically by the bursting of a quartzoid bulb filled with liquid or by the fusing of a link closing the sprinkler head. Only a few sprinkler heads are activated by the fusion of their links to emit water. The flow of water caused by the activation

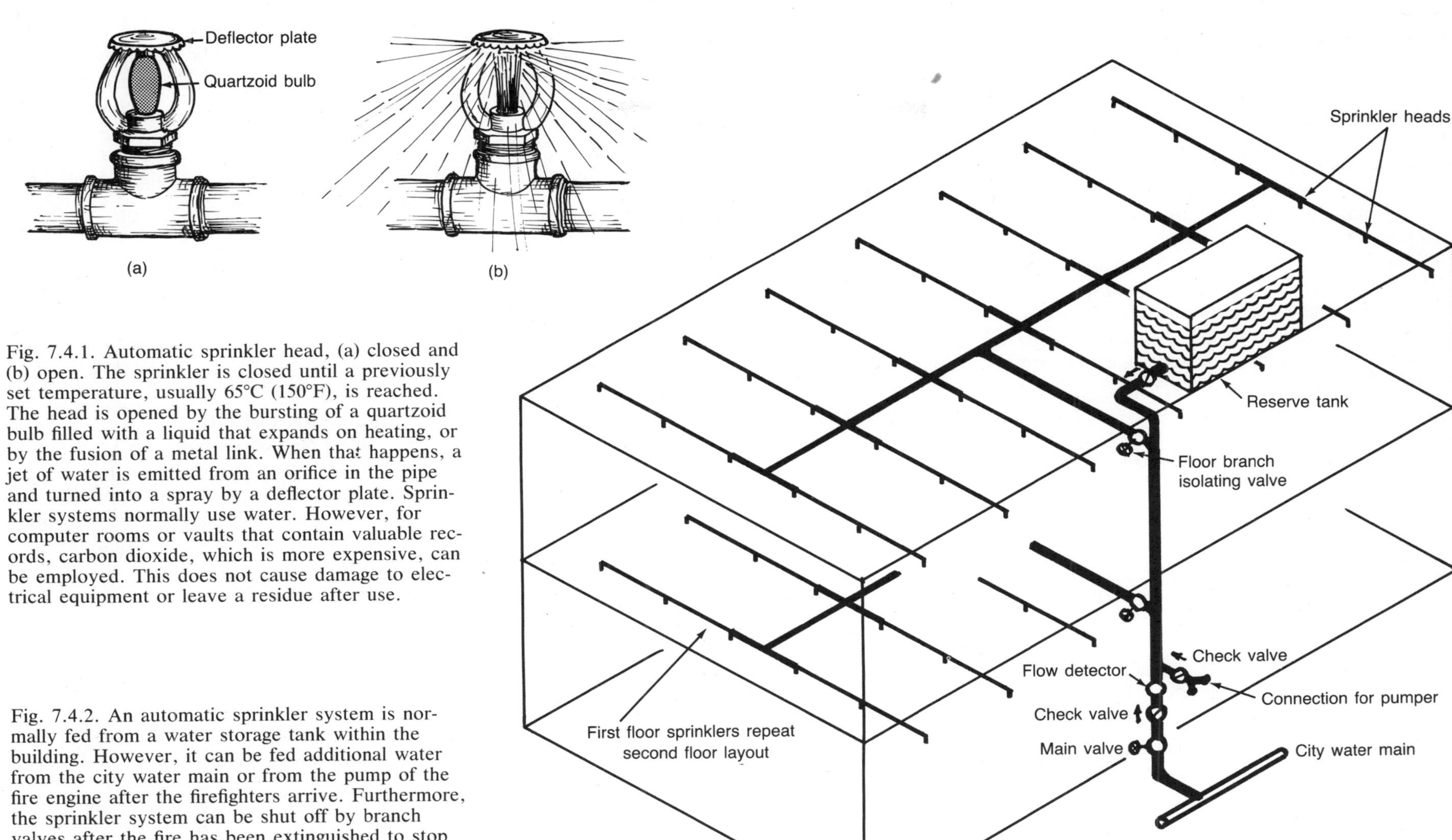

Fig. 7.4.1. Automatic sprinkler head, (a) closed and (b) open. The sprinkler is closed until a previously set temperature, usually 65°C (150°F), is reached. The head is opened by the bursting of a quartzoid bulb filled with a liquid that expands on heating, or by the fusion of a metal link. When that happens, a jet of water is emitted from an orifice in the pipe and turned into a spray by a deflector plate. Sprinkler systems normally use water. However, for computer rooms or vaults that contain valuable records, carbon dioxide, which is more expensive, can be employed. This does not cause damage to electrical equipment or leave a residue after use.

Fig. 7.4.2. An automatic sprinkler system is normally fed from a water storage tank within the building. However, it can be fed additional water from the city water main or from the pump of the fire engine after the firefighters arrive. Furthermore, the sprinkler system can be shut off by branch valves after the fire has been extinguished to stop further damage from water. The individual sprinkler heads cannot normally be closed, except by replacing the fusible link.

of an automatic sprinkler generally triggers an alarm at a central watchroom or at the headquarters of the fire department (Fig. 7.4.2).

As has already been mentioned, more people are killed or incapacitated by smoke and hot gases than by the flame of a fire. The removal of smoke and hot gases by venting is therefore desirable, provided that the venting does not produce a stack effect that increases the combustion in the remainder of the building. The practice of automatic venting was developed after a fire in 1953 at the automatic transmission factory operated by General Motors at Livonia, near Detroit. This large single-story building, covering 14 hectares (35 acres), was totally destroyed because firefighters were unable to penetrate the smoke to get near the fire.

Fig. 7.4.3. Automatic venting of a large single-story building. Each bay has a vent that is normally kept closed but is opened automatically by a smoke detector. To prevent air being drawn in from adjacent bays, each bay is surrounded within the level of the roof structure by a steel curtain.

Automatic venting is possible only in rooms that are immediately below the roof, but this includes all single-story buildings covering large areas of ground. The vents are normally closed, but each can be opened automatically by a smoke detector. To ensure that the opening of a vent does not produce air currents promoting the growth of the fire elsewhere, each vent is surrounded by a steel curtain that restricts the flow of the plume of hot air and smoke immediately below the roof (Figs. 5.10.1 and 7.4.3).

Automatic venting is also commonly used in the stage towers of theaters. These pose a particular fire risk because of the flammable nature of the scenery. In the event of a fire, a fire-resistant curtain is brought down to separate the auditorium from the stage, and any smoke in the stage tower is removed by an automatic vent operated by a smoke detector.

7.5 Lightning Conductors

Any building that is one of the tallest objects in the landscape is liable to be struck by lightning and requires protection. Lightning conductors were discovered by Benjamin Franklin in the eighteenth century, and they are now standard equipment for tall buildings. Lightning can do appreciable damage to an unprotected building. The potential discontinuity during a thunderstorm around a building 300 m high can exceed 3 000 000 volts, and the current at the core of lightning discharge may approach 200 000 amperes (Ref. 7.12). A lightning conductor consists of one or more pointed rods, usually of copper, connected to the ground by an insulated cable of low electrical resistance.

7.6 Extinguishment of Fires

Single-family residential buildings rarely have any equipment for extinguishing a fire, except the garden hose and perhaps a portable fire extinguisher. For larger buildings firefighting equipment may be required by the local building code, or its installation may reduce the cost of the fire insurance to such an extent that its installation actually saves money, even when there is no fire.

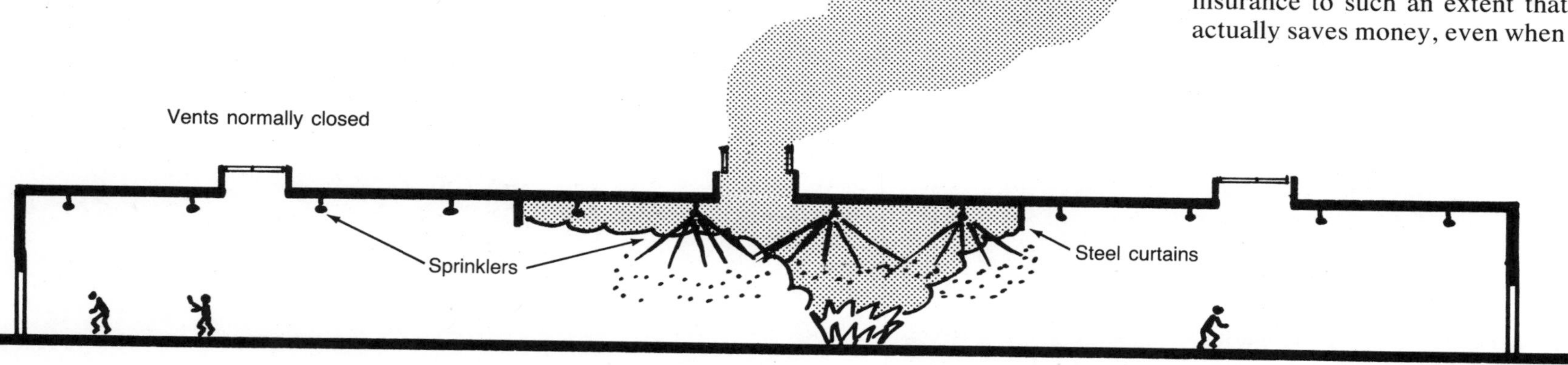

A distinction must be made between single-story buildings, multistory buildings up to 27.5 m (90 ft) high, and multistory buildings above that height. The tallest ladder of a fire department reaches to a height of approximately 30 m (100 ft), but this is the height to which a firefighter can rise and operate a hose. The limit for entering a window and rescuing a person inside is only about 27.5 m. Therefore when buildings exceed that height, it becomes necessary to rely entirely on internal means for egress. Above 30 m it also becomes necessary to rely entirely on internal means for extinguishment of the fire.

The occupants of a single-story building can leave it, provided excess smoke is removed by venting (Fig. 7.4.3), by walking outside on the same level. Consequently fire-resistant construction is not normally required.

In multistory buildings, except for those occupied by a single family, fire-resistant construction must be used, and stairs must be protected by fire-resistant walls and doors.

Buildings taller than 27.5 m, that is, about 9 stories high, must be provided with firefighting equipment inside the building. However, this is a useful provision even for lower buildings and is required by some building codes.

Furthermore, buildings less than 27.5 m high should be designed so that the fire department's ladders can in fact reach the facade of the building. This is not possible if the building is set on a podium whose roof is not accessible to the fire engine (Fig. 7.6.1).

A particular problem may arise with old buildings that have timber floors and masonry walls (Fig. 7.6.2).

Fig. 7.6.1. If the building is mounted on a podium that is not accessible to the fire department's ladder, it becomes impossible to rescue people through windows from a ladder.

Fig. 7.6.2. Collapse of a wall of the Buckingham Department Store in Sydney, Australia, in 1968. The building had timber floors and masonry walls, which depended on the timber floors for their stability. When the timber floors burned, one of the masonry walls collapsed in a dangerous manner. (*Photograph by courtesy of the Sydney Morning Herald.*)

The firefighting equipment inside the building consists of sprinklers, already discussed in Section 7.4, portable fire extinguishers, fixed-hose reels, and standpipes for the fire department. Portable fire extinguishers are of two main kinds. The first uses water for extinguishment; this is expelled by pressure, usually generated by chemical reaction between sulfuric acid and sodium bicarbonate solution. The second uses chemicals, generally in the form of a dry powder or foam, to extinguish the fire.

The fixed-hose reels are intended to enable occupants of the building to fight a fire before the fire department arrives. They are generally required for tall buildings by building regulations in Europe and Australia, but in America they are frequently regarded as an alternative to a sprinkler installation, and some regulations require only one or the other.

In addition, the fire department requires a system of standpipes within a tall building, so that it is freed from the need to bring a long line of hoses into the building. These standpipes are usually provided in or near a fire-resistant stairwell, and their supply pipes must be kept entirely separate from other water pipes. For buildings less than 60 m (200 ft) high the pipes are kept empty until they are needed for firefighting, and the water is pumped from below by the fire department's pump.

If the building is too tall to pump water from the ground, the standpipes must be kept full of water supplied from gravity tanks at the top of the building. The power of fire pumps varies in different cities, and the critical height ranges at present from 60 m to 100 m (200 ft to 330 ft). For these very tall buildings the fire brigade depends on the instant functioning of the internal standpipe system, and duplicate water supplies must be provided through separate supply mains and through the use of an on-site reservoir. All components and power supplies must be duplicated to ensure that at least one is working (Ref. 7.11, p. 328).

It would evidently not be practicable for the firefighters to walk up 60 flights of stairs, and the elevators must therefore be at the disposal of the fire department. They are generally programmed to return to the ground in the event of a fire and to remain there until the emergency is canceled. The elevators are then controllable only from within the car. At least one elevator should be operable from an emergency power supply. The firefighters can then travel to a level about two floors below the seat of the fire, and proceed from there to the fire on foot. Elevators must not be taken to the fire floor, because the occupants might be incinerated or suffocated when the doors open.

7.7 People and Fire

Most building codes include detailed requirements for fire exits from all buildings, except private homes and small apartment buildings. These vary for different codes, but generally a fire exit must be provided no further than 50 m (165 ft) from the door of any room. This must be marked by an illuminated exit sign. The corridors and stairs of each fire exit must be protected by fire-resistant floors, walls, and self-closing doors. They must have an emergency lighting system that takes over automatically if the regular power supply fails. The exit doors from the building must open outwards. In buildings used by a large number of people, such as theaters, each exit door must be fitted with a "panic bar" that opens the door when a person presses against it (Fig. 7.7.1). The sites required for doors, corridors, and stairs are discussed in Sections 13.5 and 13.6.

A special problem is posed by tall buildings, because it is not at present considered safe to use elevators for evacuating during a fire, and it is not practicable for the occupants of a large city building to walk perhaps more than 60 flights of stairs to the ground. It is also doubtful if stairs could be made large enough to evacuate up to 10,000 people from a large, tall building or kept free from smoke for the time that this would take.

Tall buildings are therefore designed to contain "refuge areas" that are kept free from smoke by an excess air pressure maintained by an emergency power supply. These can be reached within 5 minutes by a "safeguarded travel route" in which smoke is kept to a tolerable minimum, and this route can, in turn, be reached from any smoke-polluted area in $1\frac{1}{2}$ minutes (Ref. 7.11, p. 335). At the time of writing there has been no

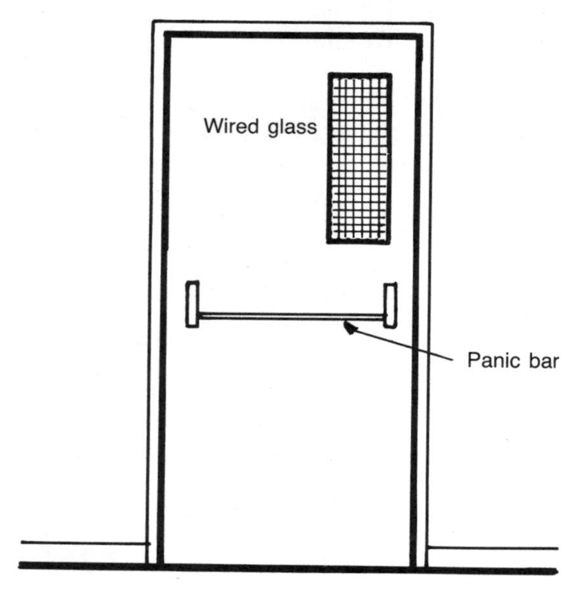

Fig. 7.7.1. Exit doors for theaters and other buildings used by a large number of persons must be fitted with "panic bars" that open the doors when people press against them.

major fire in a building designed by this concept whereby its efficacy could be tested. Few fires have occurred in tall buildings, compared with a large number in lower buildings.

In recent years building regulations have been introduced that make many buildings accessible to disabled people without assistance from other persons [see Section 13.4]. However, none of the present fire escape procedures make any allowance for their special problems. A disabled person who has entered a building unaided will for the present require help to escape from it in case of a fire.

References

7.1 E. GRANDJEAN: *Ergonomics in the Home*. Taylor and Francis, London, 1973. 344 pp.

7.2 E.W. MARCHANT (Ed.): *A Complete Guide to Fire and Buildings*. MTP Publishing Co., Lancaster, England, 1972. 268 pp.

7.3 *Encyclopaedia Brittanica*. Vol. 9, Fire Protection. Chicago, 1965.

7.4 C. ROETTER: *Fire Is Their Enemy*. Angus and Robertson, Sydney, 1962. 184 pp.

7.5 N. FITZSIMMONS (Ed.): *The Reminiscences of John B. Jervis—Engineer of Old Croton*. Syracuse University Press, Syracuse, 1971. 196 pp.

7.6 ROBERT CROMIE: *The Great Chicago Fire*. McGraw-Hill, New York, 1958. 282 pp.

7.7 A.L.A. HIMMELWRIGHT: *The San Francisco Earthquake and Fire, 1906*. Roebling Construction Company, New York, 1907.

7.8 STEPHEN BARCLAY: *Fire*. Hamish Hamilton, London, 1972. 293 pp.

7.9 *Report of the Summerland Fire Commission*. Government Office, Isle of Man, Douglas, 1974. 82 pp.

7.10 *Fire Test Performance*. Special Technical Publication 464. American Society for Testing Materials, Philadelphia, 1970. 243 pp.

7.11 Fire. Chapter CL-4. Tall Building Criteria and Loading, Volume CL. *Monograph on the Planning and Design of Tall Buildings*. American Society of Civil Engineers, New York, 1980. pp. 349–390.

7.12 *Encyclopaedia Brittanica*. Vol. 14, Lightning. Chicago, 1965.

Suggestions for Further Reading

References 7.2, 7.4, and 7.11.

7.13 M.D. EGAN: *Concepts in Building Fire Safety*. Wiley, New York, 1978. 269 pp.

7.14 T. LIE: *Fire and Buildings*. Applied Science, London, 1972. 276 pp.

7.15 G.J. LANGDON-THOMAS: *Fire Safety in Buildings*. Black, London, 1972. 296 pp.

Chapter 8 The Luminous Environment

Artificial lighting was used only to a limited extent prior to the eighteenth century; there has been a sharp increase in the level of illumination during the twentieth century. The units used for the design of lighting are explained, and methods of measuring illumination are described. The human tolerance of changes in the level of lighting is much greater than that of changes in temperature. The problem of glare and the method of classifying color are briefly considered.

8.1 Historical Note

Prior to the eighteenth century most people relied on natural light for illumination. After dark they went to sleep or performed whatever activities were possible by the light of a candle or an oil lamp. The materials used for lighting, except for beeswax (which was available only in small quantities), were all edible, and thus artificial lighting competed with the supply of food.

During the eighteenth century the new sperm whale fisheries augmented the fuel supply, and gaslight was first used in 1765. In the nineteenth century a plentiful supply of lamp oil was produced by the distillation of petroleum (Ref. 1.7, p. 255), and electric light was invented in 1878.

At the beginning of this century it was still common practice to design office buildings around light wells to provide natural light for each room, and to use large areas of glass in the roofs of department stores and shopping arcades for natural lighting (Fig. 8.1.1).

By 1950 the provision of daylight had ceased to be a major consideration in the design of office buildings, factories, and stores. Electricity had become cheap and generally available in apparently unlimited quantities. The construction of glass roofs and light wells was expensive, and it wasted a great deal of valuable space (Fig. 8.1.1). In buildings occupying entire city blocks without light wells it was impossible to provide daylight for the workplaces of most of the people, so that artificial lighting was needed even on a bright day [Section 10.1]. The windows were sometimes

Fig. 8.1.1. Strand Arcade, Sydney, a glass-roofed shopping arcade built in 1891 by John B. Spencer. It is a survival from the large number of glass-roofed buildings erected during the late nineteenth century to give good natural lighting in locations remote from the external walls.

a source of glare [see Section 8.5], and the venetian blinds were often kept permanently drawn [see Sections 5.9 and 10.1].

Since the energy crisis of 1973 there has been a return to natural lighting. Electric lighting consumes a substantial proportion of the energy used in buildings. In addition, it is a major source of heat. This is not unwelcome in winter, but it greatly adds to the air conditioning load in summer [see Section 10.6].

8.2 The Measurement of Light

The first lighting standard was produced by the (British) Metropolitan Gas Act of 1860, which defined the size and composition of a standard wax candle. This became an international standard in 1883. The light-radiating capacity, or *luminous intensity* (I), of this standard candle was a *candle power* (cp). The metric *candela* (cd), named after the Latin word for candle, differs from the candle power by only 2%. The candela is a measure of the luminous intensity in a given direction of a standard radiator defined by an international agreement.

In practice the *luminous flux* (F) is more important for the design of lighting. It is the flow or amount of light from a light source emitted through a solid angle; the unit for a solid angle is the *steradian,* explained in Fig. 8.2.1.

The unit of luminous flux is called *lumen,* from the Latin word for light source:

$$1 \text{ lumen} = 1 \text{ candela} \times \text{steradian}$$

The solid angle subtended by the surface of an entire sphere at its center is $4\pi = 12.57$ in absolute angular measure. Thus if a light source emits a luminous flux of 12.57 lumen uniformly over the surface of a sphere, its luminous intensity in any direction is 1 candela.

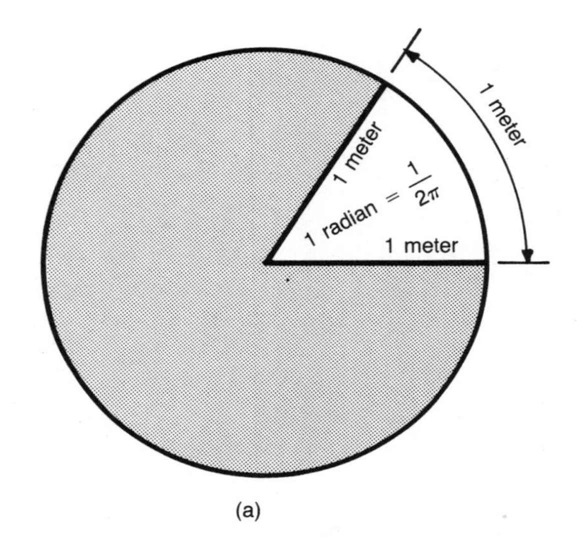

(a)

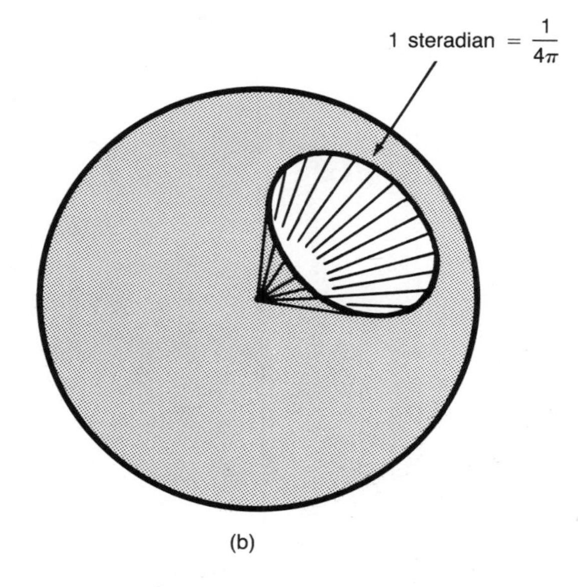

(b)

A lumen is a unit of the same kind as the watt, the unit used to measure the flow of thermal and electrical energy. The relation between the watt and the lumen varies with the wavelength of light. The eye has the greatest response to yellow-green light with a wavelength of 550 nm. For monochromatic light of this wavelength, 1 watt produces 680 lumen. All other colors produce less response from the eye; for example, 1 watt of red light produces only about 100 lumen, and for white light 1 watt produces 250 lumen.

The *illuminance* (E) is the luminous flux per unit area (Fig. 8.2.2). It is measured in *lux* (lx), from the Latin word for light, where 1 lx is 1 lumen per square meter.

The relation between the illuminance and the luminous intensity of the light source depends on the inverse square law and the cosine law.

Fig. 8.2.1. Angles are normally measured in degrees. This convention was developed by the Ancient Babylonians, who divided the quadrant of a circle into 90°, so that the full circle had 360°. It is an arbitrary system, and in light measurement the absolute system, which is independent of any units of measurement, is used.

The circumference of a circle of radius 1 m (a) is 2π m. The angle subtended at the center of the circle by 1 m of the circle's circumference is called 1 *radian*. It measures $1/2\pi$ of the circle's circumference and therefore is equal to $360/2\pi = 57°$. All angles can be expressed in radian measurement. For example,

$30° = 0.52$ radians
$45° = 0.79$ radians
$60° = 1.05$ radians
$90° = 1.57$ radians

The same system of absolute measurement can be applied to solid angles (b). The surface area of a sphere of 1 m radius is 4π m^2. The solid angle subtended at the center of a sphere by an area of 1 m^2 on the surface of the sphere is called 1 *steradian,* and it is therefore $1/4\pi = 1/12.57$ in absolute angular measure.

Figure 8.2.2 shows the relation between the illuminance, E, on an area of 1 m^2 and a light source with a luminous intensity, I, of 1 cd. If the radius of the sphere were doubled from 1 m to 2 m, the illumination would be reduced by 2^2 (Fig. 8.2.3). This may be expressed as the *inverse square law* of illuminance:

$$E = \frac{I}{D^2} \tag{8.1}$$

where E is the illuminance in lux;
I is the luminous intensity in candela;
D is the distance from the light source in meters.

If the surface under consideration is not normal to the direction of the rays of light, the illuminance is further reduced (Fig. 8.2.4). The *cosine law* enunciated by J.H. Lambert, one of the pioneers of photometry in the eighteenth century, states that

$$E_2 = E_1 \cos \theta \tag{8.2}$$

where E_1 is the illuminance on the plane normal to the direction of the rays of light:
E_2 is the illuminance on the surface tilted at an angle θ;
θ is the angle of incidence of the light.

Illuminance is measured with a photometer. Since the discovery of photoelectricity most photometers utilize photovoltaic cells [see also Section 3.1], calibrated to read illuminance directly

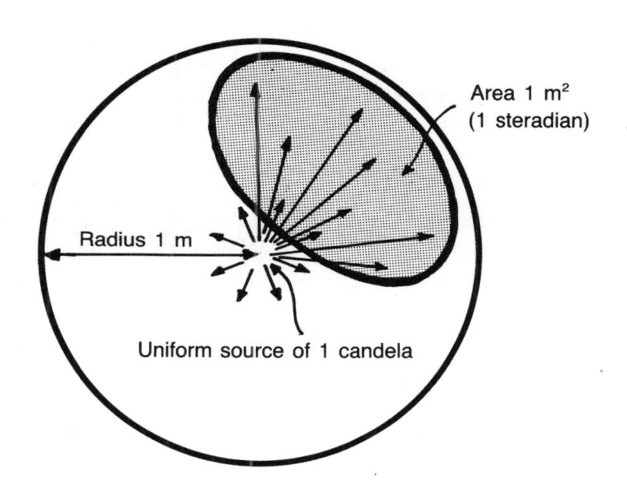

Fig. 8.2.2. A light source at the center of a sphere emitting light with a luminous intensity of 1 candela (cd) produces a luminous flux of 1 lumen for each solid angle of 1 steradian. The abbreviation for lumen is lm. The total solid angle at the center of the sphere is 4π steradian = 12.57 steradian, so that luminous flux over the entire surface of a sphere is 12.57 lumen if it has a uniform light source of 1 cd at its center.

The unit of illuminance is the lux, abbreviated lx. This is the luminous flux that falls on one square meter. In a sphere 1 m in diameter one square meter subtends a solid angle of 1 steradian at the center of the sphere. Consequently a luminous intensity of 1 cd results in a luminous flux of 1 lumen and produces an illuminance of 1 lx on the surface of that sphere.

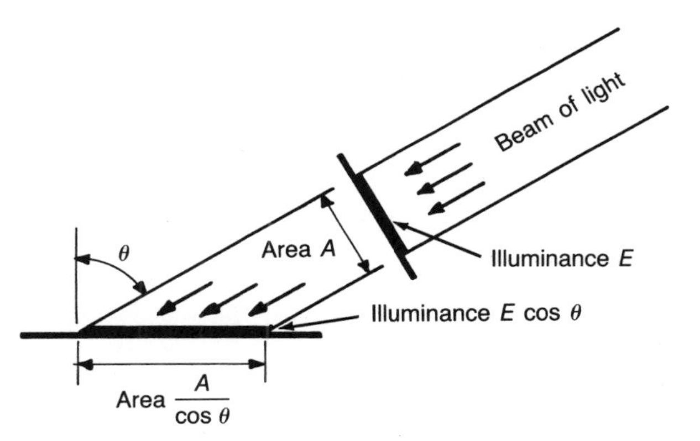

Fig. 8.2.4. The illuminance, E, is greatest if the surface under consideration is normal (that is, at right angles) to the direction of the rays of light. If the surface is tilted at an angle θ to the normal, the illumination is spread over a larger area.

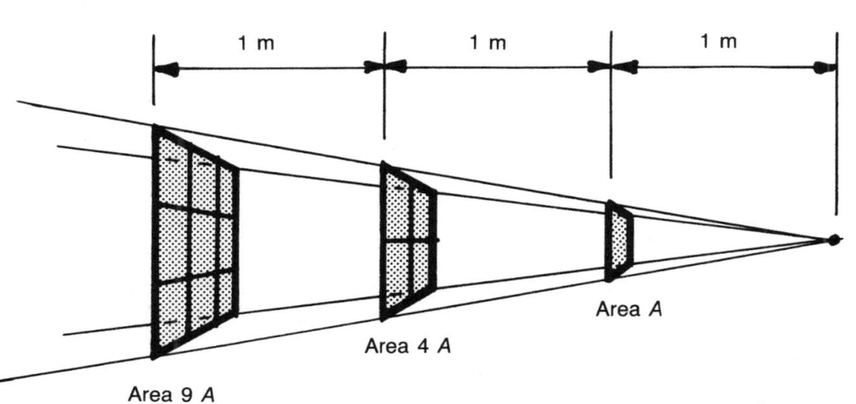

Fig. 8.2.3. As the distance from the light source increases, the illuminance, E, is spread over a greater area that is proportional to the square of the distance.

in lux. They convert light energy into electrical energy, and the current can be measured accurately and, if so desired, recorded by remote control. There are several types of photoelectric light meters. Photovoltaic cells can be made very small; for daylight studies a cell about 40 mm in diameter is convenient (Fig. 8.2.5).

Illuminance is what a light source produces on a surface reached by it. What the eye sees reflected from that surface, or what it sees by looking directly at the light source, is called *luminance* (L). When an object is illuminated by light, this produces an illuminance (E); the properties of the material determine the luminance (L) of the object's surface.

Luminance is measured in candela per square meter (cd/m^2). Several other units are still in use, and they are listed in the appendix, together with their conversion factors.

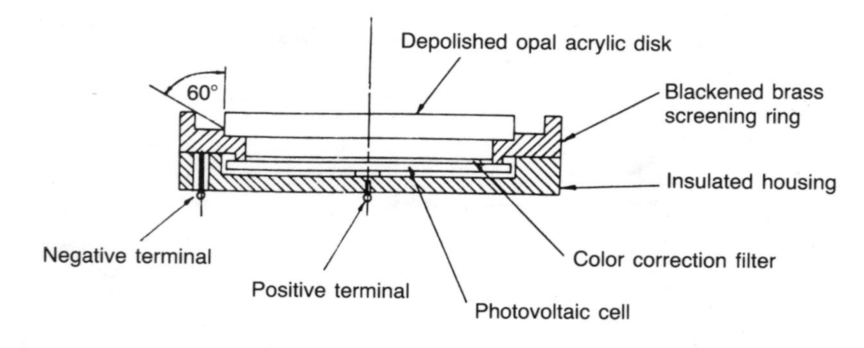

Fig. 8.2.5. Photovoltaic cell suitable for determining illuminance for daylight studies. It includes a color correction filter and a matte acrylic disk for cosine correction, to ensure that it responds proportionately to the cosine of the angle of incidence of the light in accordance with Lambert's law (Eq. 8.2). The cosine correction was devised by Pleijel and Longmore (Ref. 8.1), and it is described in several books on daylight design (for example, Ref. 8.2). The electrical circuit is not shown.

For matte reflectors the luminance of the surface is

$$L = \frac{Er}{\pi} \text{ cd/m}^2 \qquad (8.3)$$

where r is the *reflectance* of the surface, which is a ratio;
π is the circular constant (3.1416).

Seeing is the result of light sources and reflected light illuminating the retina of the eye. Two light sources that produce the same illuminance on the eye may appear to be different from each other and thus may have different luminances.

Example 8.1 *A person stands 100 m from a spotlight with a luminous intensity of 10 000 cd, and 100 m from a floodlit building whose luminous intensity is also 10 000 cd. What are the illuminances and the luminances of the spotlight and the floodlit building?*

Solution *From Eq. (8.1) the illuminances produced by both the spotlight and the floodlit building at the eye are*

$$E = \frac{I}{d^2} = \frac{10\,000}{100^2} = 1 \text{ lx}$$

The illuminances are the same, but the luminances differ greatly:

$$L = \frac{I}{A}$$

where I *is the luminous intensity and* A *is the projected area (normal to the observer) of the spotlight and the floodlit building, respectively.*

If the spotlight is circular and has a diameter of 0.5 m, its area is $A = \frac{1}{4} \times \pi \times 0.5^2 = 0.196$ *m*^{2}*, and the luminance is*

$$L = \frac{10\,000}{0.196} = 50\,900 \text{ cd/m}^2$$

If the building is 12 m high and 20 m long, its area is $A = 12 \times 20 = 240$ *m*^{2}*, and the luminance is*

$$L = \frac{10\,000}{240} = 41.7 \text{ cd/m}^2$$

The illuminances produced by both the spotlight and the floodlit building are the same, but the luminance of the spotlight is much greater than that of the floodlit building; this is what we observe *to be so.*

The comfort conditions for visual tasks and the visibility of objects depend on luminance, which is a property of an object in the direction of view. We see an object together with the other objects that constitute the *visual field*. Our subjective response to *apparent brightness* or *luminosity* is produced by the relationship between all the luminances in the visual field. This depends on the adaptation level of the eye. Thus the luminance of a car headlight is the same by night and by day, but its apparent brightness is much greater by night; if a car drives with its headlights on a sunny day, this may not even be noticed by passing cars.

The efficiency of energy conversion is normally measured as the ratio of power output to power input, and this is a dimensionless ratio smaller than 1. However, we are interested in the light output in lumen; as stated earlier in this section, lumens are watts evaluated in terms of the response of the eye. The *luminous efficacy* (sometimes wrongly called luminous efficiency) is the ratio of the luminous flux emitted by a lamp in lumen to the power input in watts. Since the watt is a much larger unit than the lumen, the numerical value of the luminous efficacy is always greater than 1. For example, a 40-W tubular fluorescent lamp with a luminous efficacy of 70 lm/W is a more efficient light source than an incandescent lamp whose efficacy is 10 lm/W.

8.3 The Eye

The photographic camera is essentially a copy of the human eye (Fig. 8.3.1), and its construction is perhaps more familiar. Like the camera, the eye has a lens that inverts the image and projects it on the retina. However, in a camera the image is focused by moving the lens, whose focal length cannot be altered, nearer or further to the photographic film. The human eye does not alter the location of the lens but uses the ciliary muscle to alter the shape of the lens, and thus the focal length.

The eye adjusts the amount of light required partly in the same way as the camera, by reducing or increasing the size of the pupil, which is an opening in the iris. The greater range of light control is exercised by a change in the sensitivity of the retinal receptors, equivalent to changing from a fast to a slow film or vice versa.

We can judge distances and perceive objects three-dimensionally by using both eyes and their muscles like a range-finder. The eyes rotate horizontally until their lines of sight intersect at the correct distance.

The retina contains two types of light-sensitive cells, the cones and the rods. The cones are concentrated in the central portion of the retina and are about 10 000 times as sensitive as the rods. They react much faster to a light stimulus, and, unlike the rods, they result in color vision [Section 8.6]. Therefore we see objects within 2° of our line of sight better than those nearer to the periphery of our visual field.

Visual acuity, the ability to distinguish fine detail, depends on the luminance of the object,

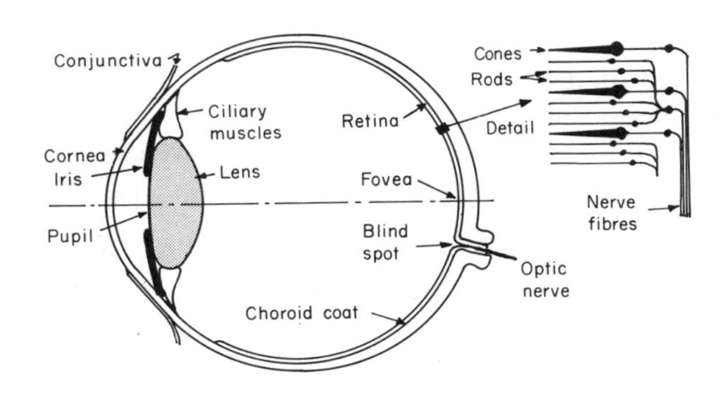

Fig. 8.3.1. The human eye. Its focal length can be changed by the ciliary muscle, which alters the shape of the lens. The image is inverted by the lens and projected on the retina, which contains light receptors, the rods and the cones. Their sensation is interpreted by the brain through the optical nerve. The amount of light is adjusted by alteration in the size of the pupil; this has the same effect as altering the f-number in a camera.

its size, the contrast with its surroundings, and the time available for looking at it. On the assumption that all these are determined by the task to be performed, except for the luminances, it is necessary to provide sufficient illumination by natural or artificial light to achieve the required standard of luminance.

The range of luminance at which the eye can function is far greater than the range of tolerance of the human body to changes in temperature. The human eye can detect brightness over a range of more than 1 to 100 million, although at the lowest levels of luminance it may require one and a half hours to adapt itself to the dark, and at the highest levels it experiences discomfort. However, the eye can perform efficiently and comfortably within a range of luminance of 1 to 1000 units without requiring time to adapt itself to this variation.

The eye becomes tired when the task it is asked to perform is too arduous, and it may require a rest. Thermal comfort can often be achieved when hard work is performed by adjusting the temperature accordingly [Section 4.6], and visual comfort can similarly be achieved by adjusting the luminance. There is no visual equivalent to the unit clo, except perhaps the use of magnifying glasses; however, the range of visual comfort is much greater than that of thermal comfort.

8.4 Light Sources and Illumination Levels

Daylight is the visible part of the radiation received from the sun. It ranges from 390 nm for violet light to 760 nm for red light. However, some animals can see solar radiation that is not visible to the human eye, and vice versa. Fig. 3.2.1 shows that the band of visible light includes the range of the highest relative energy of solar radiation. The strongest sensation of light is in the yellow-green range at a wavelength of 555 nm.

The design of buildings in relation to daylight is discussed in Chapter 9. When daylight does not suffice, the necessary luminance must be provided by artificial lighting. This happens in parts of a building that are too far from windows [Section 10.1], and in all parts of a building in the late afternoon and, of course, at night.

Although candles, oil lamps, and gas light are still used occasionally, the main sources of artificial light are electric lamps. For interior lighting the most common are incandescent and tubular fluorescent lamps. An incandescent filament lamp consists of a coil of tungsten wire that is heated to a temperature of about 2700°C in a bulb containing an inert gas.

A fluorescent lamp is a glass tube with electrodes at each end, filled with mercury vapor. It is coated internally with fluorescent powder. The electric discharge between the two electrodes of the tube produces ultraviolet radiation, mostly with a wavelength of 254 nm. This ultraviolet radiation is not visible to the human eye, but it activates the fluorescent coating of the tube to emit visible light. Lamps are discussed in Chapter 10.

The illuminance levels recommended in North America to provide adequate luminance for various tasks are set out in Figs. 2.2–2.5 of the (American) *IES Lighting Handbook* (Ref. 8.4, pp. 2.5–2.23 or Ref. 8.3, pp. A.3–A.22). They range from 30 lx for night lighting in hospital wards to 27 000 lx for the operating table in the surgical suite of a hospital.

The illuminance levels recommended in Britain are given in the (British) IES Code (Ref. 8.5, pp. 64–88) and those for Australia in the Australian Standard 1680 (Ref. 8.6, pp. 42–59). The illumination levels recommended in Britain and in Australia are lower than those recommended in North America.

There has been a steady increase in the level of illuminance since the beginning of this century, but it is doubtful if recent increases have resulted in increased efficiency at work or in increased comfort for people.

The minimum recommended illuminance levels are based on experiments during which persons performed specific tasks, but it has become common practice to provide these or higher levels for the entire area. This has some advantages, because machines or desks can be moved freely without the need for providing an electric outlet for task lighting when the layout of the building is altered. However, it greatly increases the energy consumption and the heat load on the air conditioning plant in summer. It also leads to monotonously uniform lighting, and the levels needed for specific tasks are often too high for comfort as background lighting.

8.5 Glare

Glare is caused by excessive contrast between the brightness of the task and the general brightness of the surroundings to which the eyes are adapted.

Disability glare prevents a person from performing a task. It may be caused by vehicle headlights or by a floodlight on a sports field, but it is unlikely to occur indoors.

Discomfort glare may be caused by a direct view of a sunlit window or an unshielded light fitting within the field of view (Fig. 8.5.1). It can be avoided by shielding the sources, for example, by the sunshading of windows or by placing shades or diffusers in front of light sources. It is helpful to use light-colored paint on surfaces against which glare sources will be seen, for example, the ceiling and the upper part of the walls, to reduce the contrast.

Discomfort glare may also be due to the reflection from any surface, including that on which the task is being performed (Fig. 8.5.2); the task itself may have a glossy surface and pick up reflections. One possibility is to move the light source so that the glare is not reflected into the eyes of the person performing the task; another is to move the person to a different location or turn him in a different direction; and a third is to change the surface causing the reflection and substitute a matte, nonreflecting surface.

The subject of glare is further discussed in Sections 9.6 and 10.8.

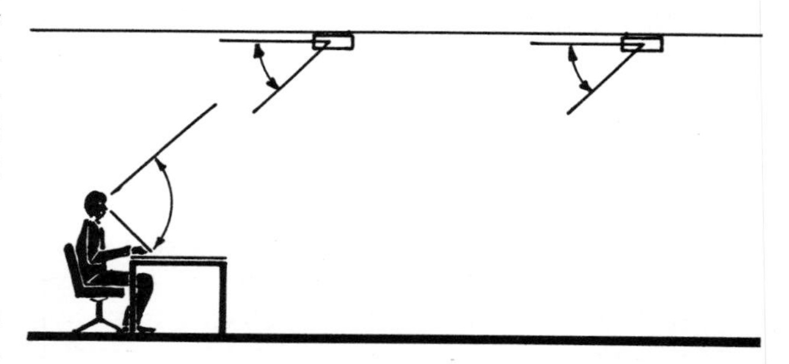

Fig. 8.5.1. Discomfort glare caused by an insufficiently shielded light fitting within the field of view.

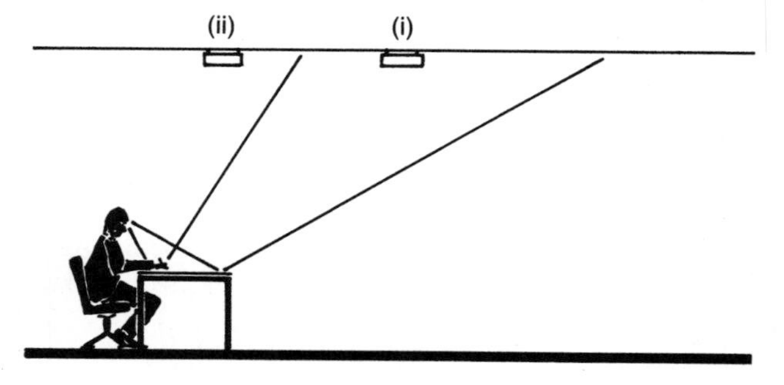

Fig. 8.5.2. Discomfort glare caused by reflection from the desk on which the task is performed. In this instance the problem can be solved by moving the light fitting from (i) to (ii), or by moving the desk.

8.6 Color

The eye perceives three primary colors, blue, green, and red, through the cones in the retina [Section 8.3], and the brain interprets the responses to identify the colors.

The color of light depends on its wavelength. It ranges from red (760 nm) through orange, yellow, green, and blue to violet (390 nm). The colors of sunlight are normally mixed together, so that it appears white. The colors can be separated by refraction through a prism; the laws of refraction are explained in most textbooks on physics (for example, Ref. 8.7, pp. 471–503).

Pigments of a primary color absorb all other colors and reflect only light of their own color. Thus an object appears blue because it reflects only blue light.

A sodium lamp emits only yellow light. If a blue object that reflects only the blue part of the spectrum is viewed in light emitted by a sodium lamp, it appears nearly black, because the blue object absorbs the yellow light and there is no other color to reflect. The sodium lamp is economical in the use of electricity, but it has never been used for interior lighting because it makes all colors but yellow look black or gray. The earliest fluorescent tubes also suffered from poor color rendering, and food lit by them sometimes looked unappetizing. A correct description of color is therefore important.

The use of a color classification scheme helps to reproduce colors accurately. The most widely used method was devised by A.H. Munsell in 1905. Colors are matched with standard colors in the *Book of Color* published by the Munsell Color Company of Baltimore.

Munsell arranged color according to hue, value, and chroma. The hue scale consists of the basic colors, red, yellow, green, blue, and purple, and five intermediate colors, identified by combinations of the letters R, Y, G, and P. Value correlates with the degree of lightness to which the color is perceived to belong, and it ranges from 0/ for ideal black to 10/ for ideal white. The chroma scale correlates the saturation of the color, and it ranges from /1 in arbitrary steps to express departure from the equivalent gray.

The power scale is the composite of Munsell value and chroma; thus 4/14 means a color slightly darker than the middle value between black and white, and 14 arbitrary steps from the equivalent gray, that is, highly saturated.

The Munsell System works well for colors viewed by daylight or other white light. The CIE System is better adapted to the classification of colored light. It is described in the *IES Lighting Handbook* (Ref. 8.3, pp. 5.4–5.8). The CIE System gives only the color, and this is specified by two chromaticity coordinates. In addition, a photometric specification of the luminance or intensity of the light is required.

References

8.1 G. PLEIJEL and J. LONGMORE: A method for correcting the cosine error of selenium rectifier photocells. *Journal of Scientific Instruments,* Vol. 29 (1952), pp. 137–38.

8.2 J.A. LYNES: *Principles of Natural Lighting.* Elsevier, London, 1968. 212 pp.

8.3 *IES Lighting Handbook.* Sixth Edition. Iluminating Engineering Society, New York, 1981. Reference Volume: 9 chapters.

8.4 *IES Lighting Handbook.* Sixth Edition. Illuminating Engineering Society, New York, 1981. Applications Volume: 22 chapters.

8.5 *The IES Code for Interior Lighting.* The Illuminating Engineering Society, London, 1977. 127 pp.

8.6 *Interior Lighting and the Visual Environment.* Australian Standard 1680. Standards Association of Australia, Sydney, 1976. 76 pp.

8.7 LLOYD W. TAYLOR: *Physics, The Pioneer Science.* Dover, New York, 1941. Two volumes. 847 pp.

Suggestions for Further Reading

Reference 8.2.

8.8 H.C. WESTON: *Sight, Light, and Work.* Second Edition. H.K. Lewis, London, 1962. 251 pp.

8.9 R.G. HOPKINSON and J.B. COLLINS: *The Ergonomics of Lighting.* Macdonald, London, 1970. 272 pp.

8.10 R.L. GREGORY: *Eye and Brain—The Psychology of Seeing.* Weidenfeld and Nicholson, London, 1967. 251 pp.

8.11 D.B. JUDD and G. WYSZECKI: *Color in Business, Science and Industry.* Wiley, New York, 1963. 500 pp.

Chapter 9 Daylight

Daylight is important for providing contact with the natural environment outdoors, and it is a source of illumination that uses solar energy.

Daylight varies with the luminance of the sky, and this depends mainly on latitude. The International Commission on Illumination recommends the use of the daylight factor for the design of natural lighting; the lumen method and model analysis with an artificial sky are alternatives.

Because of the variable nature of daylight, glare is a particular problem, especially in the tropics because of the high luminance of the sky.

9.1 Daylight for Delight and for Commodity

The artistic aspects of daylighting have been studied much longer than the technical aspects. The stained glass windows of Gothic cathedrals are early examples of the use of daylight to create a work of art inside a building. In Renaissance and Baroque buildings flat or simply shaped surfaces were sometimes painted in perspective to create an illusion of large and complex spaces. For example, a complete room extending the interior might be painted on a blank wall, or a vault on a flat ceiling. The effectiveness of the illusion frequently depended on the careful control of indirect natural lighting.

Perhaps the most extreme example is the *Trasparente,* a Baroque dormer window, added in the early eighteenth century to the ambulatory of the thirteenth-century Gothic cathedral of Toledo (Ref. 9.1, Fig. 79, p. 188), which introduced a bright light into the otherwise gloomy cathedral; in it there appeared a figure of Christ seated on clouds surrounded by a heavenly host which, although painted on the walls, has a startling three-dimensionality.

Good design can overcome the climatic limitations of daylighting. The solar radiation, and thus the level of daylight, is reduced as one moves from the tropics to the temperate zone [Section 3.2]. However, the natural light of interiors need not be affected in the same way. The windows of English Renaissance buildings are generally larger than those of the Italian buildings that served as their models (Fig. 1.2.6). On the other hand, the windows in the traditional buildings of the hot, arid countries of the Middle East are often small, so that the daylight level in the interior is lower than in English traditional houses, because control of the thermal environment is more important than lighting.

Daylight for interior rooms presents a difficult problem, sometimes solved brilliantly, as in Fig. 9.1.1. However, improvements in the technology of glass and iron in the middle of the nineteenth century made it much easier to provide adequate natural lighting for interior spaces (Fig. 8.1.1), even in the high latitudes of northwestern Europe.

By the end of the nineteenth century good daylight could be achieved in quite large buildings through the provision of light wells and skylights [Section 8.1]. The perfection of artificial lighting reduced the importance of daylight during the twentieth century [Section 10.1]; but since the mid-1970s there has been renewed interest in the technology of natural lighting because it saves energy.

We discussed the function of windows in relation to the thermal performance of buildings in Section 5.9. The requirements for good natural lighting and good thermal performance are sometimes in conflict. In northern Europe the luminosity of the sky is low in winter, and good daylighting requires large windows, which cause undesirable heat losses. Heat loss can be reduced by the use of double or triple glazing.

Overheating in summer can be controlled by the orientation of the windows and by sunshades [Sections 3.4, 3.5, 5.4, 5.15, 5.16, and 9.7]. It is caused by direct solar radiation, whereas daylight design is based on sunlight reflected from the sky or the ground. Thus windows facing north and east (south and east in the Southern Hemisphere) cause few problems. For windows on the opposite facades sunshades can be designed that control the admission of direct sunlight in

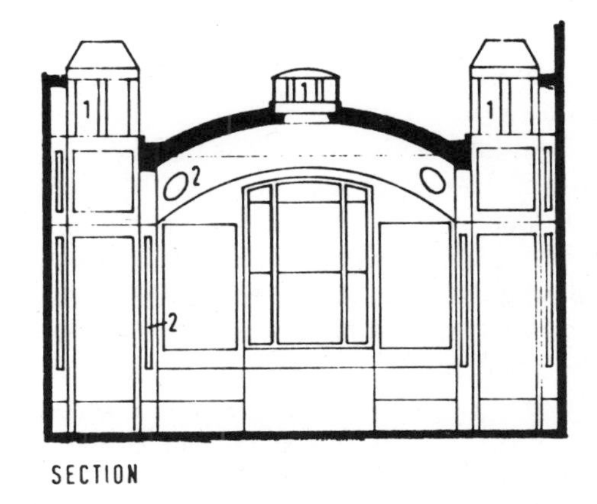

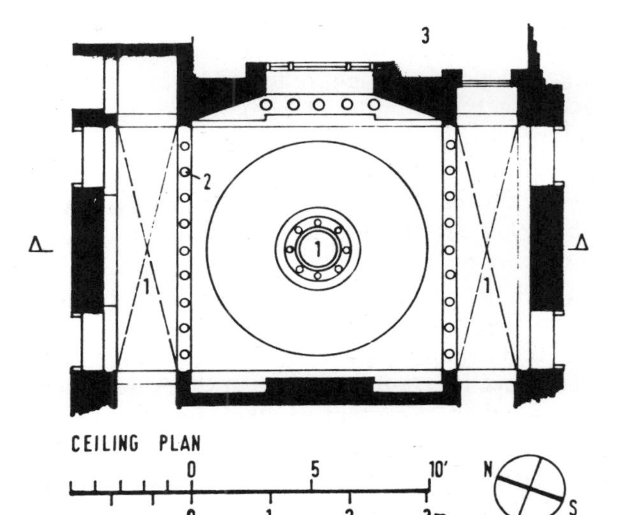

Fig. 9.1.1. The breakfast room in the house that the architect Sir John Soane built for himself in the early nineteenth century; it is now the Soane Museum at Lincoln's Inn Fields in London. This interior room obtains most of its daylight from three rooflights (1), assisted by a number of mirrors (2) and a window into a small courtyard (3). The mirrors and the courtyard window also serve to give a feeling of spaciousness by providing views beyond the immediate confines of the room. (*Reproduced from* The Lighting of Buildings, *by R.G. Hopkinson and J.D. Kay, by courtesy of the publishers, Faber and Faber.*)

summer without interfering unduly with daylight [Sections 3.4 and 9.3]. Overheating can also be controlled by the use of heat-absorbing or heat-reflecting glass [Section 5.9]. This reduces the transmittance of thermal radiation from the sun to a far greater extent than that of visible light, but some loss of daylight and distortion of color toward the blue end of the spectrum [Section 8.6] must be accepted.

While daylight enables people to see without the expenditure of energy, it also makes a contribution to the amenity and the general character of the building. This aspect has not been studied as carefully as it should be (Ref. 9.2, pp. 68–77), but there is a widespread feeling that buildings should have windows wherever possible to provide daylight and a view. We noted in Section 5.9 that windowless buildings have not become popular, and few have been built, even though there is no positive evidence that human performance is reduced by the absence of windows. This is discussed further in Section 10.1.

It is possible to have excellent daylight at a desk, table, or some other "workplane," and yet be unable to see the sky, because daylighting largely depends on light reflected by exterior surfaces, such as an area of grass or paving, or the wall of an opposite building, and by interior surfaces, such as the walls, floor, and ceiling of the room. The precise geometry of these lightpaths is complex, and for design it is generally necessary to make simplifications.

There are three principal methods for daylight design. The semiempirical *lumen method* [Section 9.4] treats the window as a luminous panel. The illuminance [Section 8.2] from the sky is multiplied by a number of factors to allow for the size of the windows, the loss of light caused by the glass and by obstructions, and the reflectances of various interior and exterior surfaces. Thus the illuminance on the workplane (the horizontal surface used for reading or other tasks) is obtained in lux, that is, lumen per square meter. This method is used mainly in North America.

The *daylight factor* [Section 9.3] is defined as the ratio of the illuminance at a point on the workplane to the exterior daylight illuminance, that is, the illuminance received at the same time on a horizontal plane exposed to the unobstructed sky. It is a dimensionless ratio that can be derived theoretically from the solid geometry of the building and its surroundings, and the optical properties of the glass and the reflecting surfaces. In practice, however, empirical simplifications must be made.

The daylight-factor method became particularly popular in England, where it was developed, at least partly because of its legal significance. English law has recognized "ancient lights" since medieval times. If a window has been used without interruption for a long time (defined in 1832 as twenty years), the owner of the building acquires the right to prevent the owner of adjoining land from obstructing the light received through this opening. This principle has not been accepted by American or Australian law. In 1922 an English judge adopted a daylight factor of 0.2 as the borderline between adequacy and inadequacy of daylight in a room. If we take the exterior daylight illuminance on an English winter day as 5000 lux [Section 9.2], this gives an illuminance on a workplane adjacent to a window of 1000 lux.

In principle the two methods are the same, but in practice each has some limitations [Section 9.4].

In addition, daylight can be investigated experimentally in an artificial sky [Section 9.5]. This is warranted only for buildings of special complexity or for research.

Whichever method is used, it is necessary to know the magnitude of the sky luminance and its distribution.

9.2 The Sky as a Source of Light

The luminance [Section 8.2] of the sky varies greatly with latitude and with cloud cover. It varies with the time of year and the time of day. The distribution of the luminance across the sky also varies according to the cloudiness of the sky. It is therefore necessary to make simplifying assumptions. As a first approximation the distribution of luminance across the sky can be assumed as uniform (Fig. 9.2.1).

In northwestern Europe, including the British Isles, daylight design is based on the assumption that the sky is overcast, as the illumination is then low; this is the critical condition for design. When the sky is partly cloudy, its luminosity is higher, but excessive daylight is not a problem in northwestern Europe, except to the extent that it may cause glare [Section 9.6].

In India, southern Africa, central Australia, and some parts of North America it may be assumed that the sky is clear blue, as fully overcast skies are comparatively rare.

Fig. 9.2.1. Sky with uniform distribution of luminance. We assume that the sky uniformly diffuses the sunlight, so that its reflectance $r = 1$. From Eq. (8.3) the illuminance, E, on a horizontal surface due to the uniform luminance, L, is

$$E = \pi L = 3.14L \qquad (9.1)$$

In this equation the illuminance, E, is measured in lux and the luminance, L, in candela per square meter.

A vertical surface is exposed to only half the sky's hemisphere, and

$$E = \frac{1}{2}\pi L = 1.57L \qquad (9.2)$$

In most parts of North America and in eastern Australia both conditions should be considered.

The overcast sky is brightest at the zenith, that is, directly overhead. The luminance decreases toward the horizon, where it is about one third of the luminance at the zenith (Fig. 9.2.2). In 1941 Professors Parry Moon and Domenica Spencer in the United States proposed Eq. (9.4), illustrated in Fig. 9.2.2.

Observations in various parts of the world have shown reasonable agreement with this formula (Ref. 9.3, pp. 41–42), and in 1955 it was adopted by the CIE (International Commission on Illumination) as the standard for the distribution of relative sky luminance for the overcast sky. At present most daylight calculations are made in locations where the overcast sky is considered the only or the principal design criterion.

The pattern of sky luminance on an overcast day is affected by the reflection from the ground. Richard Kittler, in Czechoslovakia, observed that in the mountainous regions of central and northern Europe overcast skies seldom occur except when the ground is covered with snow. Under these conditions the zenith luminance is only about twice the horizon luminance, and Kittler proposed the following formula (Ref. 9.5, p. 59):

$$L_\theta = \frac{1}{2} L_z (1 + \sin\theta) \qquad (9.7)$$

Kittler also proposed a formula for the luminance distribution of the clear blue, cloudless sky (Fig. 9.2.3), based on his analysis of obser-

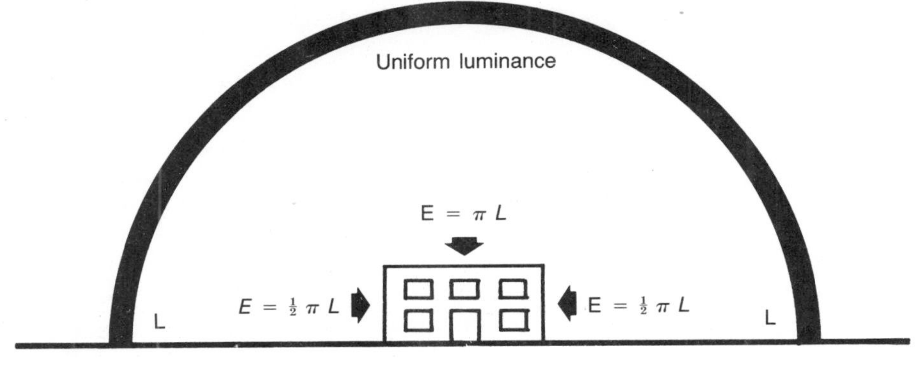

9.1.1

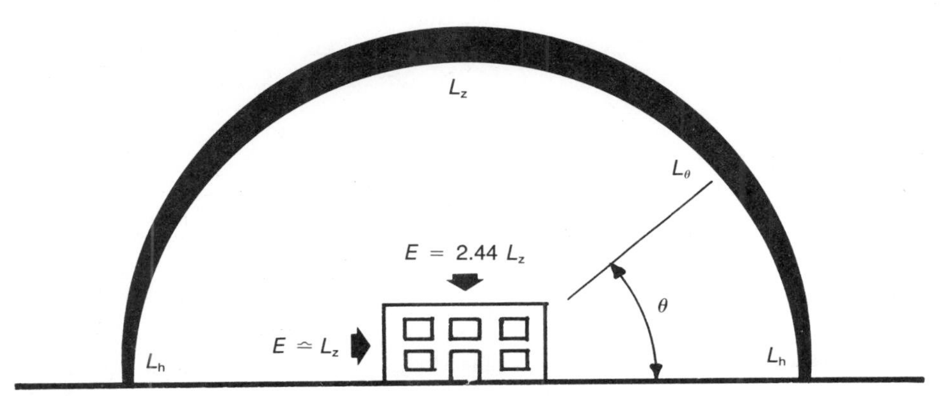

9.2.2

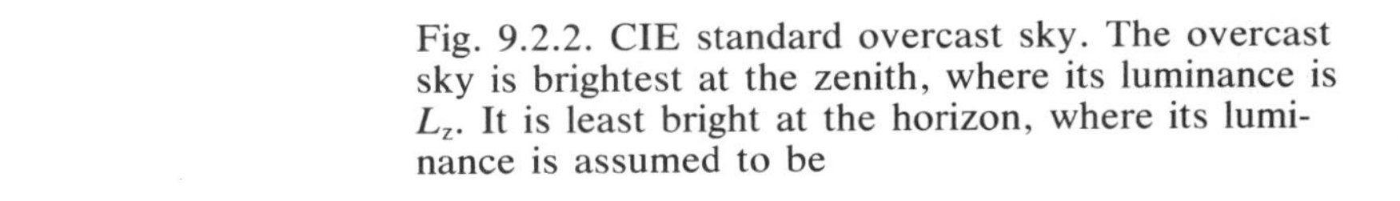

Fig. 9.2.2. CIE standard overcast sky. The overcast sky is brightest at the zenith, where its luminance is L_z. It is least bright at the horizon, where its luminance is assumed to be

$$L_h = \frac{1}{3} L_z \qquad (9.3)$$

The luminance elsewhere is assumed to vary in accordance with the formula

$$L_\theta = \frac{1}{3} L_z(1 + 2\sin\theta) \qquad (9.4)$$

where θ is the altitude of the point under consideration, that is, the angle it makes with the horizon;
L_θ is the luminance at θ;
L_z is the luminance at the zenith.

By integrating the variable luminance over the entire hemisphere of the sky (Ref. 9.4, p. 25), the illuminance of a horizontal surface is found to be

$$E = \frac{7}{9}\pi L_z = 2.44 L_z = 0.81 L_h \qquad (9.5)$$

and the illuminance on a vertical surface is approximately given by

$$E \approx L_z \qquad (9.6)$$

vations in various parts of the world (Ref. 9.3, pp. 36–40). This was adopted in 1967 as the standard clear-sky luminance distribution by the CIE:

$$L_\theta = L_z \frac{(1 - e^{-0.32/\sin\theta})(0.91 + 10e^{-3a} + 0.45\cos^2 a)}{0.274\,(0.91 + 10e^{-3z} + 0.45\cos^2 z)} \tag{9.8}$$

where θ, L_θ, and L_z have the same meaning as in Eq. (9.4);
- e is the base of the natural logarithm, that is, 2.7183;
- z is the angle that the sun makes with the zenith, that is, the difference between the solar altitude and 90°;
- a is the angle between the sun and the point in the sky under consideration, at the altitude θ.

This formula implies that the cloudless sky has a very bright aura of light around the sun and has its lowest luminance at an angle of approximately 90° from the position of the sun. The maximum luminance, for an observer facing the sun, is 20 to 30 times that at 90° to the sun. Clear, dry air and high altitudes result in skies of a deeper blue, and of greater areas of low luminance (Ref. 9.3, p. 45).

The Kittler formula for the clear blue sky is complicated. A simpler formula has been derived by V. Narasimhan and B.K. Saxena in India, who claim that it provides a better fit for the observed luminance distribution of the clear blue sky in India (Ref. 9.6). It applies only to the region of the sky away from the sun. In 1968 it was incorporated into the Indian *Standard Code of Practice for Daylighting of Buildings, IS 2440:*

$$L_\theta = \frac{L_z}{\sin\theta} \tag{9.9}$$

when θ lies between 90° and 15°; and

$$L_\theta = \text{constant} = \frac{L_z}{\sin 15^\circ}; = 3.86 L_z \tag{9.10}$$

when θ lies between 15° and 0°. L_θ and L_z have the same meaning as in Eq. (9.4).

The Indian formula is not an approximation to the Kittler formula but gives a different distribution of luminance.

The various distributions of luminance can be modeled experimentally in artificial skies [Section 9.5] or mathematically, using calculations, graphical aids, or computer programs [Sections 9.3 and 9.4].

However, the main source of error in daylight design is not due to imprecision in the distribution of luminance, but to simplistic assumptions concerning its magnitude. A substantial proportion of daylight design has been, and still is, done with a standard exterior daylight illuminance of 5000 lx. This is based on observations made by the (British) National Physical Laboratory of the average conditions near London over the greater part of winter and on rainy days in summer.

For some locations there are data for the hour-by-hour variation of daylight levels on typical days; these can be fed into a computer program to produce accurate predictions of the interior illuminance due to daylight.

However, a single figure for exterior daylight illuminance in a particular location is commonly used, irrespective of the extent of cloud cover, the time of day, or the season. This is based on the latitude, which determines the greatest altitude of the sun on the shortest days of the year (Fig. 9.2.4).

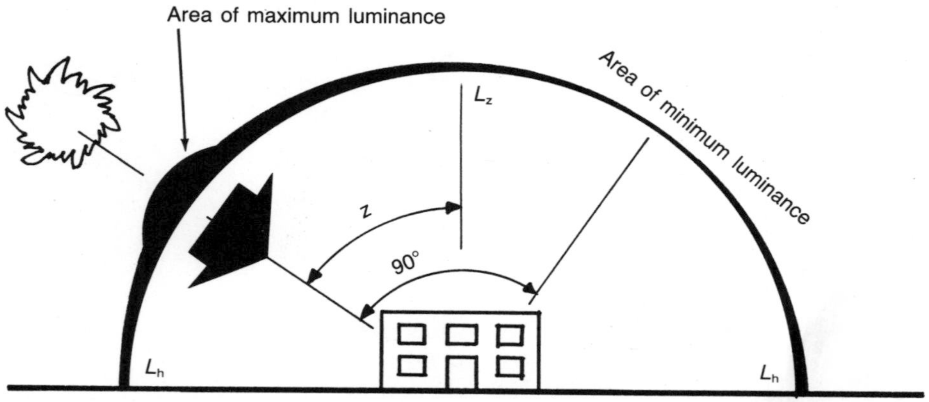

Fig. 9.2.3. Clear blue sky. The cloudless sky has its maximum luminance in the area surrounding the sun and its lowest luminance at approximately 90° to the position of the sun. Whereas the overcast sky has a horizon luminance that is lower than the zenith luminance, the reverse is true for the clear blue sky. The ratio L_z/L_h depends on the position of the sun.

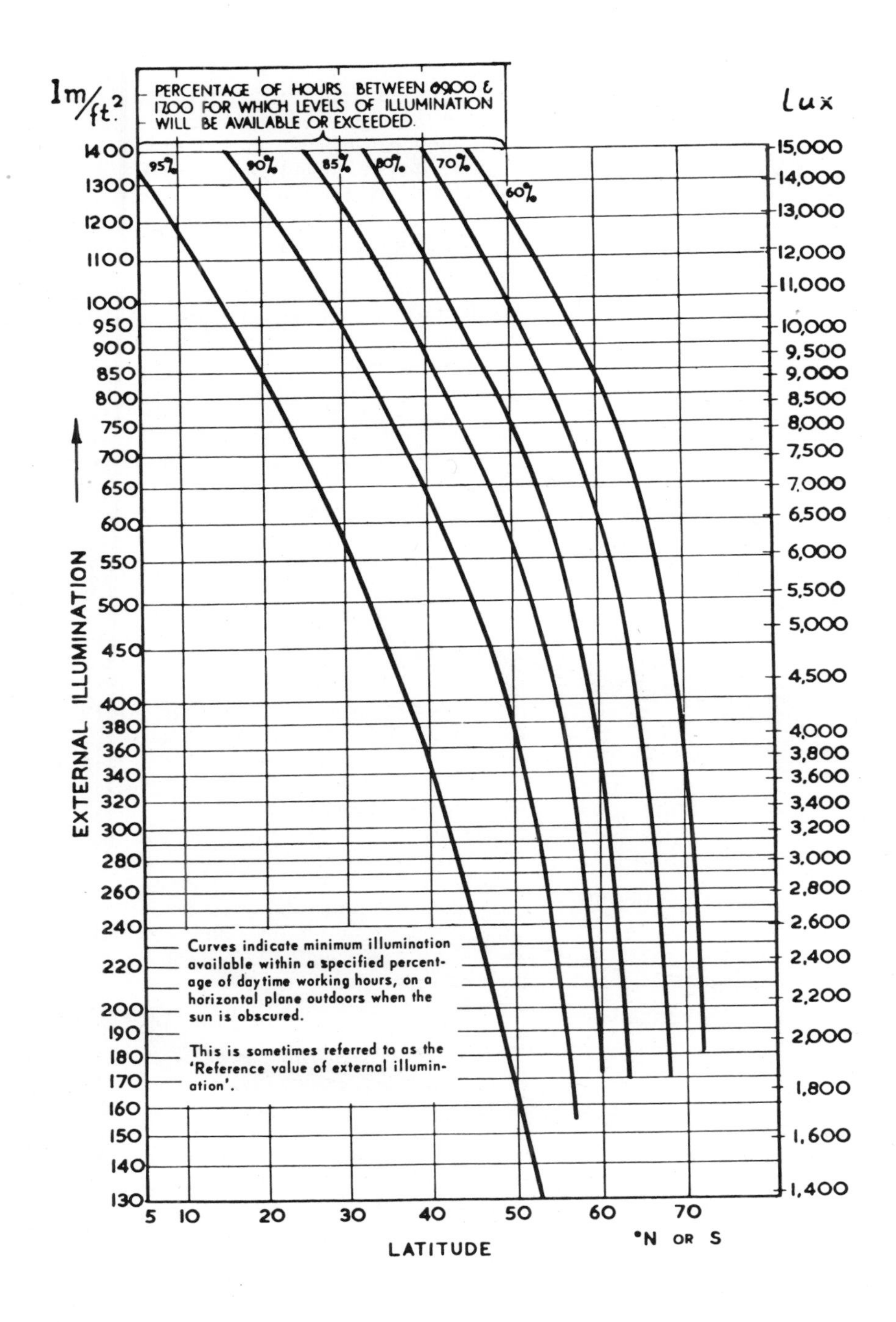

Fig. 9.2.4. Variation of daylight with latitude recommended by the International Commission of Illumination (CIE), based on an analysis of daylight data from several countries, made by A. Dresler in Australia. The designer can choose the percentage of time during which the design illumination level is to be exceeded. The chart was originally drawn in lumen per square foot; the scale in lux appears on the right of the chart.

The curves apply to working hours from 9 A.M. to 5 P.M. For different working hours the variations given in Table 9.1 should be made. (*Reproduced from Fig. 6 in Ref. 9.10.*)

Table 9.1

Percentages to Be Used When the Curves in Fig. 9.2.4 Are Applied to Periods Other Than 9 A.M. to 5 P.M.

Alternative Period	Revised percentage of alternative period for curve labeled: 95%	90%	85%	80%	70%	60%
7 A.M. – 3 P.M.	95	90	85	80	70	60
8 A.M. – 4 P.M.	100	100	95	85	70	60
7 A.M. – 5 P.M.	95	85	75	65	55	45
6 A.M. – 6 P.M.	75	70	65	60	50	40

Reproduced from Table I in Ref. 9.10.

There is no need for great precision in daylight calculations, because daylight is only one aspect of lighting. Prior to the eighteenth century most people were dependent on daylight for illumination; today most rooms have electric light. Daylight design does not, therefore, determine the illuminance on a particular workplane absolutely; it merely specifies when artificial light is required. This happens every day after the sun sets, and on other days if the cloud cover is exceptionally heavy. It is only necessary to switch on the electric light, which must be provided in any case for the hours of darkness. In some buildings this is accomplished automatically by photovoltaic cells that monitor the illuminance levels at selected points [Section 10.1].

There are exceptions to this rule. In the tropics and subtropics the length of day does not vary greatly with the season, and thus schools and some factories are used only during daylight hours. If it is acceptable to suspend normal operations on a few occasions when heavy clouds reduce daylight below the minimum design level, electric light is not needed, except in a few rooms used at night and in the circulation spaces for safety. This results in appreciable savings, which are particularly appropriate in developing countries in the tropics. Even in affluent countries, such as Australia, there are classrooms in some schools that do not have artificial lighting, and this causes slight inconvenience. When daylight provides the only illumination, however, it is necessary to ensure that it is adequate without supplementation.

9.3 The Daylight Factor

As mentioned in Section 9.1, the daylight factor is the ratio of the illuminance on a given horizontal plane in the room, called the workplane, to the illuminance at the same time on a horizontal plane exposed to the unobstructed sky. The daylight factor can be used with any exterior daylight illuminance, and indeed this is one of its advantages.

The daylight factor is partly due to the light received through the window directly from the sky, and partly due to light reflected from buildings opposite, from the ground outside, and from the internal surfaces of the building (Figs. 9.3.1 and 9.3.2). It is necessary to know the reflectances of the reflecting surfaces. It is also necessary to know the loss of light caused by the window, the blind or curtain, and the sunshade (if any) and to include a maintenance factor to allow for the frequency of window cleaning.

The remainder of the calculation consists of determining the solid angles due to the window under consideration and the surrounding building(s) obstructing the daylight. That theory (Ref. 9.3, pp. 59–76, or Ref. 9.4, pp. 102–64) can be used directly, but the calculation is lengthy. Several computer programs and tables are available. Alternatively the solution can be obtained with one of a series of daylight protractors (Ref. 9.7) produced by the (British) Building Research Station, or with the "pepperpot charts" and overlays produced by the Daylight Advisory Service of Pilkington Brothers, a British glass manufacturer (Ref. 9.8).

A. Dresler used the daylight protractors to derive charts for a number of standard building types, and these were published in 1963 by the Industrial Services Division of the Australian Department of Labour and National Service. These include single-story and multistory buildings with vertical windows, and also factories with

Fig. 9.3.1. The daylight factor. The daylight factor (*DF*) consists of three components, the sky component (*SC*), the externally reflected component (*ERC*), and the internally reflected components (*IRC*).

The reference plane inside the room is called the workplane. The *sky component* is received at a point on the workplane directly from the sky through the window. The window subtends a solid angle on a unit sphere, and the ratio of this solid angle to that of a hemisphere is called the *sky factor*. The sky factor can be used directly if the sky is assumed to have a uniform luminance; for a non-uniform luminance [see Section 9.2] a correction factor must be applied. To obtain the sky component, the corrected sky factor is reduced to allow for the loss of light caused by the window. The loss factor can be measured accurately with a light meter, or an appropriate value is taken from tables of standard properties.

The *externally reflected component* (*ERC*) is particularly important in densely built-up areas where little or no light is received directly from the sky. It depends on the *reflectance factor* (*RF*) of the reflecting building, generally on the opposite side of the street, on the *configuration factor* (*CF*) of the external obstruction caused by the other building or buildings, and on the sky factor (*SF*) of the window already mentioned; that is, $ERC = SF \times RF \times CF$.

The reflectance varies from 80% for a building recently painted white, a common practice in the tropics, to 10% for a dark building. An *RF* value of 20% is frequently used in the absence of more precise data. The configuration factor depends on the geometry of the building under consideration and the building(s) opposite.

The *internally reflected component* consists of two parts, IRC_1 and IRC_2. The first part is received through the window but, unlike the sky component (*SC*), does not reach the workplane directly. It is either reflected by the floor and the ceiling, as shown, or occasionally by the walls. In addition to the first reflectances, as shown in the figure, the second reflectances contribute to the illuminance on the workplane.

The second part, IRC_2, is generally much smaller than IRC_1 and is frequently neglected. This is partly because the light is restricted by the external obstruction and partly because the external surface has generally a low reflectance. It is customary to assume a reflectance factor of 10% for the external surface, but this would be much too low if the external reflecting surface is white or of a light color [Section 9.7].

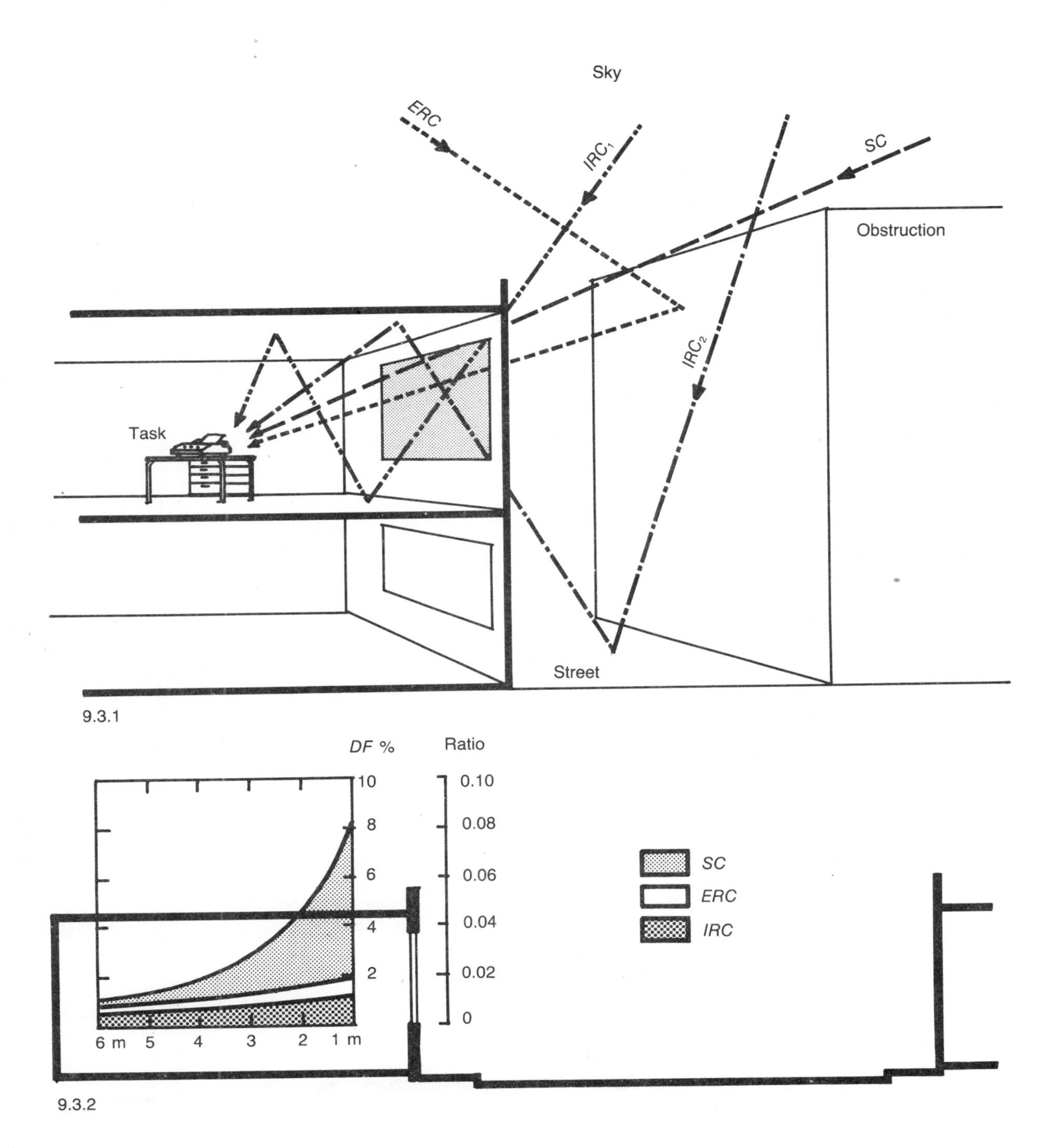

skylights, sawtooth roofs, and monitor roofs (Ref. 9.9). In 1970 the CIE adopted this method, with some modifications (Ref. 9.10), because of its simplicity.

Data needed for daylight calculations are given in several books, for example, the *IES Lighting Handbook* (Ref. 8.4) and CIE Publication No. 16 (Ref. 9.10), and in textbooks (for example, Refs. 9.3. and 9.11).

The *reflectances* of daylight by building materials range from 70% for white glazed bricks to 20% for rough, gray concrete. Smooth, clean concrete surfaces can reflect up to 40% of daylight, but dark green grass and asphalt reflect only about 7%. Snow reflects about 70%. The reflectance of building materials can be altered significantly by paint. A newly painted white surface has a daylight reflectance of about 80%, and reflectance of other colors can be obtained approximately from their Munsell value, *V* [Section 8.6], that

Fig. 9.3.2. The daylight factor falls off rapidly with distance from the window, so that supplementation of daylight by electric light is needed at the rear of deep rooms, even on a bright day [Section 10.1].

The daylight factor consists of three components (Fig. 9.3.1):

1. The light received directly, called the sky component (*SC*), which falls off most rapidly with distance from the window;
2. The light reflected from external surfaces, called the externally reflected component (*ERC*);
3. The light reflected from internal surfaces, called the internally reflected component (*IRC*), which remains almost constant with distance from the window.

The daylight factor is given by

$$DF = SC + ERC + IRC \qquad (9.11)$$

It can be expressed as a ratio or as a percentage.

is, the number between 0 and 10 that classifies the degree of lightness of a color (Ref. 9.3, p. 364):

Reflectance in percent

$$= V(V - 1) \qquad (9.12)$$

Thus a color with a Munsell value of 5 has an approximate reflectance of 20%. However, in the absence of information, the reflectance must be assumed (see caption to Fig. 9.3.1).

The light loss has three components. A window that consists of a single pane of glass has an *obstruction factor* of 1, but for wooden windows with several panes the factor may be 0.8 to allow for the obstruction caused by the frame and the glazing bars. Additional factors are needed for blinds and for curtains.

The second component of the light loss is the transmittance of the glass. The design charts in the CIE No. 16 (Figs. 9.3.9 and 9.3.13) are based on a transmittance of diffuse light of 85%. If the glass transmits a lower percentage, the daylight

Table 9.2

Correction Factor for Transmittance of Diffuse Light Through Glass That Is Less Than 85% When Clean

Transmittance of Diffuse Light Through the Glass Used	Transmittance Correction Factor
80%	0.95
70%	0.80
60%	0.70
50%	0.60
40%	0.45
30%	0.35

Table 9.3

Correction Factor for Dirt Accumulation on Windows

Location	Clean or Dirty Air	Angle of Slope of Window (90° is vertical) 90°–75°	60°–45°	30°–0°
Rural or outer suburban area	Clean	0.9	0.85	0.8
	Dirty	0.7	0.6	0.55
Built-up residential area	Clean	0.8	0.75	0.7
	Dirty	0.6	0.5	0.4
Built-up industrial area	Clean	0.7	0.6	0.55
	Dirty	0.5	0.35	0.25

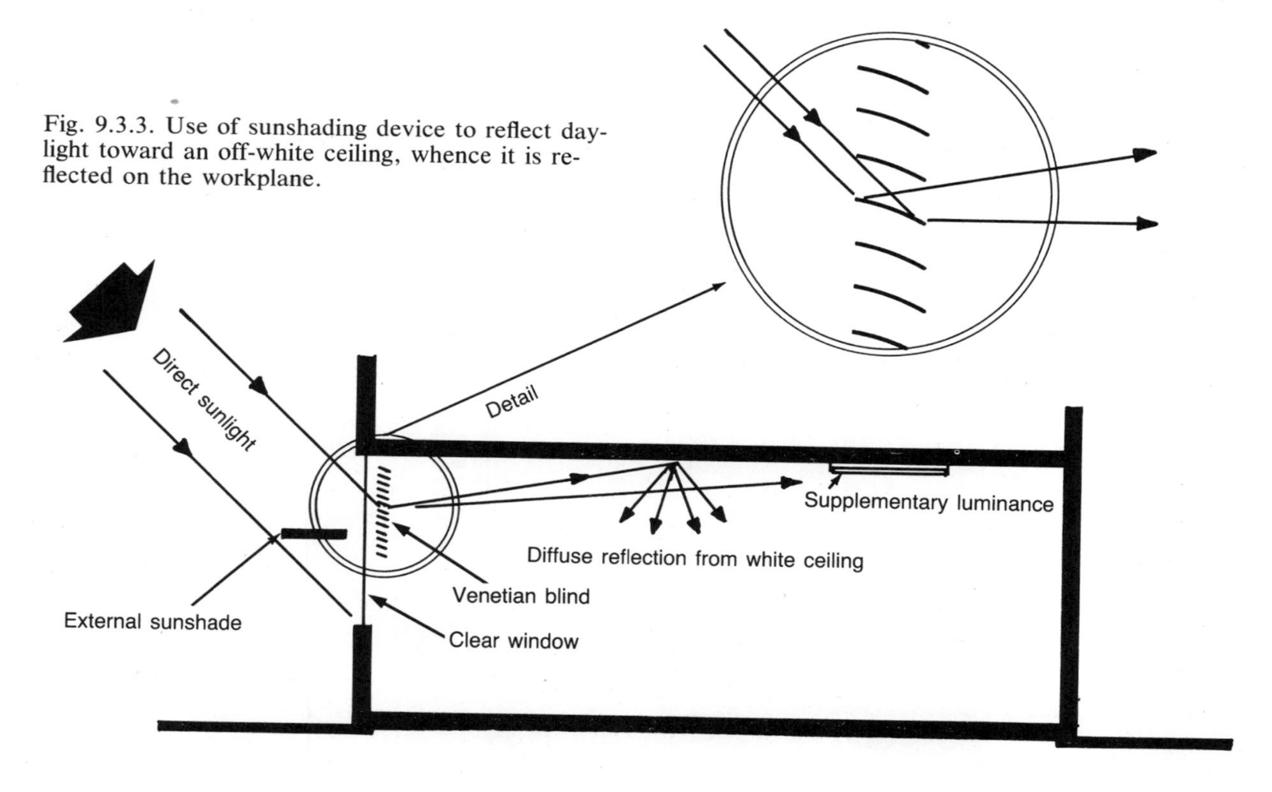

Fig. 9.3.3. Use of sunshading device to reflect daylight toward an off-white ceiling, whence it is reflected on the workplane.

factor is multiplied by the correction factor in Table 9.2. This correction is always necessary for heat-absorbing and reflective glass [Section 5.9].

The third component of light loss is the dirt accumulated on the window, which is at its worst on the day *before* it is cleaned. This is harder to predict and is beyond the control of the designer. CIE No. 16 gives Table 9.3.

A correction can also be made for light loss due to sunshading devices, although there are at present no data for their effect. Daylighting in the tropics presents special problems [Section 9.7]. However, sunshading devices do not need to result in a loss of daylight; indeed, they can be used to increase the reflected component (Fig. 9.3.3).

Example 9.1 *Determine the light-loss factor for a steel-framed vertical window with heat-absorbing glass for an office in an inner suburb.*

Solution *The obstruction factor for steel window frames is about 0.9. The dirt factor on a vertical window in an inner suburb is about 0.6. Heat-absorbing glass is tinted, and its transmittance is about 70%.*

Therefore the light-loss factor is

$$0.9 \times 0.6 \times 0.7 = 0.38$$

This means that only 38% of the daylight falling on the window is transmitted by it, and the remaining 62% is lost.

The CIE charts include an allowance for the transmittance of plain window glass, so that a 70% transmittance requires a correction factor (Table 9.1) of only 0.8. The combined correction factor is

$$0.9 \times 0.6 \times 0.8 = 0.43$$

The CIE charts for vertical windows (Fig. 9.3.12) are calculated for rooms with off-white ceilings (reflectance 70%), light colored walls (reflectance 50%), and dark floors (reflectance 15%). The CIE charts for skylights (Fig. 9.3.9) are worked out for an average reflectance of 20%, that is,

$$\frac{\Sigma\, rA}{\Sigma\, A} = 0.20$$

where A is the area of each part of the room and r is its reflectance. Table 9.4 gives a correction factor for rooms whose average reflectance is higher or lower.

Skylights provide the most effective use of daylight, because they give approximately uniform illumination; but they are limited to the top story of multistory buildings and to single-story buildings.

Skylights can be placed in a sloping roof by replacing some corrugated iron sheets, shingles, or roof tiles with transparent plastic. This is a simple and cheap method, but it can be used only in climates where overheating of the building in summer is unlikely. Flat or domed skylights (Fig. 9.3.4) are subject to the same limitation.

Sawtooth roofs (Figs. 9.3.5 and 9.3.6) facing north (south in the Southern Hemisphere) and monitor roofs (Fig. 9.3.7) are suitable for warmer climates.

We will first determine the amount of glass needed for a monitor roof.

Table 9.4

Correction Factor for Rooms with Skylights Whose Average Reflection Differs from 20%

Average Reflection	Correction Factor
15%	0.95
20%	1.00
25%	1.05
30%	1.10
35%	1.15
40%	1.20

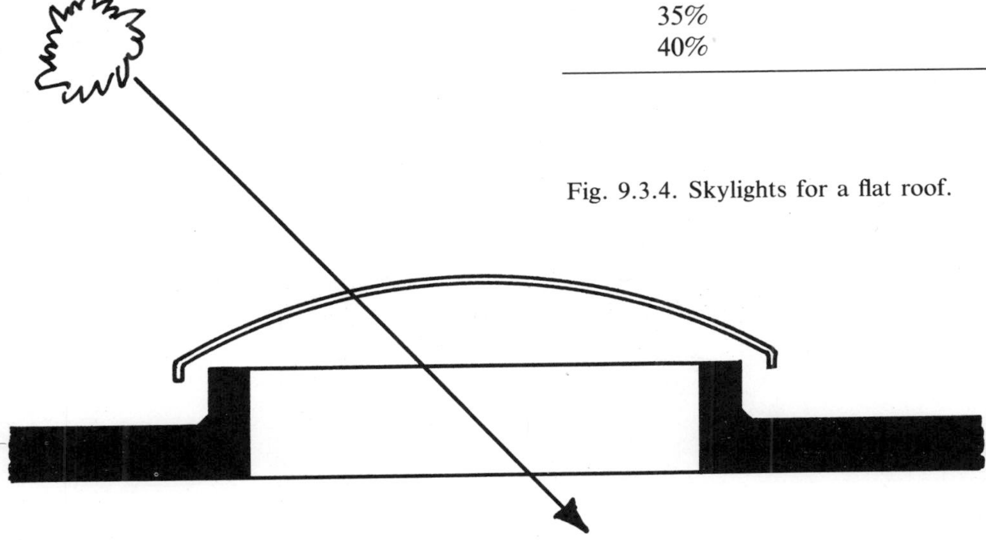

Fig. 9.3.4. Skylights for a flat roof.

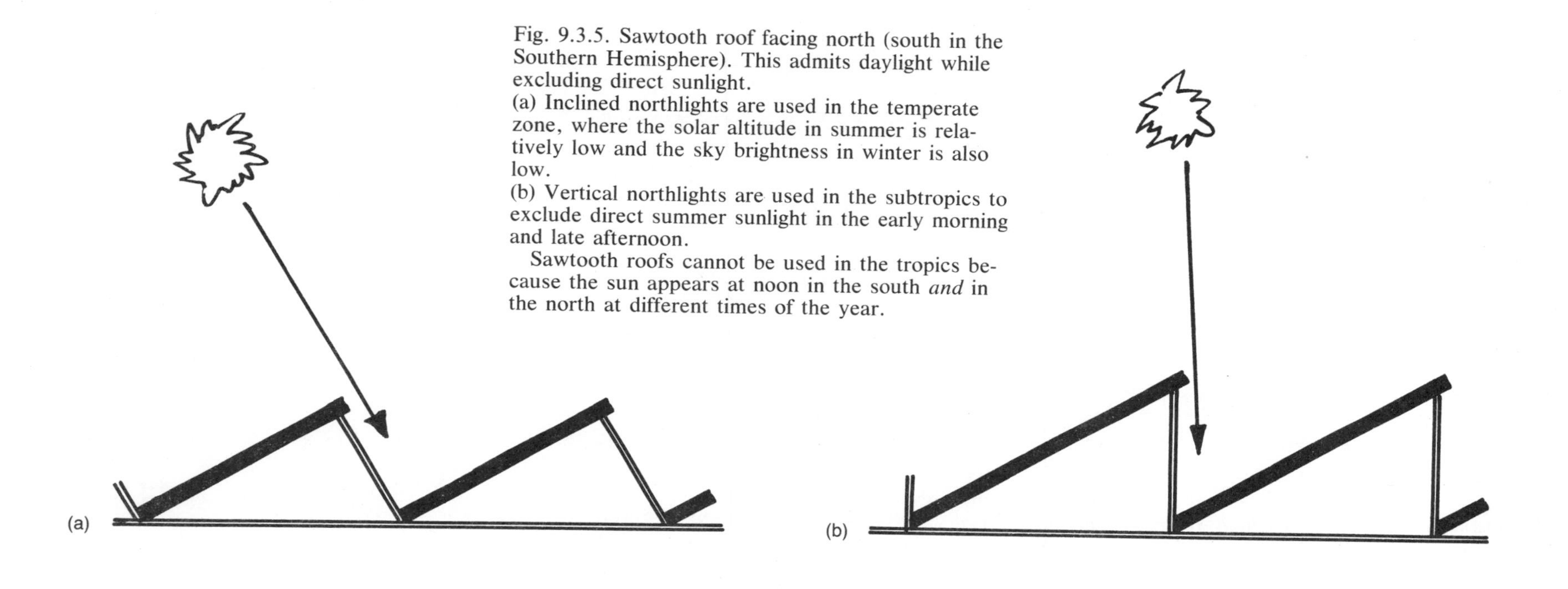

Fig. 9.3.5. Sawtooth roof facing north (south in the Southern Hemisphere). This admits daylight while excluding direct sunlight.
(a) Inclined northlights are used in the temperate zone, where the solar altitude in summer is relatively low and the sky brightness in winter is also low.
(b) Vertical northlights are used in the subtropics to exclude direct summer sunlight in the early morning and late afternoon.

Sawtooth roofs cannot be used in the tropics because the sun appears at noon in the south *and* in the north at different times of the year.

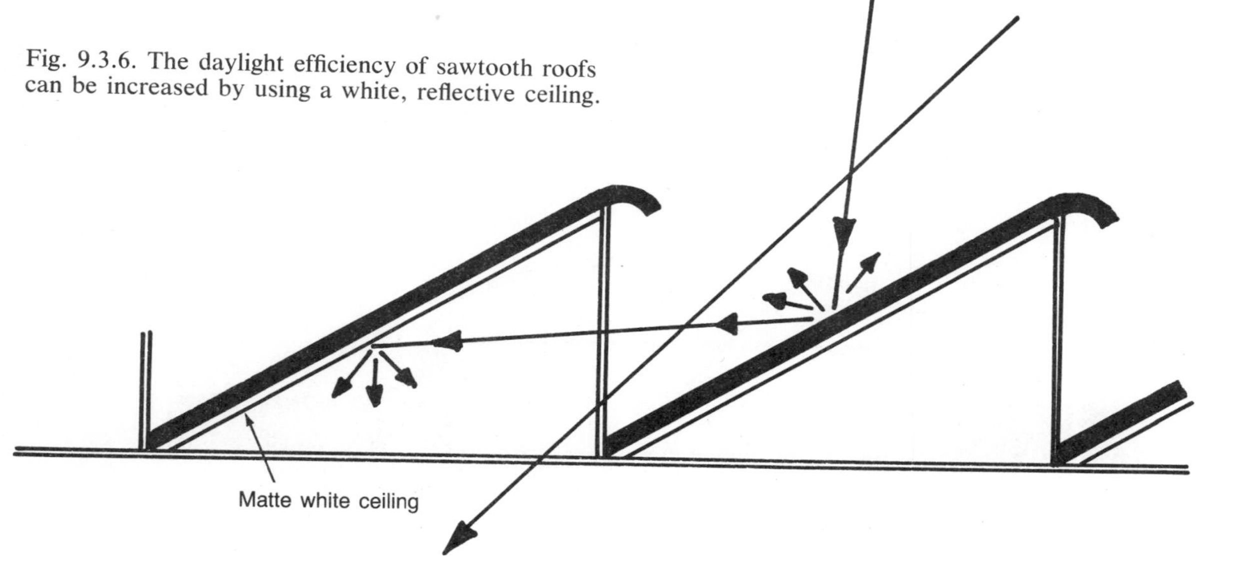

Fig. 9.3.6. The daylight efficiency of sawtooth roofs can be increased by using a white, reflective ceiling.

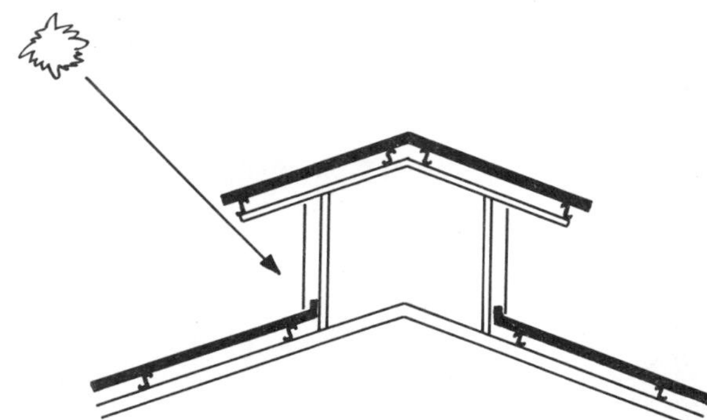

Fig. 9.3.7. Monitor roof to admit daylight while excluding direct sunlight. It is necessary to provide a roof overhang for the monitor windows on one side (on both sides in the tropics) to prevent sunlight penetration.

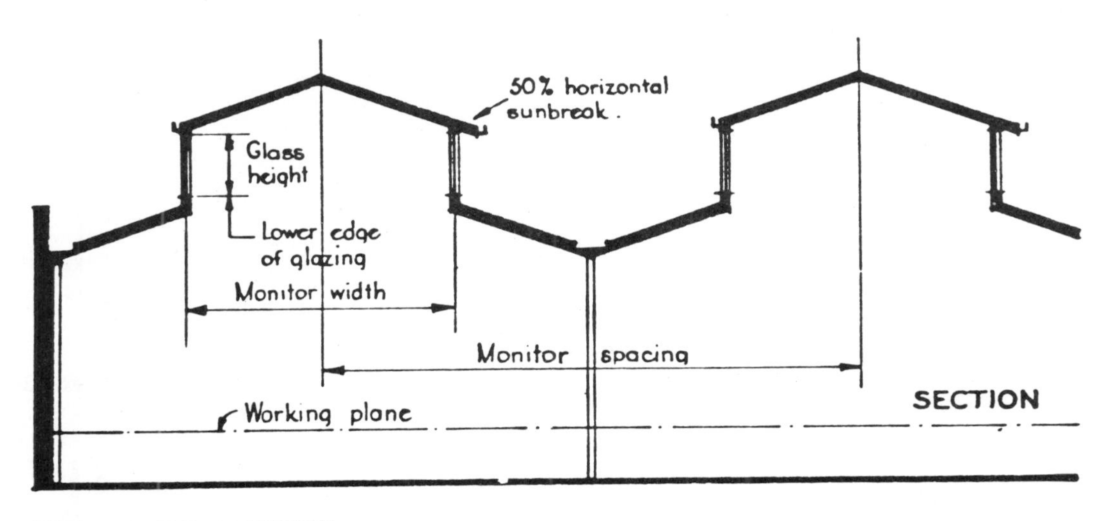

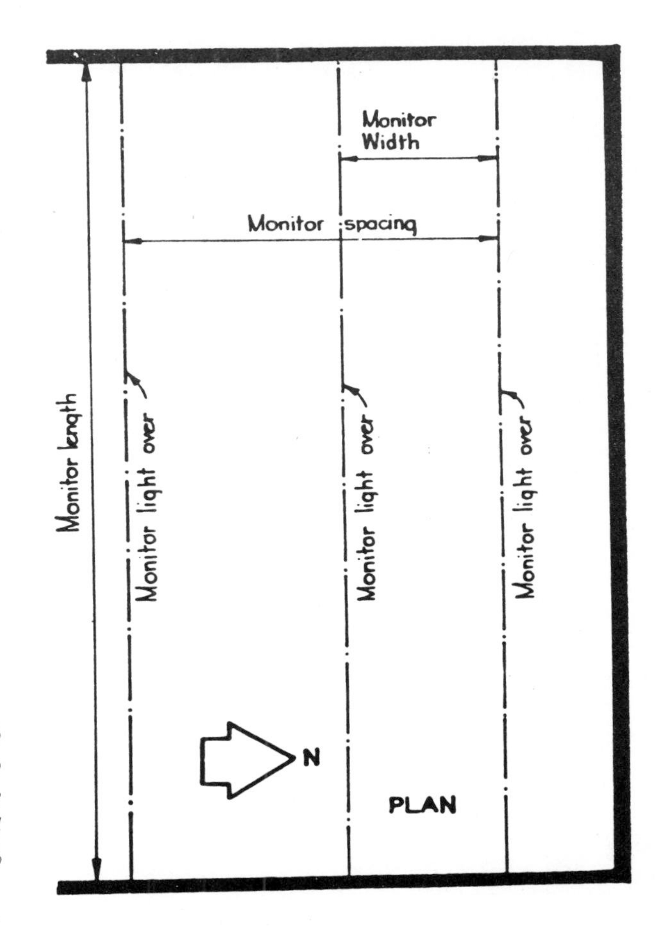

Fig. 9.3.8. Principal dimensions of a ridge-type monitor roof. (*Reproduced from Ref. 9.10.*)

Example 9.2 *A factory in an outer suburb of Sydney, Australia, is housed in a building with a multiple, ridge-type monitor roof oriented north and south, with 50% horizontal sunbreaks (Fig. 9.3.8). Determine the amount of glazing required.*

Working hours are from 8 A.M. to 4 P.M., and electric light will be used for maintenance and for work outside these hours.

An average interior illuminance of 250 lx is required, and electric task lighting is provided in addition where needed.

The building is 100 m long, the lower edge of the glazing is 9 m above floor level, and the working plane is 1 m above floor level. The monitors are spaced at intervals of 15 m.

The average reflectance of the interior surfaces is 25%.

The windows are of clear glass. Obstruction by glazing bars, by the roof structure, and by pipes and ducts amounts to 25%.

Solution *From Table 9.1 the curve labeled 90% for the hours 9 A.M. to 5 P.M. corresponds to 100% for the hours 8 A.M. to 4 P.M. The latitude of Sydney is 34°. The 90% curve in Fig. 9.2.4 intersects the 34° latitude line at 8700 lx for 100% of normal working hours.*

From Tables 9.2, 9.3, and 9.4 the transmittance correction factor is 1.0, the dirt correction factor is 0.7, and the correction factor for reflection is 1.05. The correction factor for obstructions is 1 − 0.25 = 0.75.

To achieve an interior illuminance of 250 lx, we require a daylight factor of

$$\frac{250}{1.0 \times 0.7 \times 1.05 \times 0.75 \times 8700} = 0.052, \text{ or } 5.2\%$$

The ratio

$$\frac{\text{building length}}{\text{height of lower edge of glass above working plane}} = \frac{100}{9-1} = 12.5$$

From Fig. 9.3.9 the required ratio of glass height to monitor spacing is 0.15.

Since the monitors are spaced at intervals of 15 m, we require glazing 15 × 0.15 = 2.25 m deep on each side of the monitors.

The distribution of daylight is not uniform. It is greatest under the ridge, where both monitors contribute equally to the illumination, and lowest at the columns. However, the variation of daylight would be much greater if it were admitted through windows in the walls of the factory.

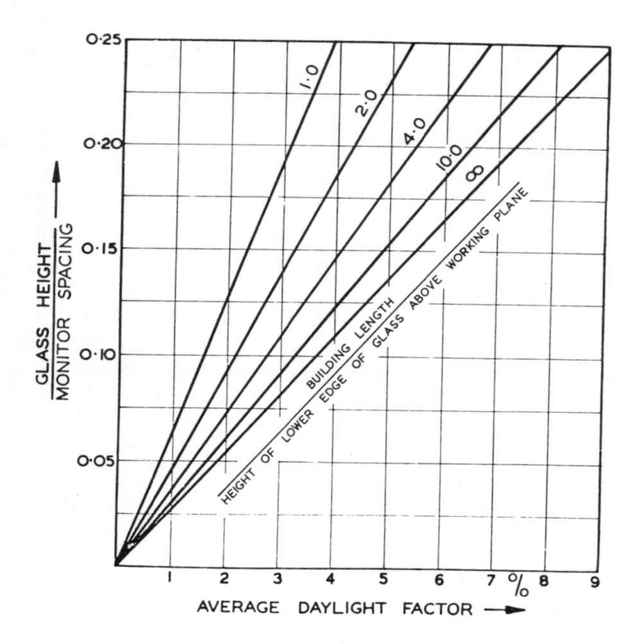

Fig. 9.3.9. The average daylight factor for a multiple ridge-type monitor roof, with a 50% sunbreak (see Fig. 9.3.8). (*Reproduced from Ref. 9.10.*)

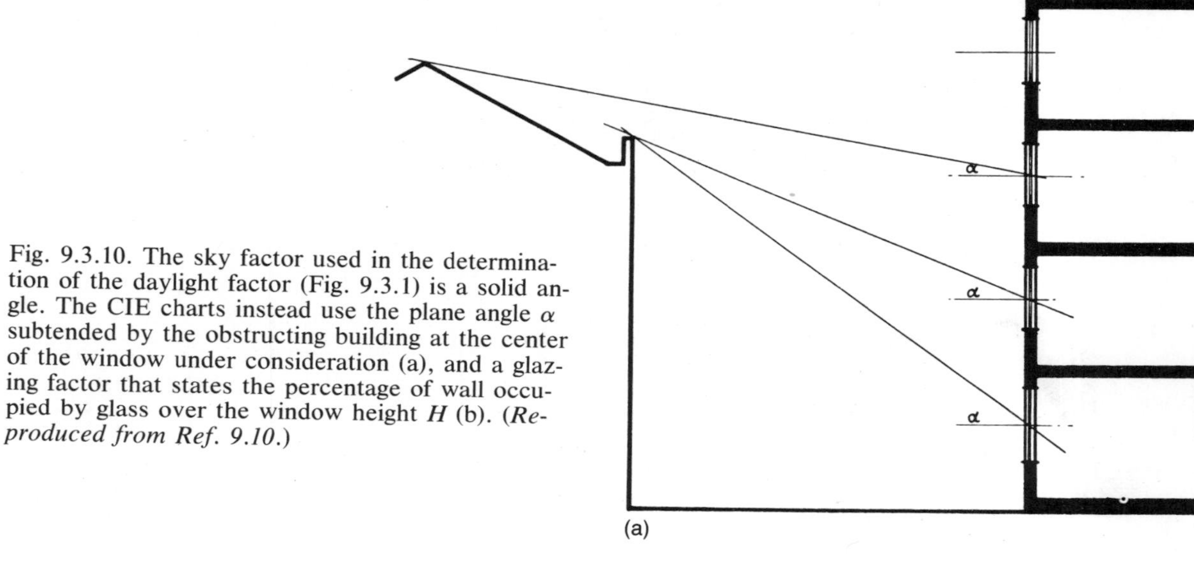

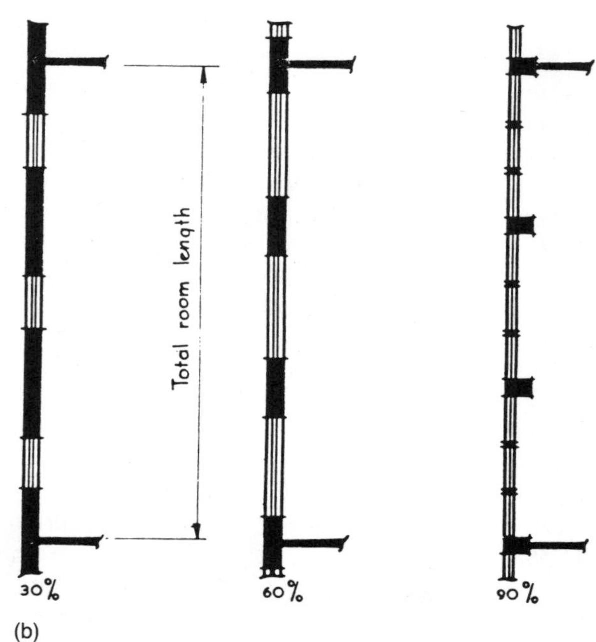

Fig. 9.3.10. The sky factor used in the determination of the daylight factor (Fig. 9.3.1) is a solid angle. The CIE charts instead use the plane angle α subtended by the obstructing building at the center of the window under consideration (a), and a glazing factor that states the percentage of wall occupied by glass over the window height H (b). (*Reproduced from Ref. 9.10.*)

Since the monitors are sunshaded, there should be no overheating in summer, if the building is adequately insulated and ventilated.

Vertical windows are used more frequently than skylights, because they provide a view, because they are more accessible, and because they are the only windows possible in the lower stories of a multistory building. However, they are less efficient for daylighting.

The question of the adequacy of daylight arises most frequently in built-up areas where nearby buildings obstruct the access of daylight. CIE No. 16 (Ref. 9.10) defines a plane angle of obstruction, α, and a glazing factor (Fig. 9.3.10) from which the sky factor can be determined.

The daylight factor is influenced by the height of the workplane, the height of the window, H (which determines the amount of daylight admitted), the height of the ceiling (which influences the internal reflection), and the ratio of the length to the depth of the room (which also influences the internal reflection).

Example 9.3 *The daylight of an apartment building in London is obstructed by a building on the opposite side of the street. The angle* $\alpha = 10°$ *(Fig. 9.3.10) for the third-story windows. Determine the maximum room depth for daylighting.*

The room height is 2.7 m. The length of the room is 5 m. The other dimensions are as shown in Fig. 9.3.11. The room is to have 75% glazing as shown in Fig. 9.3.10.b, which means that 42% of the wall is glass.

An average interior illuminance of 100 lx is required for 85% of the time between the hours of 8 A.M. and 4 P.M.

Solution *From Table 9.1 the curve labeled 80% for the hours 9 A.M. to 5 P.M. corresponds to 85% for the hours 8 A.M. to 4 P.M. The latitude of London is 51°. The 80% curve in Fig. 9.2.4 intersects the 51° latitude at 7500 lx. Therefore the exterior daylight illuminance exceeds 7500 lx for 85% of the time between the hours of 8 A.M. and 4 P.M. At latitude 51° the summer days are much longer than those in winter. Ample daylight is available in summer from early morning until 8 P.M. For plain glass there is no transmittance correction factor, and the obstruction factor is included in the 75% glazing statement. From Table 9.3 the dirt factor for a built-up residential area is 0.8.*

To achieve an interior illuminance of 100 lx, we require a daylight factor of

$$\frac{100}{0.8 \times 7500} = 0.0167, \text{ or } 1.67\%$$

From Fig. 9.3.12 the room depth corresponding to this factor and 75% glazing is 2.9 H.

The window height H *is the difference between the room height (2.7 m), the sill height (0.9 m), and the distance of the window head from the ceiling (0.3 m). This is 1.5 m.*

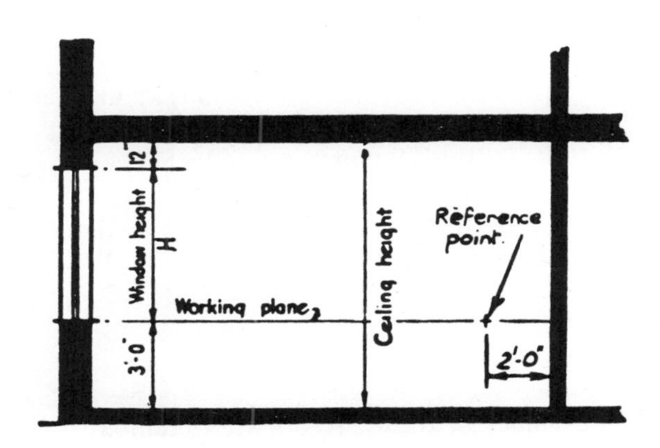

Fig. 9.3.11. The CIE charts are computed for a standard height of the workplane above the floor of 0.9 m (3 ft) and a standard location of the window head of 0.3 m (1 ft) below the ceiling. The height *H* of the windows accordingly increases with room height.

The minimum daylight factor is determined for a distance 0.6 m (2 ft) from the rear wall, that is, the position of a person seated at a table as close as possible to the rear wall. (*Reproduced from Ref. 9.10.*)

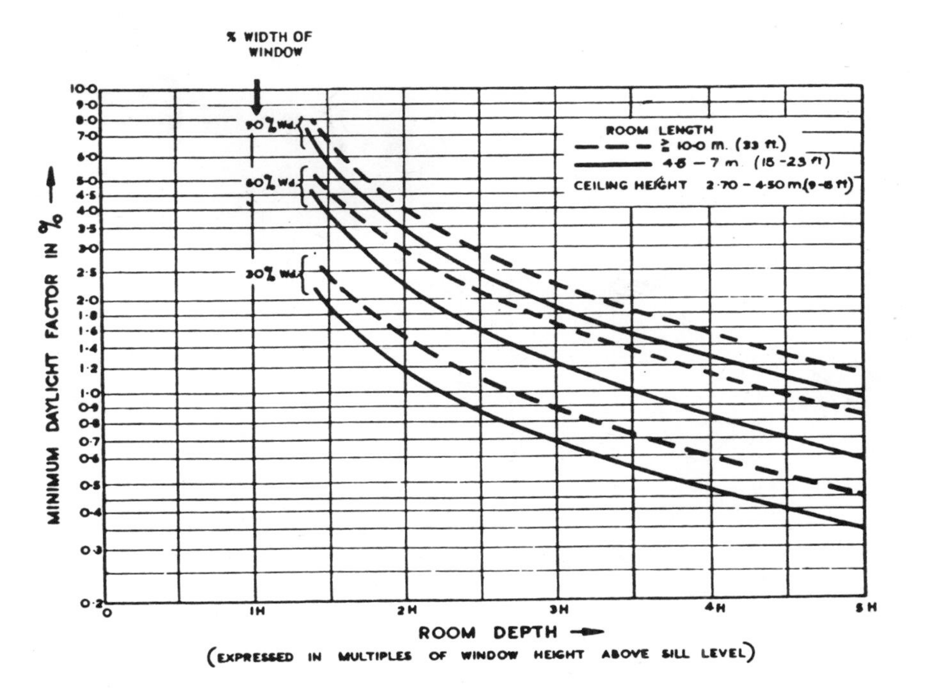

Fig. 9.3.12. The minimum daylight factor at a distance of 0.6 m (2 ft) from the rear wall for a room with windows on the opposite wall only, when there are no external obstructions. (*Reproduced from Ref. 9.10.*)

Fig. 9.3.13 gives the correction factor for the angle of obstruction α *= 10° as 0.89.*

Therefore the maximum permissible room depth is

$$0.89 \times 2.9 \times 1.5 = 3.9 \text{ m}$$

Fig. 9.3.14 gives the depth of the room at which the daylight factor doubles. This is 1.7 H, *or 2.6 m. Thus the minimum daylight illuminance is 200 lx or more over two thirds of the room.*

London is quite far from the equator. Its latitude corresponds in North America to the southern tip of the Hudson Bay, and in South America to the Straits of Magellan. Evidently daylighting of houses and apartment buildings is practicable in most countries, provided that daylight is not blocked too much by surrounding buildings.

However, daylighting of commercial buildings presents greater problems, which are discussed further in Section 10.1.

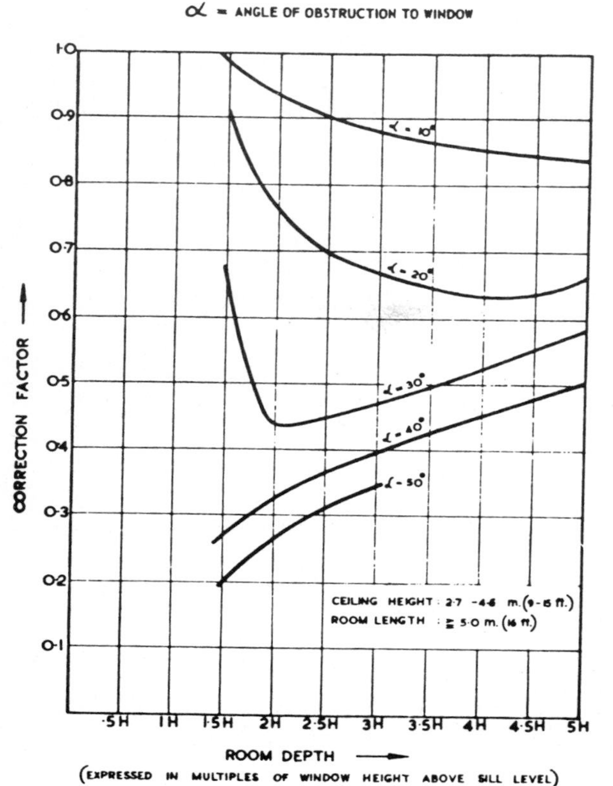

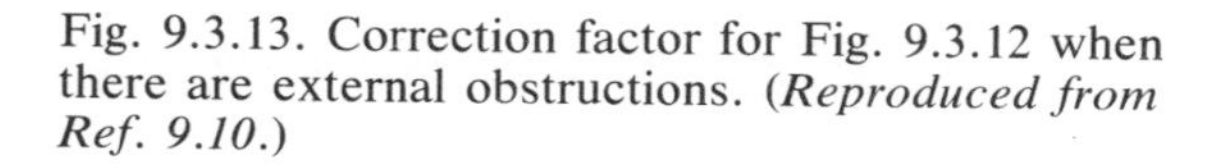

Fig. 9.3.13. Correction factor for Fig. 9.3.12 when there are external obstructions. (*Reproduced from Ref. 9.10.*)

Fig. 9.3.14. Distance from window at which the daylight factor is double the minimum daylight factor at a distance of 0.6 m from the rear wall. (*Reproduced from Ref. 9.10.*)

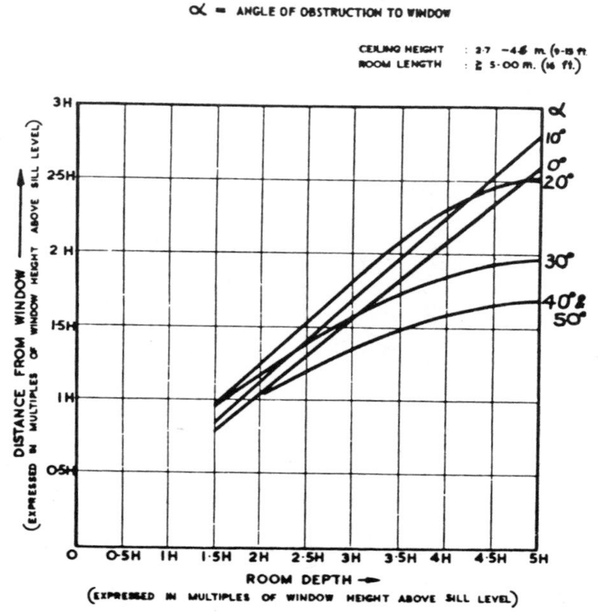

9.4 The Lumen Method

The lumen method, developed by J.W. Griffith at the Southern Methodist University in Dallas (Ref. 9.12), treats the window as a luminous panel, as if it were a source of artificial light. Its illuminance is derived by multiplying the exterior daylight illuminance by a number of factors. The method is also described in the *IES Lighting Handbook* (Ref. 8.4).

As in the daylight factor method, the illuminance on a horizontal workplane transmitted through a vertical window consists of two parts, one received directly from the sky and one reflected by exterior surfaces:

$$E_p = E_i A_w K_m K_u + E_r A_w K_m K_u \qquad (9.13)$$

where E_p is the illuminance on the horizontal workplane, measured in lx; that is, lumen per square meter;
E_i is the illuminance from the sky incident on the window;
E_r is the illuminance reflected by the ground outside;
A_w is the gross area of the window, in square meters;
K_m is the light-loss factor;
K_u is the utilization factor.

The values of these factors are obtained from tables (Ref. 9.12 and Ref. 8.4, pp. 9.84–9.92). The light-loss factor allows for transmission loss due to glazing and any dirt on it, for obstruction by window frames, glazing bars, blinds, and curtains. The utilization factor allows for the interior reflections. It depends on the reflectances of the ceiling, the walls, and the floor and on the relative dimensions of the room.

The lumen method can also be used for skylights. The second part of Eq. (9.13) then becomes inoperative because there is no additional daylight due to external reflections.

The CIE daylight-factor method and the lumen method are, in principle, identical. In practice the tables and diagrams have in each case been designed for some problems but not for others. Therefore each method has some advantages and disadvantages compared to the other:

1. Both methods are limited to a number of standard cases, but the range is wider for the CIE method.
2. The CIE method, like other daylight-factor methods, makes allowance for the obstruction to daylight caused by nearby buildings. It is thus particularly useful for buildings in the central business district and in other closely built-up areas. The lumen method ignores obstructions.
3. The CIE method cannot be used for buildings with sunshades or overhangs, except for monitor windows; the lumen method can be used for buildings with roof overhangs.
4. The CIE method can be used for design, whereas the lumen method can only be used to check a room whose dimensions are known.
5. The lumen method is based on the uniformly lit sky and is thus only approximately correct for the overcast and the clear sky, with direct sunlight excluded. The CIE method is corrected for the CIE overcast sky. It is therefore accurate for the overcast sky, and it is not intended to be used for the clear blue sky; but when the approximate nature of sky luminance figures are considered [Section 9.2], the error is not serious provided that direct sunlight is excluded.

9.5 Artificial Skies and Experimental Techniques

While the methods described in the two previous sections are, in practice, restricted to a limited number of standard cases, the daylighting of any building, however complex, can be investigated in an artificial sky. In particular the effect of sunshading devices on daylighting can be examined.

There are two types of artificial sky. One consists of a hemisphere illuminated to reproduce the luminance of the real sky (Figs. 9.5.1 and 9.5.2). The simplest luminance distribution is uniform, but it is not difficult to reproduce the luminance distribution of the CIE standard overcast sky given in Eq. (9.4).

The clear sky condition is more difficult to simulate, particularly if the effect of direct sunlight

Fig. 9.5.1. An artificial sky consists of a hemisphere painted white on the inside and lit to simulate the luminance of the sky. A model is placed at the center of the hemisphere in line with its "equator," which forms the horizon plane of the model. The equipment shown in Figs. 9.5.1 through 9.5.4 was built by staff of the Department of Architectural Science, University of Sydney.

is included. Richard Kittler in Bratislava (Czechoslovakia) built an artificial sky that includes a movable sun with the correct high luminance relative to the clear blue portion of the sky, and a series of small lamps surrounding the sun to reproduce the aura of light around the sun; but that is the only one of its kind.

The model inside an artificial sky must be large enough to accommodate photovoltaic cells for illuminance measurements on differently located

Fig. 9.5.2. Model of a sunshaded schoolroom in the artificial sky shown in Fig. 9.5.1. The distribution of the luminance in this sky can be altered by using different combinations of the lights at the base of the hemisphere, which are shielded from the model. Thus it is possible to reproduce both the uniformly lit and the CIE standard overcast sky. The horizontal exterior illuminance from the sky and the interior illuminance on different workplanes within the model are measured with an illuminance meter. Alternatively, photovoltaic cells can be used.

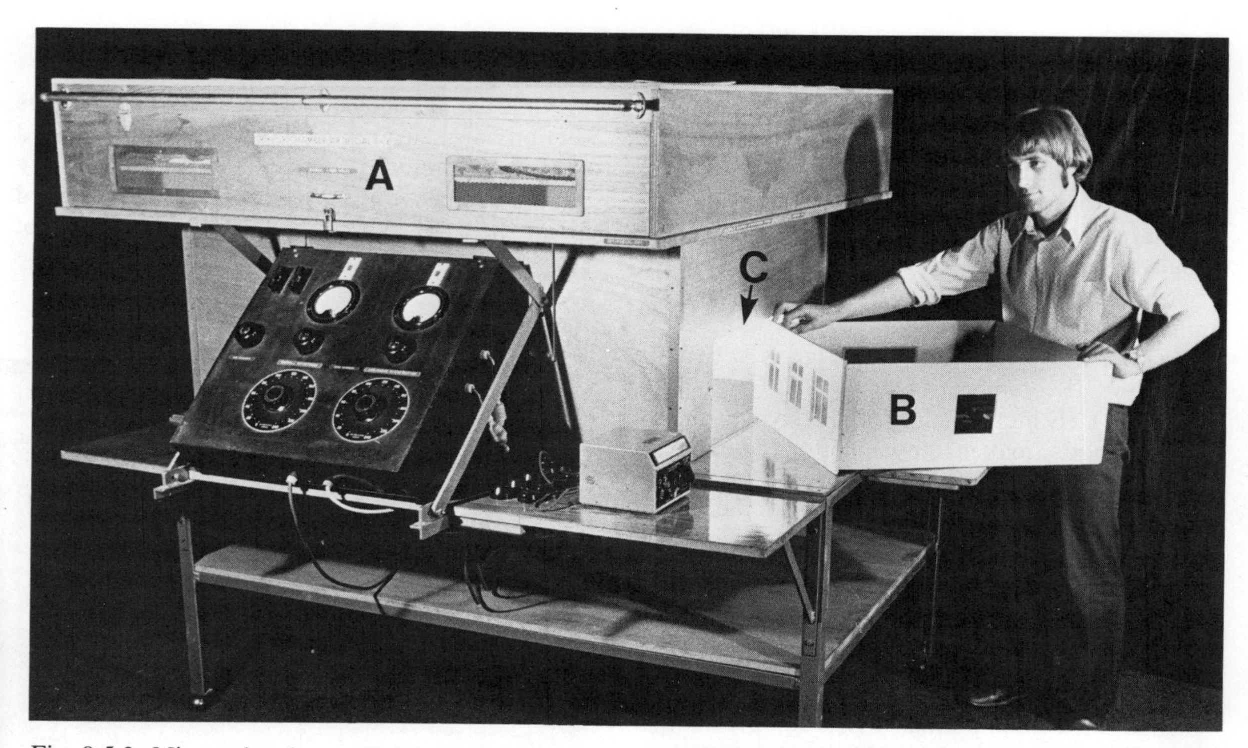

Fig. 9.5.3. Mirror-chamber artificial sky, which can be used only for vertical windows. The upper part (A) of the chamber contains an array of fluorescent lamps and/or cold cathode lamps to match daylight. A dimmer is used to give varying proportions of light from the two types of lamp, so that it is possible to simulate different sky luminances for a CIE standard overcast sky. The lower part of the chamber is lined with mirrors.

The model B is placed in front of the opening marked C. Various window arrangements and shading devices can be placed in this opening.

Fig. 9.5.4. Scale model of a school classroom for use with the mirror-chamber sky shown in Fig. 9.5.3. The desks are made to scale; for horizontal illuminance measurements they are replaced by cosine-corrected photovoltaic cells mounted at desk height. The vertical illuminance on the chalkboard is recorded with a photovoltaic cell mounted within the board.

workplanes (Fig. 8.2.5). Furthermore, there must be enough room for a person to work inside the artificial sky. As a result hemispherical artificial skies cannot be made smaller than 6 m in diameter, so that they require a large space and height.

Vertical windows can be investigated with the simpler and smaller mirror chamber sky (Figs. 9.5.3 and 9.5.4).

Artificial skies can be used for objective illuminance measurements and for more subjective observations of the quality of lighting, particularly the likelihood of glare. Photographs are a useful aid to this subjective investigation.

Photography can also be used for evaluating solid angles for complex configurations. A 35-mm camera is normally fitted with a 50-mm lens that has an angle of view of 45°. The camera records what can be seen within that angle. A conventional "wide-angle" lens with a focal length of 28 mm has an angle of view of about 75°. Some cameras can be fitted with a "fish-eye" lens with an angle of view of 180°. The full-field picture so obtained inevitably contains gross distortions of the objects recorded, but it becomes possible to see all the obstructions to daylight from a window on a single flat picture, or to see at the same time all the windows that provide daylight for a plane (Fig. 9.5.5).

Fig. 9.5.5. Photograph of a narrow city street, taken with a fish-eye lens to show all the obstructions to daylighting on a single, if distorted, photograph. (*Photograph by Mr. Frank Morand.*)

With suitable optical systems, full-field photographs that are in accordance with a spherical projection can be produced. Solid angles that could be calculated only with difficulty can then be measured in steradians [Section 8.2] by means of a suitable overlay. The *globoscope*, developed by G. Pleijel in Sweden, produces a photograph that conforms to a stereographic projection, while the full-field camera, developed by Robin Hill in England, produces a photograph that conforms to an equidistant projection (Ref. 9.3, pp. 366–71).

9.6 Glare

We noted in Section 8.5 that glare can be produced by daylight or by artificial light. R.G. Hopkinson (Ref. 9.2, p. 93) considered that people are less disturbed by a mild degree of glare from daylight than from artificial light and suggested that this greater tolerance was due to the more diverse nature of the light admitted by windows. The view of the sky through most windows is surrounded by trees, or an open space, or other buildings, and this provides a transition between the bright sky and the much darker interior of the room. In addition to this physical aid Hopkinson suggested that the psychological effect of a pleasant view would distract attention from the excessive brightness of the sky and would modify an overall judgment of discomfort.

On the other hand, glare from daylight is far more variable than glare from electric light. We design natural light for the minimal condition, which assumes an overcast sky or a clear blue sky without direct sunlight. Since the sun moves across the sky, it is difficult and expensive to ensure that it cannot at any time be seen through a window. Even when the sun is not visible, reflection from a bright surface outside can create glare at certain times of the day.

An excess of *indirect* daylight can produce glare. The relatively large windows needed for the minimal conditions of daylight discussed in Sections 9.3 and 9.4 are liable to admit too much light for visual comfort when conditions are much better than minimal. Discomfort glare can be caused either by sheer excess of light to which the eye cannot adapt or by excessive contrast between the bright window and a dark interior. Disability glare [Section 8.5] can also be caused by daylight; for example, a brightly lit window at the end of an otherwise dark passage may cause disability to vision due to glare.

Glare from sunlight is most likely through skylights, clearstory windows, and vertical windows facing west. Windows facing east present fewer problems because the sun enters through them at a time when it disturbs few people.

Discomfort glare caused by the temporary admission of direct sunlight can be corrected by drawing curtains or blinds. They are particularly needed on windows facing west, where the sun is low in the sky even in summer (Fig. 5.4.1). Blinds also provide a solution for clearstory windows and for skylights, but these generally require a mechanism for operation.

Blinds cannot be drawn when direct sunlight is deliberately admitted for solar heating in winter [Sections 3.4, 3.5, and 5.14 through 5.16]. Glare thus needs particular attention in passive solar design.

Glare becomes a problem only to persons who *look* at a window and *see* a bright light. Thus glare can be avoided by suitable arrangement of the "workplane," for example, by turning a desk through a right angle or moving it elsewhere.

A graded contrast between the window and its surroundings helps to reduce glare. Windows liable to glare should be set in light-colored walls, and their frames should be of aluminum or painted white. A translucent white curtain that can be drawn back, or an adjustable venetian blind with white slats, is useful for controlling glare without shutting out the daylight.

Tinted, heat-absorbing glass and heat-reflecting mirror glass [Section 5.9] are also effective in reducing glare, but they also reduce the amount of daylight. Unlike a curtain or a blind, they cannot be removed on a dull day.

The placement of windows should be related to the use of the room. Windows should not be put behind a focus of attention. Thus the chairman at a committee meeting, the teacher in a classroom, or a lecturer at a public meeting should not stand or be seated in front of a window, because the contrast between the person and the much brighter window would make it uncomfortable to look at the speaker.

When the function of a room is particularly sensitive to glare, as is an art gallery, daylight can be controlled by motor-driven venetian blinds operated by photovoltaic cells, and supplemented by electric light also controlled by photovoltaic cells (Ref. 9.2, plates 40–43).

Apart from direct glare due to daylight, discomfort and disability glare can be caused by reflection. However, this problem is the same for daylight and for electric light, and it is discussed in Sections 8.5 and 10.8. It is advisable to avoid shiny surfaces on "workplanes," such as desks, machines, and typewriters, and to use matte finishes instead.

9.7 Daylighting for the Tropics

Daylight design was developed in the countries of the temperate zone, notably in England, where daylight levels in winter are low and solar radiation rarely causes overheating in summer.

The design criterion is the exterior daylight illuminance available for, say, 90% of the year. This depends mainly on the altitude of the sun during the shortest days of the year, and that depends on latitude (Fig. 9.2.4).

The requirements for interior illuminance levels are generally more modest in developing countries. Thus the Indian Standard (Ref. 9.6) specifies illuminances that are lower than those customary in Britain and Australia, and appreciably lower than those recommended in the United States.

The higher daylight levels and lower illuminance requirements for developing countries in the tropics suggest that daylighting is easier than in the temperate zone, and there is scope for different solutions.

In some hot-arid regions the living rooms in traditional houses are kept in semidarkness, which provides a welcome contrast to the bright light outside; they are adequately lit for talking but not for reading. On the other hand, many tropical buildings are designed without any sunshading, and the daylight-design methods discussed in Sections 9.3 and 9.4 can be applied without modifications. For comfort these buildings need a large air conditioning plant; but some are used without air conditioning and become very hot in the afternoon.

Sunshading devices designed for good thermal performance are common, but daylighting is not always considered in their design and many exclude too much daylight. It is possible to produce

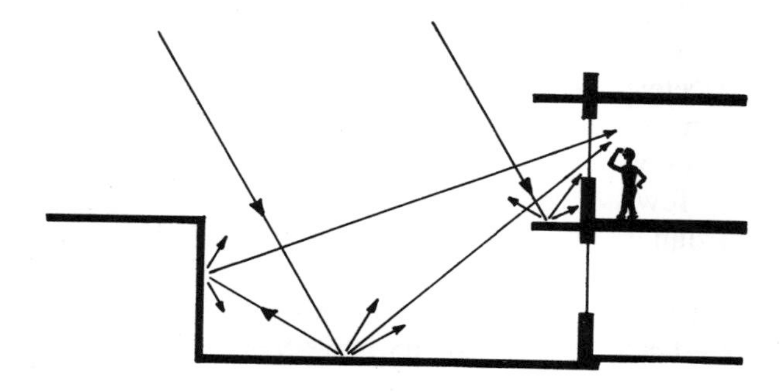

Fig. 9.7.1. In the hot-arid tropics glare results from external reflections rather than from the clear blue sky, which has a low luminance, provided the window is shaded from direct sunlight. Because of the high luminance of the sun and the aura surrounding it, reflected light is very bright if the ground and nearby buildings have a high reflectance. Buildings are often painted white or made of concrete. The ground also may have a high reflectance because the soil is dry and there is little vegetation.

Fig. 9.7.2. Glare can be avoided in the hot-arid tropics if the daylight is reflected by an interior matte surface and if people inside have no direct view of the sun. However, the daylight may be insufficient, and the absence of a view except for some blue sky may be disconcerting.

good daylight with sunshades by using them as reflecting surfaces (Fig. 9.3.3).

Determination of the daylight factor of a building with a complex sunshading device from first principles would be a lengthy operation, and there are few design aids [Section 9.4]. However, the design can be examined experimentally in an artificial sky [Section 9.5].

It is customary to assume a clear blue sky for the hot-arid tropics; its luminance distribution was discussed in Section 9.2. In the hot-humid tropics the sky is *partly* overcast for most of the year. The luminance distribution of this type of sky is very variable, and it is customary to design for a uniformly bright sky.

Glare [Section 9.6] is a particular problem in the tropics. The conditions in the hot-humid tropics are similar to those in the temperate zone, except that the sky luminance is much higher. Vegetation grows easily, and green plants have a low reflectance. Buildings are generally of light construction and are not normally painted white, which is a common practice in the hot-arid tropics. Therefore external reflectance rarely presents a problem, but direct glare from the sky should be avoided through shielding of the windows.

In the hot-arid tropics the clear blue sky has a low luminance, but the sun is very bright, and reflected light from the ground can be very glaring (Figs. 9.7.1 and 9.7.2).

The traditional architecture of the Middle East made extensive use of small courtyard gardens [Section 5.12], mainly to improve the thermal performance of the building. However, the low reflectance of green plants also assists the daylighting of buildings in a hot-arid region by eliminating the glare without eliminating the view.

Both in the hot-arid and the hot-humid tropics the main technical problem is the thermal performance of the building [Chapter 5]. The daylight problem should be solved within that context.

References

9.1 N. PEVSNER: *An Outline of European Architecture*. Seventh Edition, Penguin, Harmondsworth, 1972. 446 pp.

9.2 R.G. HOPKINSON and J.D. KAY: *The Lighting of Buildings*. Faber, London, 1972. 320 pp.

9.3 R.G. HOPKINSON, P. PETHERBRIDGE, and J. LONGMORE: *Daylighting*. Heinemann, London, 1966. 606 pp.

9.4 J.W.T. WALSH: *The Science of Daylight*. Macdonald, London, 1961. 285 pp.

9.5 J.A. LYNES: *Principles of Natural Lighting*. Elsevier, London, 1968. 212 pp.

9.6 V. NARASIMHAN and B.K. SAXENA: Measurement of the luminance distribution of the clear blue sky in India. *Indian Journal of Pure and Applied Physics,* Vol. 5 (1967), pp. 83–86.

9.7 J. LONGMORE: *BRS Daylight Protractors*. H.M. Stationery Office, London, 1968. 25 pp. + 10 protractors.

9.8 W. BURT et al.: *Windows and Environment*. Pilkington Environment Advisory Service, St. Helens, England, 1969. 202 pp. + 45 transparencies.

9.9 *Industrial Data Sheets. L2—Natural Lighting*. Commonwealth of Australia, Department of Labour and National Service, Industrial Services, Division, Sydney, 1963. 136 pp.

9.10 *Daylight—International Recommendations for the Calculation of Natural Daylight*. Publication No. 16 (E-3.2). Commission Internationale de l'Éclairage (CIE), Paris, 1970. 79 pp.

9.11 DEREK PHILLIPS: *Lighting in Architectural Design*. McGraw-Hill, New York, 1964. 310 pp.

9.12 *Predicting Daylight as Interior Illumination*. Libby-Owens-Ford Glass Company, Toledo, Ohio, 1960. 27 pp.

Suggestions for Further Reading

References 9.2 and 9.11.

9.13 B.P. LIM, K.R. RAO, K. THARMARATNAM, and A.M. MATTAR: *Environmental Factors in the Design of Building Fenestration*. Applied Science, London, 1979. 273 pp.

Chapter 10 Artificial Light

During daylight hours most rooms make use of daylight, at least near the windows. The integration of daylight and electric light poses special problems because of glare from windows.

Incandescent lamps, fluorescent lamps, and their luminaires are briefly described, and their maintenance, depreciation, and replacement are discussed. The color temperature of lamps is defined. All lamps produce heat, and some of it can be extracted before it enters the room when integrated ceilings are used.

The number of lamps required for a given level of illuminance is calculated by the lumen method. Glare from artificial light can be avoided by suitable design.

10.1 Integrated Daylight and Electric Light

We noted in Sections 9.3 and 9.4 that it is possible in almost any latitude to use daylight to illuminate houses, apartments, small offices, and single-story factories during daylight hours, although for factories electric light may be more efficient. However, there are problems in daylighting deep office spaces and department stores. A glance at Fig. 9.3.12 shows that the daylight factor decreases rapidly with the depth of the office. It would not be economical to increase the ceiling height merely for the purpose of extending the use of daylight. In fact, the room height of offices has been consistently reduced during the last century.

In practice the potential use of daylight is therefore restricted to an outer zone which extends $1\frac{1}{2}$ D to $2\frac{1}{2}$ D from the windows, where D is the room height.

We mentioned in Section 5.9 that the thermal performance of windowless buildings can be superior to that of buildings with windows. Nevertheless, few windowless office buildings have been erected. It is generally accepted that people feel a need for access to daylight and a view of the world outside.

It remains a matter of opinion whether it is worthwhile to utilize daylight for the illumination of deep office spaces, or whether daylight should be treated merely as an amenity for the people in the building, to give them a feeling of contact with the natural environment outside and with the passage of time and of the seasons. The office would then be permanently lit by artificial light (PAL). The alternative approach is to utilize daylight as far as possible in that part of the office that is close enough to the windows, and to use permanent supplementary artificial light for the interior (PSALI). The lights forming part of the PSALI system would remain switched on during the whole of the time when the office is in use (Fig. 10.1.1). Additional artificial lights would be switched on when it gets dark.

The success of a combination of natural and artificial light depends to a large extent on the illuminance due to the artificial light. A high illuminance level wastes energy, and a low illuminance level may cause glare from the windows [Section 9.6]. The problem disappears when there are no windows, for example, in interior offices, or in theaters or department stores where artificial light is used exclusively.

Prior to 1950 offices, even in America, were rarely designed for illuminance levels in excess of 250 lx. The office worker furthest from the window received no illumination from daylight, but if he looked at the much brighter windows and then back at his work he would perceive the windows as glaring. As a result blinds were often kept permanently drawn, at least on the sunny side of the building, but sometimes on all sides, because working conditions were otherwise intolerable. This negated the very purpose for which windows were provided.

By 1970 most American offices were designed for illuminances of 500 lx or more; windows were frequently made from heat-absorbing or heat-reflecting glass to reduce solar heat gain. The combined effect of tinted windows and eyes

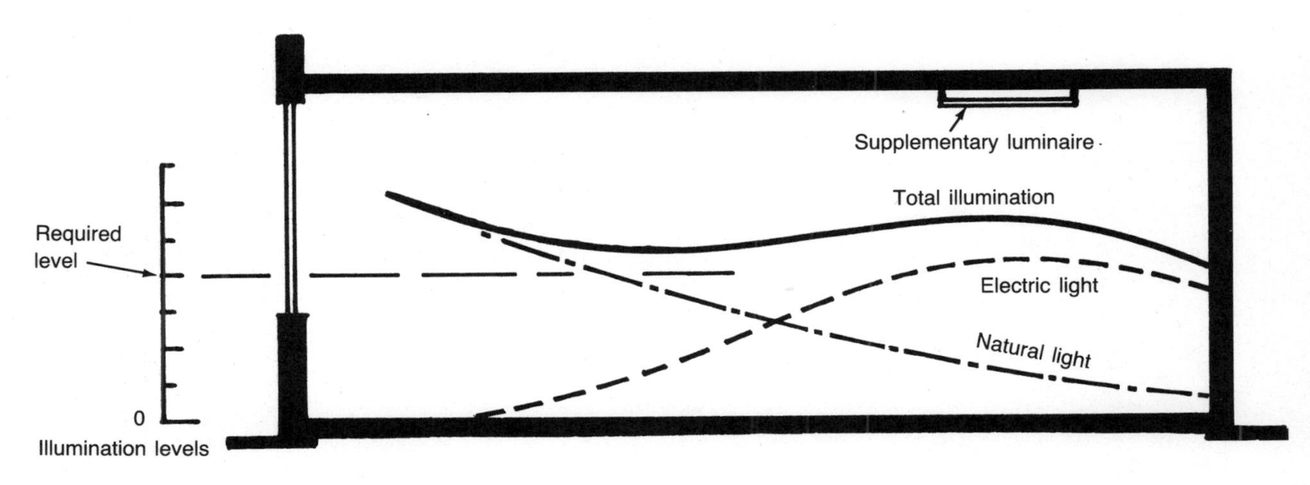

Fig. 10.1.1. Use of permanent supplementary artificial lighting for the interior of a room (PSALI) when the room is too deep to be lit by daylight alone. The portion of the room near the window is lit mainly by daylight, the rear of the room mainly by artificial light, and both contribute to the illumination of the central portion of the room. Additional electric light is switched on at nighttime to light the entire room.

adapted to higher illuminance levels eliminated the glare from the windows, although the artificial light itself sometimes caused glare [Section 10.8].

However, it is open to question whether this can still be considered a satisfactory solution since energy conservation has become a major design criterion. Electric light installations with illuminance levels higher than necessary are wasteful in themselves; they also generate heat. This is not a severe disadvantage in winter, although electric light is not an efficient source of heat. In summer the additional load on the air conditioning plant is a major source of energy consumption.

The original PSALI system developed by the (British) Building Research Station (Ref. 9.3, Chapter 19) also had limitations. It was appreciated that glare would be a problem with the lower illuminance levels used in Britain. In the ABC System devised by R.G. Hopkinson (Fig. 10.1.2) additional electric lights are provided during daylight hours at the rear of the room to overcome glare, and these are switched off at nighttime.

British PSALI were generally installed with only three switching positions: off, daytime, and nighttime. This proved a limitation because of the ever-changing nature of daylight. A better result could be obtained by dimming the electric light than by switching it on or off.

J.B. de Boer and D. Fischer (Ref. 10.1, p. 179) describe the available methods:

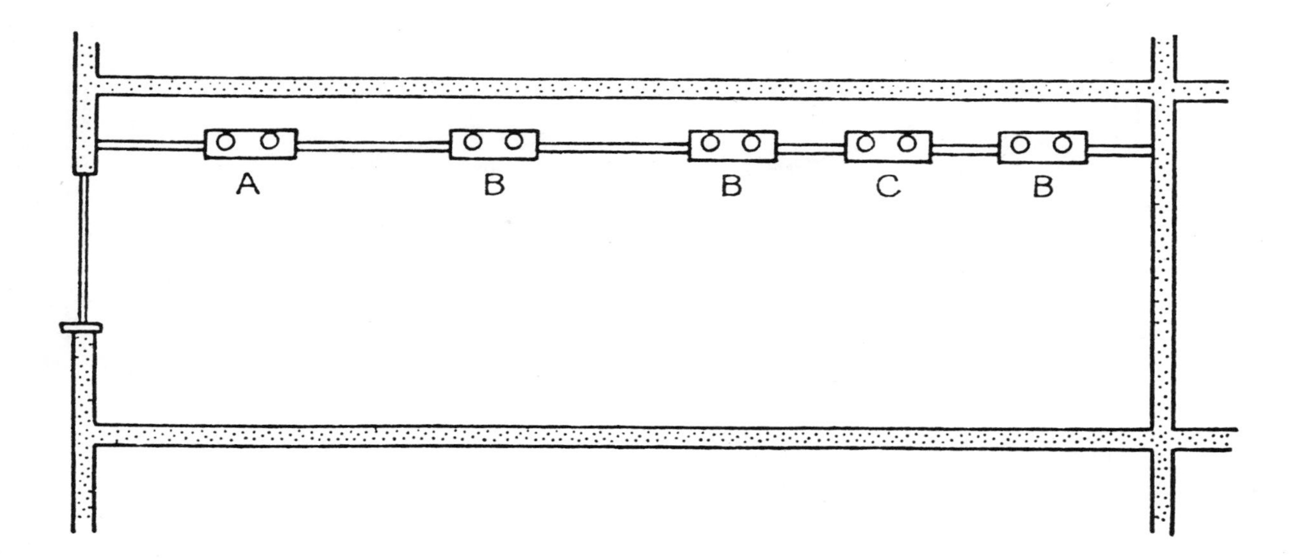

Fig. 10.1.2. The ABC System of PSALI (permanent supplementary artificial lighting of interiors), with three individually switched sections of electric light. During daytime sections B and C provide the PSALI. At nighttime the electric lights A are switched on, and the electric lights C are switched off, to provide a uniform, and lower, level of illuminance. (*From Ref. 9.3, p. 477.*).

The most practical method of controlling light output is, of course, to control the electrical input to the light source rather than employ filters or shutters. The methods by which this may be accomplished are based on one of two principles: the input current can be varied by changing its amplitude, or it can be varied by changing the length of time during a cycle that it is permitted to flow.

Change in amplitude—a principle that, where full control is required, can only be employed for incandescent lamps [Section 10.2] can be accomplished by means of a resistance dimmer or a variable transformer. The resistance dimmer is connected in series with the lamp circuit and the voltage that appears across the lamp is equal to the supply voltage, less the voltage dropped across the dimmer. The dimmer itself is relatively inexpensive, but it is also wasteful in the use of energy because at settings to give low light outputs an appreciable percentage of the power fed to the circuit is dissipated in the dimmer itself.

This waste is avoided with the continuously variable (auto) transformer. Here one pays mainly for the energy consumed by the lamp. Dimmers of this type are frequently used in places such as cinemas, where a heavy lighting load must be operated dimmed for long periods of time.

Regulators working on the timed-flow principle can be used to control both incandescent lamps and, with appropriate special ballasts [Section 10.4], tubular fluorescent lamps. The basic element in modern electronic light regulators of this type is the thyristor (a silicon controlled rectifier). A typical regulator consists of a pair of thyristors connected in inverse parallel so that each one may conduct during each half cycle of the alternating current supply voltage. Drive circuits control the switching on of the thyristors at a chosen instant in the supply voltage waveform, thereby increasing or decreasing the mean power supplied to the load. The drive circuits may derive their input signal from a variety of controllers, for example, manual or electronic potentiometers and light sensitive cells.

Installations of this type may in due course eliminate the problem of glare from windows, give more agreeable lighting with lower illuminance levels, and save some electric power.

10.2 Incandescent Lamps and Luminaires

The electric light bulb was invented independently by Joseph Swan at Newcastle-upon-Tyne, England, in 1878 and by Thomas Alva Edison at Menlo Park, New Jersey, in 1879, and they combined their efforts. In a poster distributed in 1882 Edison pointed out the advantages over electric-arc lights, gas light, oil lamps, and candles:

"This room is equipped with Edison Electric Light. Do not attempt to light with a match. Simply turn key on wall by the door. The use of electricity is in no way harmful to health, nor does it affect the soundness of sleep."

The original Swan and Edison lamps were made with carbonized threads placed in a glass tube from which the air had been evacuated, so that the carbon could not burn. The higher the temperature of the filament, the greater the amount of light for a given amount of electricity. In 1905 the tantalum filament was introduced, and in 1908 D. Coolidge developed the tungsten filament at the laboratories of the General Electric Company; this is still in use today. Tungsten has a melting point of 3655 K (3382°C), which permits high operating temperatures; it is a ductile and strong material, so that thin wire can be produced.

Because the hot filament evaporated in a vacuum and blackened the glass, Irving Langmuir, also at General Electric, suggested in 1913 the replacement of the vacuum by an inert gas. This solved one problem but created another, because heat was now lost by convection. Langmuir then coiled the filament, and later coiled the coil. After many years of experiments, the coiled-coil filament, still in use, was introduced in 1934 (Fig. 10.2.1).

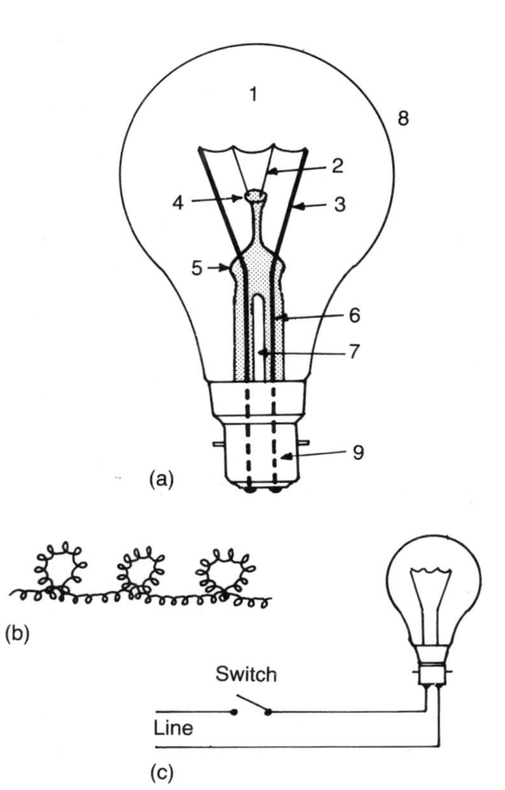

Fig. 10.2.1. Incandescent electric lamp for general lighting purposes, with bayonet fitting.
(a) Components of lamp: 1 = inert gas; 2 = molybdenum support; 3 = lead-in wires; 4 = glass button; 5 = glass stem; 6 = nickel fuse; 7 = exhaust tube for evacuating bulb and filling it with gas; 8 = glass bulb, which may be clear or "pearl"; 9 = base with bayonet fitting normally used in Britain and Australia; in America screw fittings are used instead.
(b) The filament of the lamp is a coiled coil of tungsten wire.
(c) Unlike the fluorescent tube, the incandescent lamp does not require any special control gear, merely a switch.

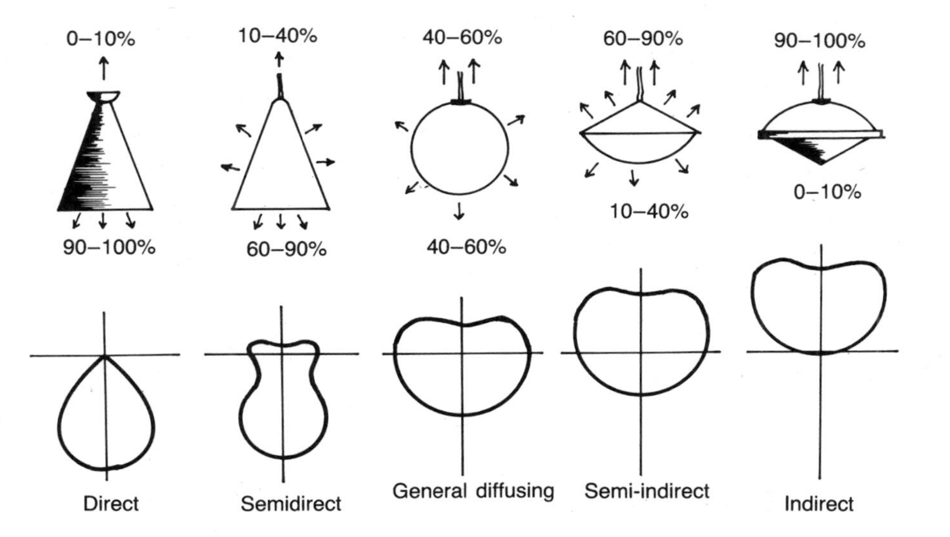

Fig. 10.2.2. Classification of luminaires according to the direction of the reflected light. They range from a direct luminaire, which reflects 90% or more of the light downwards, through a diffusing luminaire to an indirect luminaire, which directs 90% or more of the light onto the ceiling. The actual light fitting may be plain and purely functional, or highly decorative.

The light distribution of each luminaire is shown by the polar curve below it. The luminous intensity at any angle is proportional to the distance of the curve from the intersection of the axes.

A wide range of incandescent lamps is produced, ranging from 15 W to 2000 W. They are made with glass of several types: clear, "pearl" (that is, etched on the inside to diffuse the light), and "inside white," which gives even better diffusion but results in some loss of light. Bulbs can also be made with internal reflectors to provide a concentrated spotlight or a somewhat wider floodlight.

The naked light bulb is now rarely seen even in low-cost houses. Light fixtures may range from elaborate hand-cut glass chandeliers to simple paper shades; however, apart from making a room more attractive, they reflect the light downwards to provide direct lighting, upwards to provide indirect lighting, or diffuse it in all directions (Fig. 10.2.2).

One limiting factor for the life of an incandescent lamp is the evaporation of the tungsten filament; the evaporated tungsten is deposited on the glass wall of the bulb. The introduction of a halogen (generally, bromine or iodine) into the lamp reduces the filament loss, because the tungsten combines with the halogen and is eventually recombined with the filament. This *tungsten-halogen* lamp can be operated at higher temperatures at a higher efficacy and with a longer life.

Details of commercially available incandescent lamps are listed in handbooks (for example, Ref. 8.3, Chapter 8).

The luminous efficacy and life expectancy of incandescent and fluorescent lamps are compared in Section 10.4.

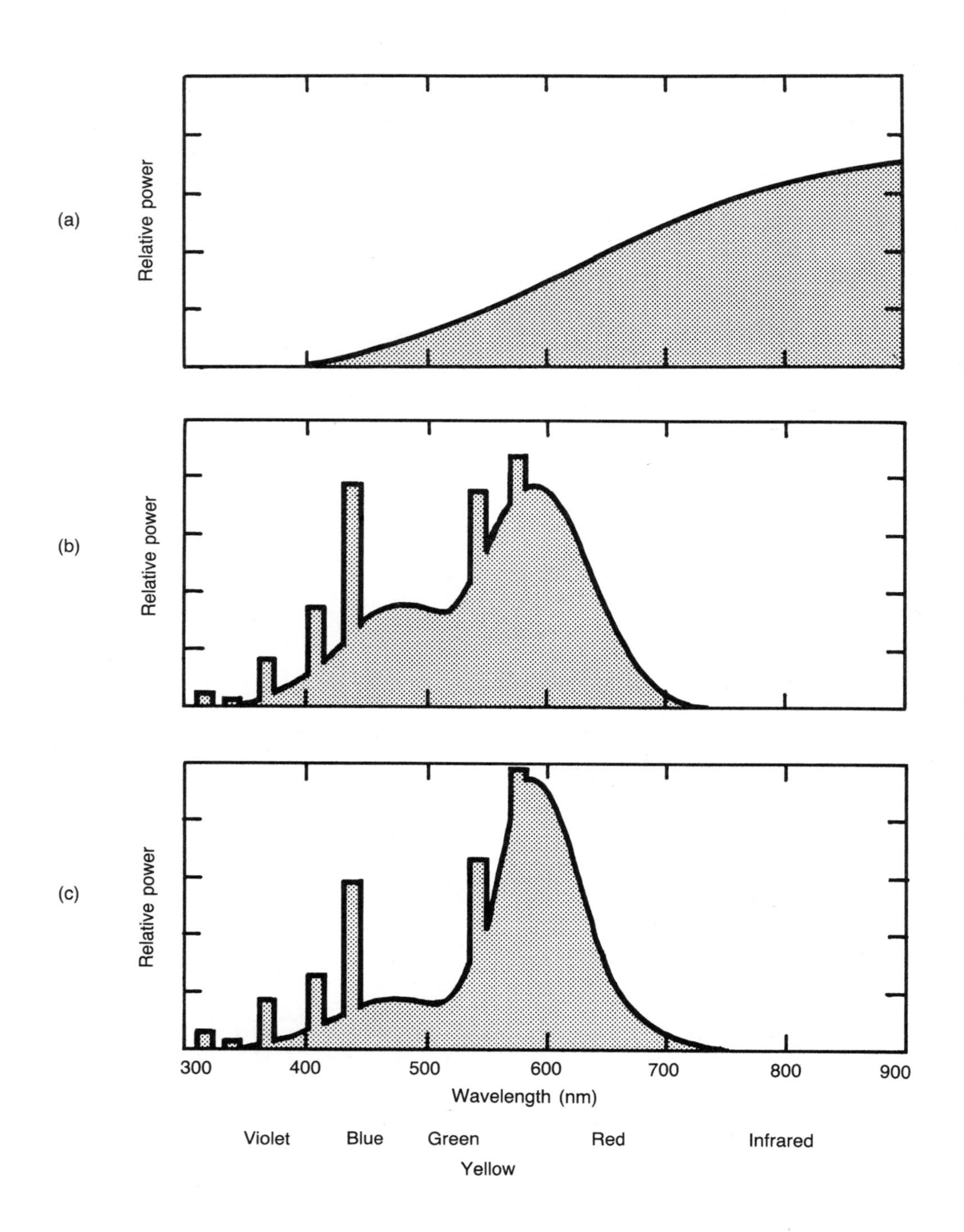

Fig. 10.3.1. Variation of power with wavelength (a) for a general service incandescent lamp, (b) for a "cool white" fluorescent tube, and (c) for a "warm white" fluorescent tube. The variation of power with wavelength for sunlight is shown in Fig. 3.2.1.

10.3 Color of Lamps

We noted in Section 4.2 that a "black body" is a body that has no reflecting power, and is thus a standard radiator. As the temperature of a black body is raised, it emits more radiation, and a greater portion of that radiation is emitted at the shorter wavelengths, which include visible light (Fig. 10.3.1).

The color temperature of a lamp is the temperature of the black body that most closely resembles its color distribution.

There is a correlation between the temperature of the light source and its color temperature. For example, daylight has a color temperature of about 6000 K, which is approximately the temperature of the sun's surface. The tungsten filament of an incandescent lamp operates at a temperature of about 3000 K, and the color temperature of the lamp is about that. To make an incandescent lamp emit light with a color distribution resembling that of daylight would require raising the temperature of the filament to 6000 K, which is not possible because tungsten melts at 3655 K. Thus an incandescent tungsten lamp with a color temperature of 3000 K contains more red and yellow and less blue than daylight at 6000 K (Fig. 10.3.1). It consequently gives a warm appearance and enhances the pink to brown color of skin, and the red, brown, and yellow of brick, natural stone, and timber. For that reason many people prefer the light of incandescent lamps to that of most fluorescent lamps, which is closer to daylight. Higher-wattage incandescent bulbs are bluer, and low-wattage bulbs are yellower in appearance. Dimmed lamps give a yellow-red light.

The other principal type of lamp, described in Section 10.4, is the fluorescent tube. This operates at a temperature of about 40°C (about 310 K), and the color of the lamp is produced by the phosphor, or chemical coating of the tube. There is therefore no relation between the temperature inside the tube and its color temperature. By a combination of suitable phosphors it is possible to produce a wide range of color mixes; however, while the color spectrum of incandescent lamps is continuous, there are discrete peaks due to mercury vapor [see Section 10.4] superimposed on the color due to the chemical composition of the phosphors.

Fluorescent tubes can be made "cool white" (color temperature 4500 K) with a high proportion of blue and green light, "white" (3500 K), or "warm white" (2900 K) with a lower proportion of blue and green light (Fig. 10.3.1).

10.4 Fluorescent Lamps and Other Discharge Lamps

It has been known since the early eighteenth century that an electric discharge produces a glow in a rarefied gas in a glass tube. The fluorescent tube, which was introduced only in the 1940s, is a special kind of discharge tube, because most of the light is produced not by the discharge but by the coating on the tube (Fig. 10.4.1).

The electric discharge produces a very small amount of visible light; most of its radiant energy is in the form of ultraviolet radiation with a wavelength of 254 nm. This is not visible to the human

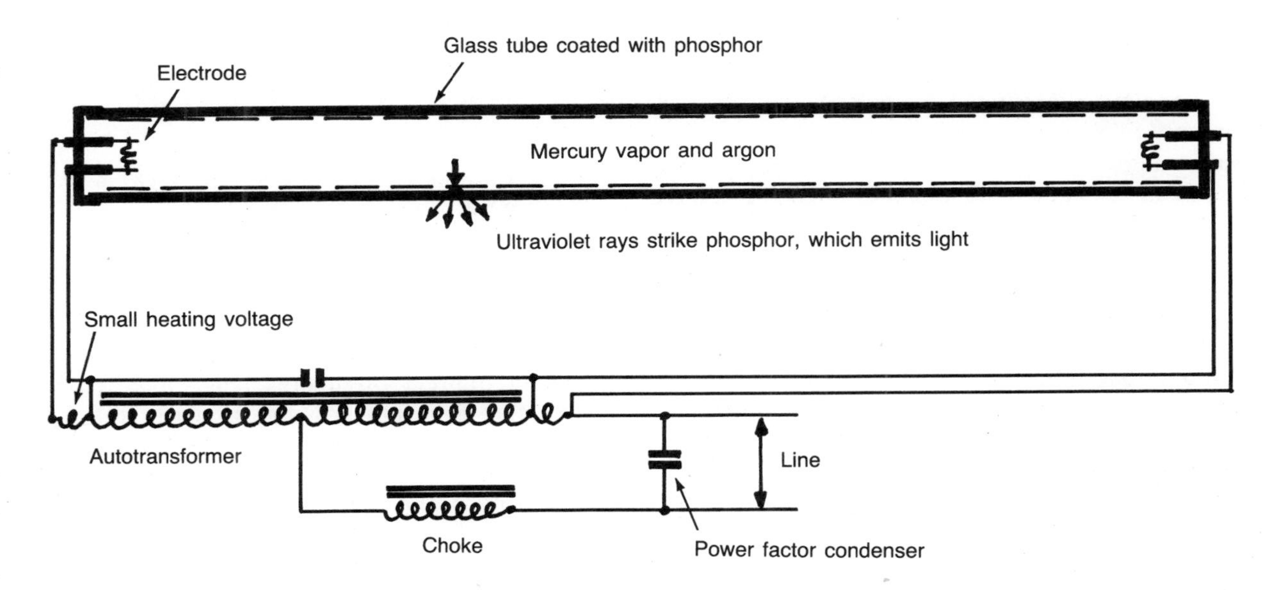

Fig. 10.4.1. Tubular fluorescent lamp. The lamp itself, shown at the top of this figure, consists of a glass tube coated inside with a phosphor that emits visible light when struck by invisible ultraviolet rays. The rays are produced inside the tube by an electric discharge between the two electrodes, set up in the column of mercury vapor and argon gas inside the tube.

The lamp needs a starter to ionize the gas along the entire length of the tube and to develop an arc within the lamp. It then needs a ballast to limit current flow. It further requires a heating circuit for the two electrodes.

eye, but it activates the fluorescent coating inside the tube, which is called a phosphor, to emit visible light. The nature of this light depends on the phosphor. It can emit light of various colors and in color mixes that closely resemble daylight (Fig. 10.3.1.b, c). Lamps that emphasize a particular part of the spectrum can be produced for enhancing certain merchandise or for improving the growth of plants.

Fluorescent tubes require a starter to initiate the arc along the tube, a ballast to limit current flow thereafter, and in some cases also a circuit for heating the electrodes. This adds to the prime cost.

While a large part of the radiation produced by incandescent lamps (Fig. 10.3.1.a) is in the infrared range, which produces heat but no light, the radiation produced by fluorescent tubes is mostly in the visible range so that they produce more lumens for the same wattage; they have a luminous efficacy of about 40–85 lm/W, compared to 10–20 lm/W for incandescent lamps.

Fluorescent tubes are a little more expensive to install than incandescent lamps but cheaper to operate. They have a longer life (5000 to 7500 hours of operation) than incandescent lamps (about 1000 hours).

Fluorescent tubes have a smaller wattage range than incandescent lamps and a more restricted range of luminaires.

For large commercial spaces fluorescent tubes provide a given level of illumination at a much lower cost, particularly if the light is switched on continuously for long periods. Incandescent lamps are better suited for locations where lights are switched frequently on and off, for decorative lighting, for directional lighting, and for floodlighting.

There are many other types of discharge lamp. They are described in the trade literature and in specialized books (for example, Ref. 8.3, Chapter 8). Many have a luminous efficacy better than that of fluorescent lamps, and much better than that of incandescent lamps. Most have a longer lamp life. All need additional electrical equipment for limiting the current, and most also for starting. Some give poor color rendering; for example, low-pressure sodium vapor lamps give almost monochromatic yellow light [see Section 8.6]. However, their high efficacy of 150 lm/W makes them useful for outdoor lighting when color is not important. The specific bright colors of some discharge tubes are useful for display lighting (for example, neon tubes). Some discharge lamps, such as high-pressure sodium vapor lamps, produce a near-white light with reasonably good color rendering (2900 K), so that they can be used indoors where high-powered lights with long life and good luminous efficacy are needed; however, they take several minutes to warm up, and some cannot be relit immediately when they are switched off.

10.5 Maintenance, Depreciation, and Replacement of Lamps

Dust accumulation is a special problem with lamps in indirect luminaires (Fig. 10.2.2) that reflect the light upwards, particularly for fluorescent lamps that present long horizontal tubes to the dust particles. The loss of light can be appreciable if the lamps are not cleaned (Fig. 10.5.1). If ventilated luminaires are used (Fig. 10.6.1), convection currents carry the dust through holes in the canopy or reflector, so that less of the dust is deposited on the lamps themselves.

We noted [Section 9.3] that a light-loss factor due to dirt accumulation must be included in the design of windows. Similarly, a light-loss factor due to dirt accumulation must be included when determining the number of lamps required for a room (Ref. 8.3, pp. 9.3–9.5). When lamps are cleaned once a year, the light loss varies from 20% to 40%.

Lamps deteriorate as they get older: They produce less light, and eventually they fail. The average life of incandescent lamps is approximately 1000 hours of operating time. The average life of fluorescent lamps is 5000 to 7500 hours, and it is influenced by the frequency of switching the lamps on and off, by the ambient temperature, by the adjustment of the control gear, and by the type of ballast. Even if all lamps are installed at the same time, they will fail at different times, and there are at any time some lamps that have failed (Fig. 10.5.2).

In private houses and small commercial undertakings individual lamps are replaced as they fail, but for large offices and factories it is more economical to allow a few failed lamps to remain in position, and to replace all lamps together on a regular schedule. Fig. 10.5.2 shows that for the first 40% of the anticipated average lamp life few failures occur; the failure rate then increases sharply.

10.6 Heat Production by Lamps, and Integrated Ceilings

The heat load from a lighting system is approximately equal to the combined wattage of the lamps in the system. The infrared radiation produced in large measure by incandescent lamps is already in the form of heat, and the radiation in the range of visible light is partly in the form of heat. The visible light and any ultraviolet radiation are eventually transformed into heat. This represents a large heat load for brightly lit office buildings and factories, and it is unnecessary to heat some buildings that would require heating if daylit. On the other hand, when the building requires cooling, the lighting system can greatly increase the cooling load.

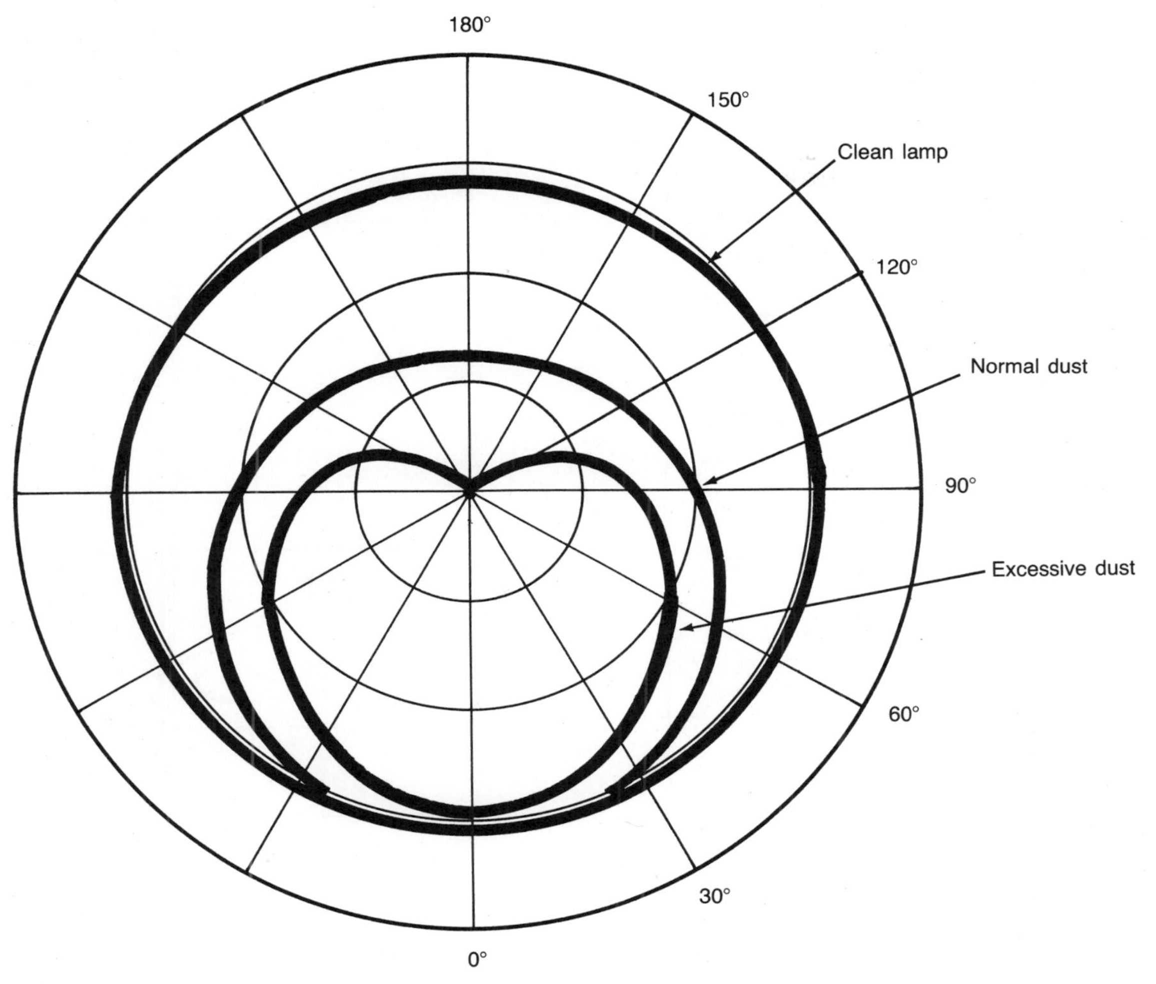

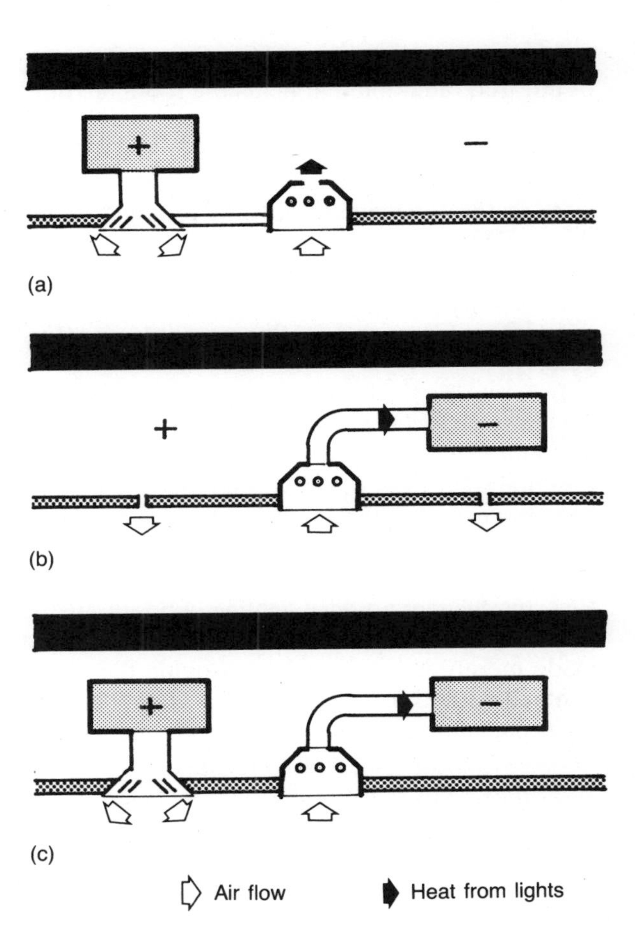

Fig. 10.5.1. Polar diagram showing the loss of light due to the accumulation of dust on standard tubular lamps mounted in a twin fitting. The loss of light is most severe on top of the tubes, where most of the dust accumulates.

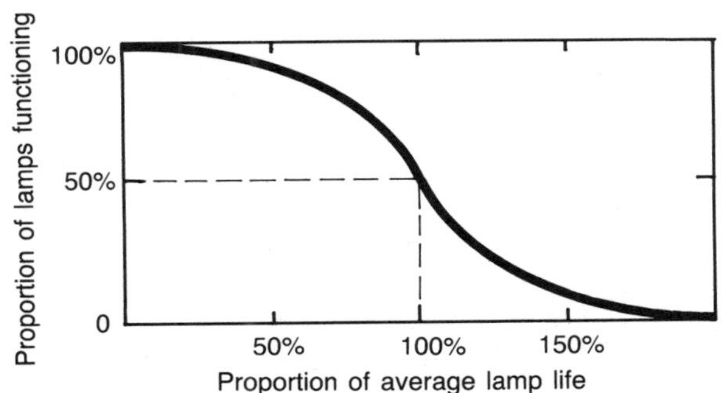

Fig. 10.5.2. Tubular fluorescent lamps still functioning in a lighting system, expressed as a percentage of their average lamp life. By definition, 50% of all lamps are still functioning when the *average* expected life expectancy is reached.

Fig. 10.6.1. Integrated ceiling systems that provide automatic cooling of the lamps by extracting the return air through slots in the luminaires. The air is supplied through ceiling diffusers.
(a) Air supplied from a pressurized ducted supply, air exhausted through plenum that is at a lower pressure than the air in the room.
(b) Air supplied from pressurized plenum, air exhausted through duct at a pressure below that of the air in the room.
(c) Air supplied and exhausted through ducts.

Incandescent light bulbs produce more heat for a given level of illumination because their lower luminous efficacy is due to the generation of infrared radiation that heats the room but does not light it (Fig. 10.3.1.a).

In an integrated ceiling it is possible to extract the return air [Sections 6.8 and 13.11] through slots in the luminaires and thus remove a large proportion of the heat generated before it enters the room. Luminaires that act as air exhausts can be used in integrated ceilings both for incandescent lamps and for fluorescent lamps (Fig. 10.6.1).

In assessing the heat load generated by lighting systems, allowance should be made for the heat that enters with the visible light through windows [Sections 3.5, 5.9, 9.3, 9.6, and 10.1]. This can be appreciable if the windows are inadequately designed and poorly shaded. Hence many architects and lighting engineers favor the use of PAL (permanent artificial lighting only), discussed in Section 10.1. In that case the windows are made merely large enough to provide the visual contact with the world outside. However, the dimming of the artificial light with increase in the level of the daylight, discussed in Section 10.1, may be a better solution.

10.7 The Lumen Method

The number of lamps is determined by the lumen method for artificial lighting, which is based on the same principle as the lumen method for daylighting [Section 9.4].

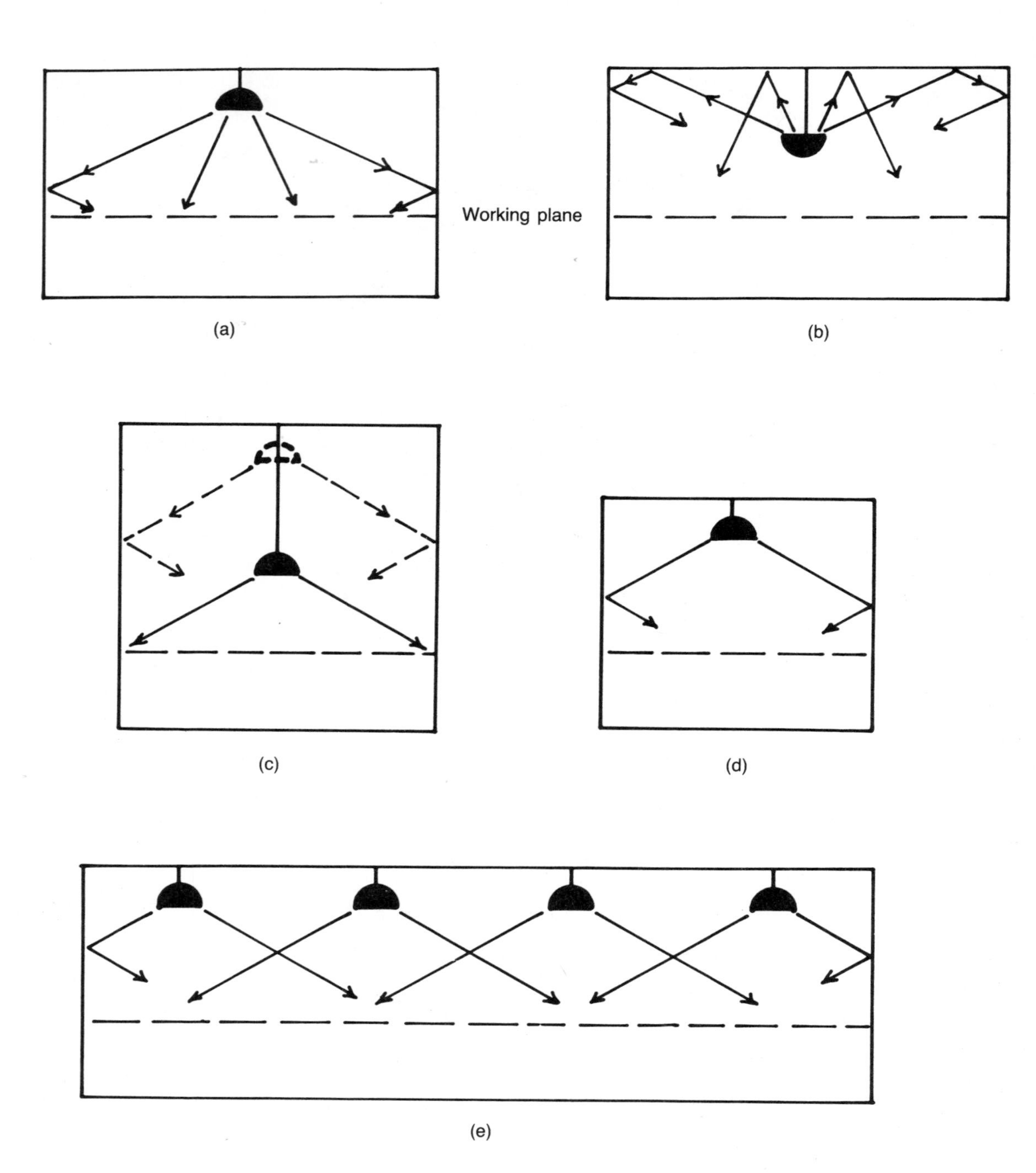

Fig. 10.7.1. The coefficient of utilization is influenced (a) by the reflection from the walls, (b) by the reflection from the ceiling, (c) by the height of the luminaire, and (d) and (e) by the width of the room. The *IES Handbook* quantifies these influences in the *cavity ratio.*

$$N = \frac{E_p \times A}{F \times CU \times LLF} \quad (10.1)$$

where N is the number of luminaires required;
E_p is the illuminance required on the horizontal workplane in lux lumen per square meter [discussed in Section 8.4];
A is the total horizontal area of the workplane in square meters;
F is the output (luminous flux) in lumen of the lamps installed in one luminaire, obtained from a table;
CU is the coefficient of utilization;
LLF is the light-loss factor, which allows for deterioration of the lamp with time [Section 10.5] and accumulation of dirt.

The coefficients CU and LLF are obtained from tables in textbooks and handbooks (for example, Ref. 8.3, pp. 9.3–9.33).

The coefficient of of utilization depends on the type of the lamp and the luminaire, the reflectance of the ceiling, walls, and floor (Fig. 10.7.1), and on the room cavity ratio:

$$\text{Room cavity ratio} = \frac{5H(\text{Room length} + \text{Room width})}{\text{Room length} \times \text{Room width}} \quad (10.2)$$

where H is the height of the luminaire plane above the workplane (Fig. 10.7.2).

A slightly different method is used in Britain and in Australia.

Example 10.1 *Determine the number of luminaires required for a room measuring 6 m by 8 m, 3.5 m high, to give an illuminance on the workplane of 500 lx.*

We will use fluorescent tubes mounted in pairs in porcelain-enameled fittings (Type 27 in Fig. 9.12 of the Reference Volume of the IES Handbook, *Ref. 8.3). From Fig. 8.114 in the same handbook we note that tubes when new produce about 2870 lumen each.*

Solution *Let us assume that the tubes are suspended 0.6 m from the ceiling and that the workplane is 0.75 m above the floor. The height of the luminaire plane above the workplane is then*

$$H = 3.5 - 0.6 - 0.75 = 2.15 \text{ m}$$

and the room cavity ratio is

$$\frac{5 \times 2.15\,(6 + 8)}{6 \times 8} = 3.14$$

We will assume that the reflectance of the floor is 20%, the reflectance of the walls is 50%, and the average reflectance of the ceiling and the walls above the luminaire plane is 70%. From Fig. 9.12 of the Reference Volume of the IES Handbook *(Ref. 8.3), the coefficient of utilization is 0.61.*

The light-loss factor has a number of components. The surface depreciation of a porcelain-enameled luminaire is very slow, and 1.0 is an appropriate factor. We will assume a 10% lamp lumen depreciation (factor 0.9); a factor for dirt on the luminaires of 0.85; a factor for dirt on the room surfaces of 0.9; and a factor of 1.0 for lamp burnout [Section 10.5]. Thus the light-loss factor is

$$\text{LLF} = 1.0 \times 0.9 \times 0.85 \times 0.9 \times 1.0 = 0.69$$

If we want to provide approximately even illumination for the entire room, the area of the workplane is equal to the floor surface of the room:

$$A = 6 \times 8 = 48 \text{ m}^2$$

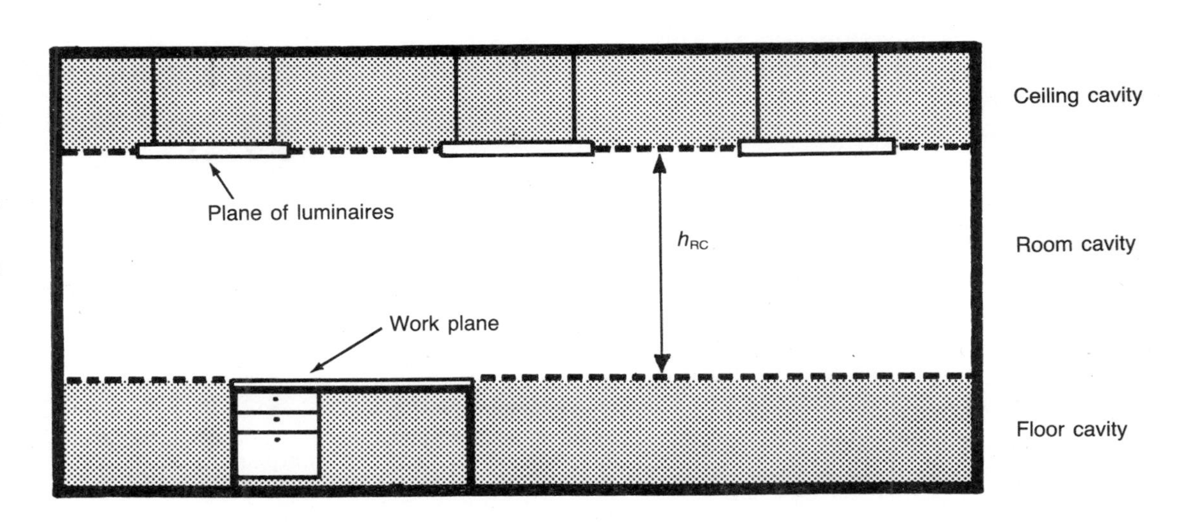

Fig. 10.7.2. The dimensions used in the calculation of the room cavity ratio in Eq. (10.2).

From Eq. (10.1) the number of luminaires required is

$$N = \frac{500 \times 48}{2 \times 2870 \times 0.61 \times 0.69} = 9.94$$

We require 10 luminaires, each with 2 tubes. For a room measuring 6 m by 8 m it would be appropriate to arrange the luminaires in 3 rows to give uniform illumination. In that case 12 luminaires are needed to give 3 rows of 4 luminaires each.

10.8 Glare from Artificial Light, and Visual Comfort

We noted in Sections 8.5, 9.6, and 10.1 that brightly lit windows produce glare from daylight and that one cure is to use higher levels of artificial light. However, illuminances due to electric light that are higher than necessary do not merely waste energy but also cause glare of a different kind.

One type of glare caused by electric light, and also by daylight, is due to the reflection of a bright light from a glossy or semimatte surface into the eyes of the observer (Fig. 10.8.1). This causes veiling reflections that reduce the contrast and result in loss of detail. Normally this produces no more than a mild distraction, but it can in extreme cases cause appreciable loss of task visibility and appreciable discomfort. We noted that this problem can be solved by using matte surfaces or by reducing the illuminance on the task. However, it can also be remedied by moving the light source so that the glare is not reflected into the eyes of the observer or by moving the person to a different location or turning him in a different direction (Fig. 8.5.2).

Direct glare is caused by a bright light source appearing in the normal field of view of the observer (Fig. 8.5.1). In extreme cases this may manifest itself as disability glare [Section 8.5]; however, it is more likely to cause discomfort. Usually, if we avoid discomfort glare from artificial light, we automatically avoid disability glare.

Direct glare is a greater problem with daylight. People do look at windows, and if windows are shielded too much to avoid glare, the admission of daylight is limited. This is one argument for relying entirely on artificial light for lighting offices.

Lamps, on the other hand, can be shielded by suitably designed luminaires, and one important reason for using luminaires instead of naked lamps is the avoidance of glare. People rarely look vertically upwards at luminaires that give a direct light. Seen from an angle, glare can be obviated by screening the light source from direct view.

Glare can be reduced by using more luminaires of lower luminance, or by mounting them higher, so that people are less likely to look at them directly from nearby. It can also be reduced by surrounding light sources with light-colored surfaces to reduce the contrast in brightness. A completely luminous ceiling produces less glare than a few widely spaced luminaires.

Derek Phillips (Ref. 9.11, p. 36) considers that visual comfort can be improved by providing visual rest centers:

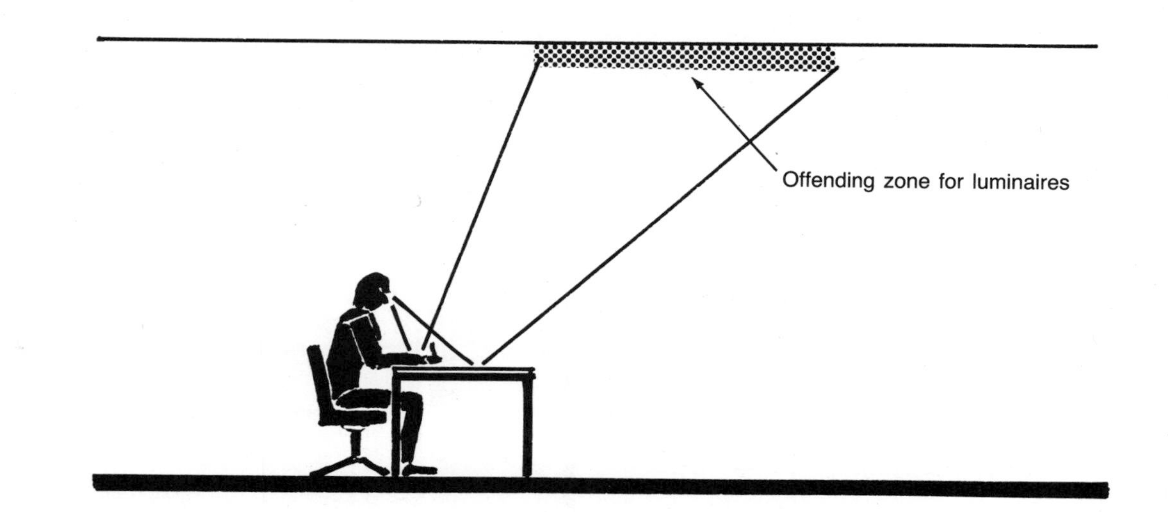

Fig. 10.8.1. Veiling reflections in the visual task are caused by a bright luminaire that is situated at a mirror angle relative to the eye, so that the workplane acts as a reflector. This happens if the angle between the eye and the workplane is equal to the angle between the luminaire and the workplane.

These are areas where the eye may wander after some minutes of concentration to release tension and recuperate for the next sequence of concentrated effort. This process takes place automatically, without in any way disturbing whatever thought processes are at work as a result of the visual task concerned.

In daylit buildings a window generally acts as a visual rest center, and the extension of view that this creates requires a change of focus, as well as a change of adaptation. . . . Where there is no view, either at night or in offices without daylight, the need for a visual rest center remains. This may be achieved by areas of lower brightness than the visual task that have no strong phototropic attraction. The ideal area is one that provides no conspicuous pattern to be perceived by the eye since its object is one of relaxation. The view is also held that pictures in a wall can act as visual rest centers, since after the picture is familiar it is treated purely as an area of color, which may give the necessary readaptation and focusing.

For most rooms the problem of glare can be solved without calculation; however, for critical cases a numerical assessment is necessary.

In the 1950s R.G. Hopkinson developed from theory the *glare index* at the (British) Building Research Station (Ref. 9.3, Chapter 12) and established limiting values for this index by asking people exposed to various degrees of glare to express an opinion whether the glare was imperceptible, acceptable, uncomfortable, or intolerable. The Glare Index is used in the (British) IES Code (Ref. 8.5). While this is an accurate method for assessing glare, it is also complicated.

The (American) *IES Lighting Handbook* (Ref. 8.3, pp. 3.12–3.14 and 9.71–9.80, Ref. 8.4, pp. 2.24–2.26) gives an alternative method for determining the *visual comfort probability* (VCP). The Australian Interior Lighting Code (Ref. 8.6) gives a number of tables for the control of direct glare and of reflected glare.

References

10.1 J.B. de BOER and D. FISCHER: *Interior Lighting*. Philips Technical Library—Kluwer Technische Boeken B.V., Deventer, 1978. 336 pp.

Suggestions for Further Reading

References 8.9, 8.10, 9.11, and 10.1.

10.2 W.T. O'DEA: *A Short History of Lighting*. H.M. Stationery Office, London, 1958. 40 pp.

10.3 H. HEWITT and A.S. VAUSE (Editors); *Lamps and Lighting*. Edward Arnold, London, 1966. 566 pp.

Chapter 11 Noise Control

This chapter deals with the control of undesirable sound. People have complained about noise at least since the days of Ancient Rome, but the scientific study of the subject dates only from the 1920s. We examine first the nature of sound and its measurement. The logarithmic decibel scale is used not only for sound pressure but also for sound power and for sound intensity.

The level of noise can be reduced by absorption in the room where it originates, or by insulation of the floor, walls, and roof. Airborne and impact sound require different treatment. Particular attention should be given to the design of doors, false ceilings, and ducts. A masking noise is often useful to protect the privacy of conversations.

11.1 Unwanted Sound

Architectural acoustics has a long and distinguished prehistory; but as this is concerned with the acoustics of theaters, it will be discussed in Chapter 12. The acoustics of noise control is essentially a development of the twentieth century.

Noise is not a new problem. Several Roman authors complained about it two thousand years ago, and edicts restricting traffic noise at night were enacted in Ancient Rome. The same happened again when Paris and London became big cities in the sixteenth century. Today's street noises are different, but probably no louder.

Prior to this century attempts at noise control were limited to the suppression of the noise at its source, by restricting traffic at night, reserving some streets for pedestrians and litter bearers, and forbidding street calls by itinerant vendors. Important buildings were constructed with thick walls of heavy materials that provided good insulation against airborne sound, and the heavy curtains and upholstery favored in the eighteenth and nineteenth centuries in the wealthier houses absorbed sound generated within the room.

During the twentieth century noise became more ubiquitous. Airplanes, automobiles, radios, and tape recorders create noise in suburbs and rural areas where high noise levels were previously unknown. On the other hand, noise control is used more widely, and the sound-insulating construction and the sound-absorbing materials developed during the twentieth century can be incorporated at a moderate extra cost into low-cost housing.

Since the 1950s there has been increasing pressure in the industrialized countries to treat noise as "pollution," to be restricted by legislation, like air pollution and water pollution.

11.2 The Nature of Sound and Its Measurement

Aristotle suggested as early as the fourth century B.C. that sound is conveyed from one point to another by motion of the air, but this proposition remained contentious until the seventeenth century.

Athanasius Kircher published in 1650 *Musurgia Universalis,* one of the first treatises on acoustics. In it he described an experiment he conducted in Rome with a bell that was struck in a vacuum; he reported that the bell could be heard, and that air was not therefore necessary for the transmission of sound. It is difficult to determine today what went wrong with the experiment, but in 1660 Robert Boyle repeated it in London with a better vacuum pump, and found that the sound diminished as the air was evacuated.

Sound is caused by vibrations or waves in any elastic substance. It travels at a velocity of 343 m/s (meters per second) in air at sea level, but much faster in solids. For example, the velocity of sound in structural steel is about 5000 m/s.

$$\text{Frequency} \times \text{Wavelength} = \text{Velocity} \qquad (11.1)$$

In this equation the frequency is measured in hertz (abbreviated Hz), which is a single unit for cycles per second, the wavelength is measured in meters (m), and the velocity in meters per second (m/s).

This equation applies to sound as it does to light, but the velocity of light (299 720 000 m/s, or approximately 300×10^6 m/s) cannot be exceeded, and it is much higher than the velocity of sound (343 m/s in air), which can be exceeded, for example, by supersonic airplanes.

Light and solar heat radiation are usually identified by wavelength; for example, green light has a wavelength of 500 nm, which is equal to 0.5×10^{-6} m. Consequently its frequency is

$$\frac{300 \times 10^6 \text{ m/s}}{0.5 \times 10^{-6} \text{ m}} = 600 \times 10^{12} \text{ Hz} = 600 \text{ THz}$$

Sound is usually identified by its frequency; for example, middle C, at the center of a piano keyboard, has a frequency of 262 Hz and a wavelength of

$$\frac{343}{262} = 1.31 \text{ m}$$

The human ear responds to frequencies ranging from about 20 Hz to about 20 000 Hz, corresponding to wavelengths ranging from 17 m to 17 mm. The fundamental notes on a modern piano range in frequency from 28 Hz to 4186 Hz. Although the ear can hear higher sounds, and some animals can hear sounds of higher pitch than the human ear, frequencies above 10 000 Hz are not significant for building design. Frequencies above 20 000 Hz are classified as ultrasonic.

The vibration that causes the sound produces a change in air pressure. Our hearing is the response of the ear (Fig. 11.2.1) to that change in pressure. The ear can respond to a very large range of sound pressure. The threshold of audibility for most people between the ages of 3 and 40 occurs at about 20×10^{-6} Pa (pascal), and the ear begins to hurt at a sound pressure of about 20 Pa. This is still a very small pressure; for comparison, standard atmospheric pressure is 101 325 Pa.

Sound pressure can be measured in pascals, but because the range of audible sound pressure is about 1 million to 1, it is more convenient to use a logarithmic scale with units called *decibels*.

A sound pressure measurement expressed in decibels is usually called a sound pressure level, to distinguish it from the sound pressure measurement in pascals.

The bel, named after Alexander Graham Bell, was originally introduced in 1923 in America by the Bell Telephone System to measure *sound power*. If the powers of two sounds are P_1 and P_0, the difference between them is

$$\log_{10} \frac{P_1}{P_0} \text{ bel} = 10 \log_{10} \frac{P_1}{P_0} \text{ decibel}$$

The alternative unit for sound power, like any other form of power, is the watt. By comparison with the wattage required for common sources of heat, the wattage of sound is small. The sound power of a person shouting is about 1×10^{-3} W (one thousandth of a watt). A full symphony orchestra during a loud passage produces about 10 W, and the engines of a large passenger jet generate a sound power of only about 10 000 W. By comparison 1000 W of electricity is required for a small portable electric heater.

The decibel scale compares two sounds. To turn it into an absolute scale, we must refer it to a standard sound power, and this is 1×10^{-12} W.

Example 11.1 *Compare the sound power of a person shouting (1×10^{-3} W) and a symphony orchestra (10 W) on the decibel scale.*

Solution *The comparative sound power of 1×10^{-3} W and 10 W is*

$$10 \log_{10} \frac{10}{1 \times 10^{-3}} = 40 \text{ decibel}$$

Example 11.2 *The sound power of a machine is 0.008 W. Determine its sound power in decibels, referred to the standard of 1×10^{-12} W.*

Solution *The sound power is*

$$10 \log_{10} \frac{0.008}{1 \times 10^{-12}}$$

$$= 99 \text{ decibel referred to } 10^{-12} \text{ W}$$

The waves from the sound source in a free space spread in all directions, and the intensity of sound is reduced with the distance from the source of the sound in accordance with the inverse square of the distance (Fig. 11.2.2). The relation is similar to the inverse square law for illuminance [Section 8.2, Eq. (8.1)].

In practice it is easier to measure sound pressure than sound intensity:

$$\text{Sound intensity} = \frac{(\text{Sound pressure})^2}{\text{Impedance of air}} \tag{11.2}$$

Consequently the sound pressure is reduced proportionately with the distance from the sound source (Ref. 11.1, pp. 3–8).

The decibel scale is also used to measure sound pressure. Since sound power is proportional to

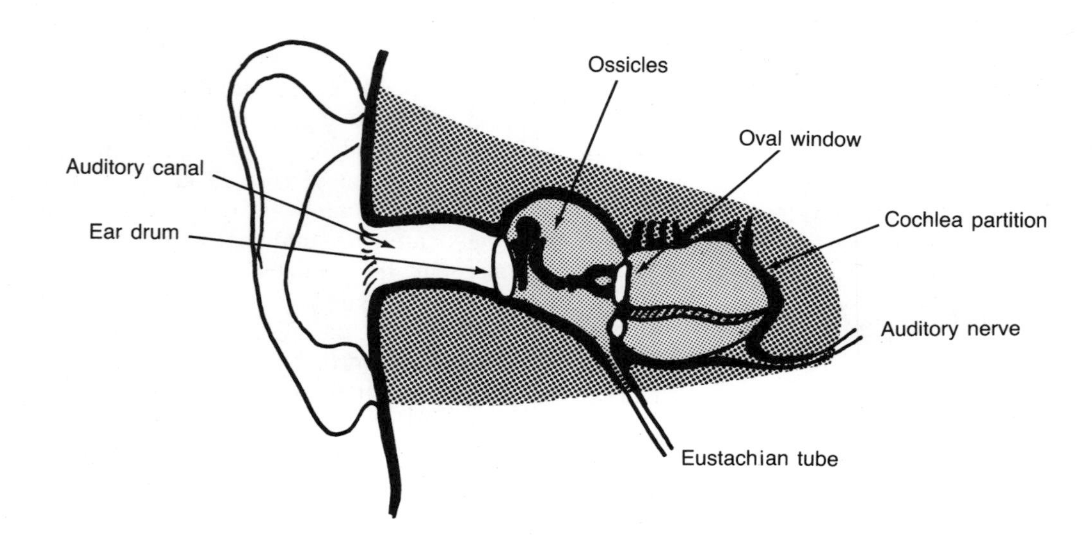

Fig. 11.2.1. The human ear. The outer ear is terminated by a membrane called the ear drum. Behind the ear drum is the middle ear, which connects with the mouth. Thus the ear is internally subject to atmospheric pressure, and the ear records only the excess pressure on the outside of the ear drum. This excess pressure is enlarged through a lever system formed by three small bones, called the ossicles, and transferred to a smaller membrane, the oval window, which closes the cochlea, a hollow bone filled with liquid. The cochlea is divided down the middle by the cochlea partition on which terminate about 25 000 ends of the auditory nerve.

sound intensity, which is proportional to the square of sound pressure, two sound pressures p_1 and p_0 differ by

$$\log_{10}\left(\frac{p_1}{p_0}\right)^2 \text{ bel} = 2\log_{10}\frac{p_1}{p_0}\text{ bel}$$

$$= 20\log_{10}\frac{p_1}{p_0}\text{ decibel}$$

The decibel scale becomes an absolute scale when it is referred to a standard value, which is taken as the threshold of normal hearing, $p_0 = 20 \times 10^{-6}$ Pa. The statement "the sound pressure level is 20 decibel," without further reference, means that the sound pressure is 20 db (decibel) higher than 20×10^{-6} Pa.

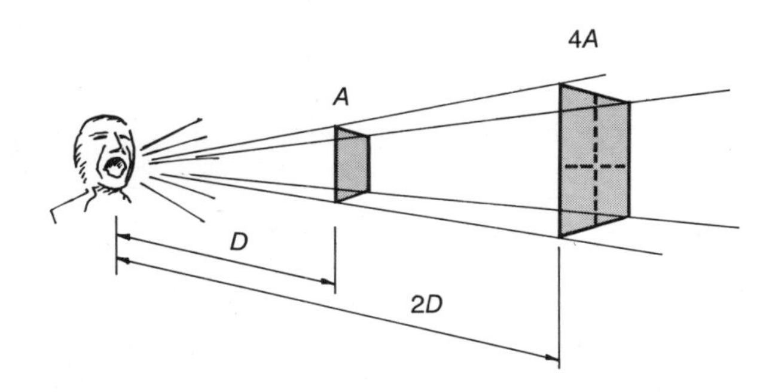

Fig. 11.2.2. The inverse square law for sound *intensity* states that it is inversely proportional to D^2, where D is the distance from the sound source. Since

$$(\text{Sound } intensity) = \frac{(\text{Sound } pressure)^2}{\text{Impedance of air}}$$

the sound *pressure* is inversely proportional to D.

Example 11.3 *Determine the increase in sound pressure due to a 3-decibel increase in the sound pressure level.*

Solution

$$3 = 20\log_{10}\frac{p_1}{p_0}, \text{ and therefore}$$

$$\frac{p_1}{p_0} = \log^{-1} 0.15 = 1.4125$$

Therefore an increase in sound pressure level of 3 decibel means an increase of 41% in the sound pressure.

If the measurement is referred to the threshold of normal hearing, which is $p_0 = 20 \times 10^{-6}$ Pa, then 3 decibel means a sound pressure of $1.4125 \times 20 \times 10^{-6} = 28 \times 10^{-6}$ Pa.

Example 11.4 *Determine the number of decibels required to double, treble, and quadruple the sound pressure level.*

Solution *If* $p_1/p_0 = 2$, *the number of decibels required for this pressure differential is*

$$20\log_{10} 2 = 6.020 \text{ db}$$

Similarly,

$$20\log_{10} 3 = 9.542 \text{ db}$$

and

$$20\log_{10} 4 = 12.041 \text{ db}$$

Example 11.5 *Determine the number of decibels corresponding to the upper limit of 20 Pa when the ear begins to hurt.*

Solution *If the measurements are referred to the threshold of normal hearing,* 20×10^{-6} *Pa, then*

$$\frac{p_1}{p_0} = \frac{20}{20 \times 10^{-6}} = 10^6$$

The number of decibels corresponding to this pressure differential is

$$20\log_{10} 10^6 = 20 \times 6 = 120 \text{ db}$$

Example 11.6 *The sound pressure level of a noisy machine is 90 db in the open air at a distance of 3 m. Determine the distance required to reduce the sound pressure level to 70 db.*

Solution *A drop in sound pressure level of 20 db means that the ratio of sound pressures*

$$\frac{p_1}{p_0} = \log_{10}^{-1}\frac{20}{20} = 10$$

Since the reduction in sound pressure is directly proportional to distance, the sound pressure level is reduced to 70 db at a distance of 30 m from the machine.

It is useful to have a table showing changes in sound pressure, sound power, and sound intensity in decibels (Table 11.1).

Table 11.2 shows the decibel ratings of some typical noise sources.

Sound pressures cannot be added arithmetically. The addition is performed on the corresponding sound powers. Since sound power is proportional to the square of the sound pressure, a doubling in sound power produces only a $\sqrt{2}$ increase in sound pressure.

We noted in Example 11.4 that a doubling of sound pressure corresponds on the logarithmic decibel scale to 6 decibels. However, the addition of a background noise and of an additional equal noise due to a new sound source produces a doubling of sound power, and thus an increase of only 3 db (Table 11.1).

Table 11.1

Increase in Sound Pressure, Sound Power, and Sound Intensity on the Decibel Scale

Increase in Decibel	Factor by Which Sound Pressure Is Increased	Factor by Which Sound Power or Sound Intensity Is Increased
0	1.00	1.00
1	1.12	1.26
2	1.26	1.58
3	1.41	2.00
4	1.58	2.51
5	1.78	3.16
6	2.00	3.98
7	2.24	5.01
8	2.51	6.31
9	2.82	7.94
10	3.16	10.00
11	3.55	12.59
12	3.98	15.85

Example 11.7 *The background noise in a room is 70 db. A machine is started up that would, by itself, produce a sound pressure also of 70 db. What is the combined sound pressure due to the background noise and the machine?*

Solution *The sound pressure* p_l *due to the background noise of 70 decibel is given by*

$$70 = 20 \log_{10} \frac{p_l}{20 \times 10^{-6}}$$

which gives

$$p_l = 20 \times 10^{-6} \log_{10}^{-1} \frac{70}{20}$$

$$= 3162 \times 20 \times 10^{-6} \text{ Pa}$$

When the sound pressure of 70 decibel due to the machine is added, the new sound pressure becomes

$$\sqrt{2} \times 3162 \times 20 \times 10^{-6} \text{ Pa}$$

and this corresponds to

$$20 \log_{10} \frac{\sqrt{2} \times 3162 \times 20 \times 10^{-6}}{20 \times 10^{-6}} = 73.0 \text{ decibel}$$

To obtain the combined effect of two sound pressures, it is evidently only necessary to add a number of 3 or less to the higher. This can be done with the aid of the curve in Fig. 11.2.3.

The sound pressure level of a composite sound depends on the sound pressure level of the individual frequencies of its components. This is called a frequency analysis, in which the instrument for measuring the sound pressure level is adjusted to measure only one frequency at a time.

Example 11.8 *The frequency analysis of a noise produced by a machine is as follows:*

Frequency (Hz)	125	250	500	1000	2000	4000
Sound pressure level (db)	54	60	67	68	60	55

Determine the combined sound pressure level due to all frequencies.

Table 11.2

Approximate Sound Pressure Level in Decibels

Threshold of hearing, by international standardization	0
Rustle of leaves	10
Background noise in TV studio	20
Rustle of paper	30
Background noise in a public library	40
Background noise in a quiet office	50
Normal conversation heard at a distance of 1 m	60
Telephone	70
Road traffic	80
Circular wood saw at a distance of 3 m	90
Discotheque	100
Pneumatic hammer at a distance of 1 m	110
Threshold of pain; jet airplane takeoff at 100 m	120

Solution *We will start with the two highest numbers. The difference between 68 and 67 is 1.0. From Fig. 11.2.3 we read 2.5 and add this to 68, which gives 70.5. The entire calculation is as follows:*

68 − 67 = 1; add 2.5: 68 + 2.5 = 70.5 db
70.5 − 60 = 10.5; add 0.4: 70.5 + 0.4 = 70.9 db
70.9 − 60 = 10.9; add 0.4: 70.9 + 0.4 = 71.3 db
71.3 − 55 = 16.3; add 0.1: 71.3 + 0.1 = 71.4 db
71.4 − 54 = 17.4; add nothing.

We round off upwards to allow for the lower and higher frequencies not included in the analysis. Thus the sound pressure level due to the noise is 72 db.

We noted that the decibel is the unit of sound power, sound intensity, and sound pressure. It was originally intended that there should be another unit to measure loudness, because the loudness of a sound as perceived by the human ear varies appreciably with the frequency of the sound. The phon was adopted by the First International Acoustical Conference, held in Paris in 1937, as the unit of loudness, but it is rarely used because phons can only be measured under laboratory conditions. Instead loudness is generally measured in dbA with a sound level meter using the A-weighted scale.

A sound level meter contains a microphone whose output is amplified and measured with a voltmeter calibrated in decibels (Fig. 11.2.4). It can be set to a "fast" response to measure the maximum value of a fluctuating noise, or a "slow" response to measure the average noise level.

Because the response of the ear varies with the frequency of sound, the better sound level meters have built-in weighting networks, that is, electrical circuits to attenuate different frequencies by different amounts. The internationally standardized networks are called A, B, and C. It was intended that the A scale would be used for sound pressure levels below 55 db, and the other two scales for the higher levels. In practice, the A-weighted scale has given the best correlation with loudness, which is stated in dbA, even if the sound pressure level is above 55 db.

Response to noise varies greatly. It is affected by personality and by tradition. As a rough guide, people feel disturbed if for more than 10% of the time the noise level in their home exceeds 35 dbA at night, or 45 dbA at day.

Higher noise levels are accepted in offices, and even higher levels in factories. A continuous noise above 80 dbA may endanger hearing, and staff should be provided with ear protectors.

Several more sophisticated and complex noise criteria have been developed and are described in recently published acoustics textbooks (for example, Ref. 11.2, pp. 156–79). Some are intended to serve as general noise criteria, while others deal specifically with speech intelligibility and with the politically sensitive problem of aircraft noise. However, the dbA scale remains the most generally useful and the one most frequently used.

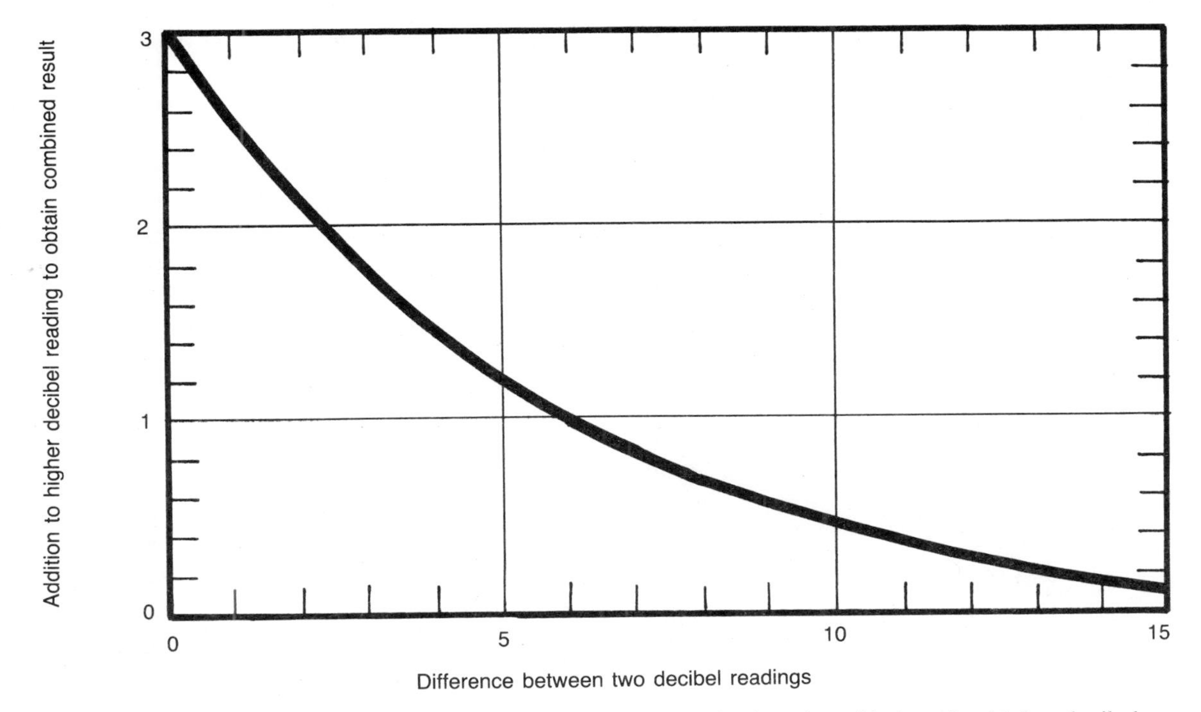

Fig. 11.2.3. Addition of two decibel readings. First, determine the difference between the two decibel readings. The curve then gives the number (always 3.0 or less) to be added to the *higher* decibel reading.

The principal other criteria are:

1. The Noise Criterion (NC) curves developed by L.L. Beranek (Ref. 11.3) and used particularly in the United States, which give a closer specification of sound pressure limits at different frequencies.
2. The Preferred Noise Criterion (PNC) curves, an improved version of the NC curves.
3. The Noise Rating (NR) curves, a European alternative to the NC curves.
4. The L_{10} dbA and L_{90} dbA Rating, which states the dbA level exceeded, respectively, 10% and 90% of the time.
5. The Speech Interference Level (SIL), which gives the arithmetic mean of the sound pressure levels in the octave bands centered on 500, 1000, and 2000 Hz.
6. The Articulation Index (AI), which measures the proportion of syllables intelligible.
7. The Perceived Noise Level in db (PNdb), which is used mainly to measure aircraft noise.
8. The Noise and Number Index (NNI), which is a measure of both aircraft noise and the number of aircraft flying over.

Some of these indices have a rough correlation with one another. For example, the sound level in PNdb is approximately equal to the sound level in dbA, plus 14.

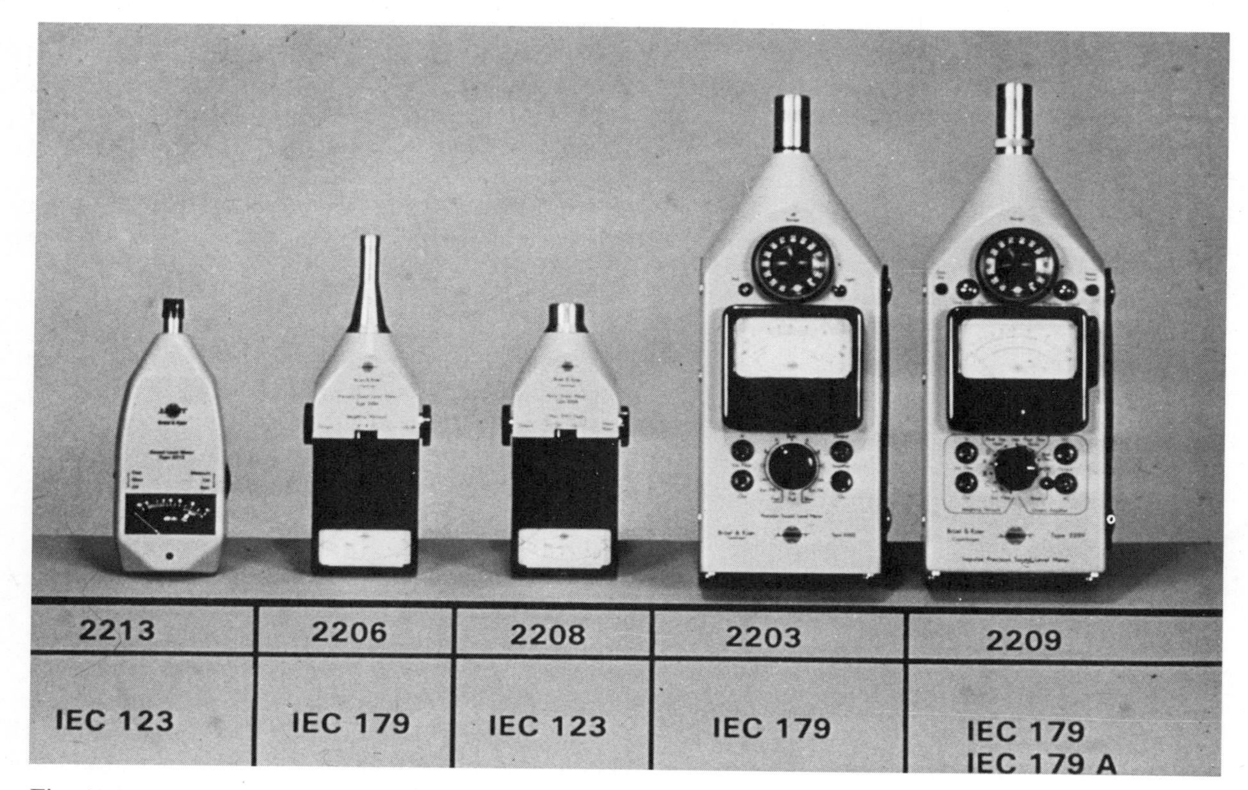

Fig. 11.2.4. Five different sound level meters of varying complexity and precision. (*By courtesy of Bruel & Kjaer, Copenhagen.*)

11.3 Sound Absorption

The most effective method of noise reduction is to absorb the sound at its source. Hard, rigid materials reflect sound. Soft, porous surfaces whose particles can vibrate easily absorb sound and convert the sound into heat. The rise in temperature is negligible because sound power is small; we noted in Section 11.2 that a person shouting generates about 0.001 W, while a small electric heater requires 1000 W, or about 1 million times as much power.

We noted in Section 5.9 that heat-absorbing glass was only partly effective, because it reradiated the heat absorbed in both directions. This is not a problem with sound-absorbing materials, which turn the sound absorbed into heat.

The sound absorption coefficient is the proportion of sound absorbed. Thus a material with a coefficient of 0.1 absorbs 10% of the sound, and another with 0.9 absorbs 90%. The sound absorption of materials varies appreciably with frequency. The coefficients are listed in the trade literature supplied by manufacturers and in many textbooks (for example, Ref. 11.2, pp. 279–83, and Ref. 11.3, pp. 349–95). A few are reproduced in Table 11.3.

The sound absorption coefficient given in the tables, and used in the calculations for the sound absorption of buildings, is that for sound arriving at all angles of incidence. This coefficient is determined by experiments in a reverberation chamber (Fig. 11.3.1).

Fig. 11.3.1. Determination of the sound absorption coefficient of a material, using a loudspeaker and a microphone in the reverberant chamber of the Department of Architectural Science, University of Sydney. (*By courtesy of Dr. Fergus Fricke.*)

Table 11.3

Sound Absorption Coefficients of a Few Typical Building Materials

Material	Sound Absorption Coefficient at a Frequency of					
	125 Hz	250 Hz	500 Hz	1000 Hz	2000 Hz	4000 Hz
Glazed tiles	0.01	0.01	0.01	0.01	0.02	0.02
Concrete with roughened surface	0.01	0.02	0.02	0.03	0.04	0.04
Timber floor on timber joists	0.15	0.15	0.15	0.10	0.10	0.08
Cork tiles, 22 mm thick, on solid backing	0.05	0.10	0.20	0.55	0.60	0.55
Draped curtains over solid backing	0.05	0.25	0.40	0.50	0.60	0.50
Carpet, 9-mm pile, on felt underlay	0.08	0.08	0.30	0.60	0.75	0.80
Expanded polystyrene, 25 mm thick, spaced 50 mm from solid backing	0.10	0.25	0.55	0.20	0.10	0.15
Acoustic spray plaster, 12 mm thick, sprayed on solid backing	0.03	0.15	0.50	0.80	0.85	0.60
Metal tiles with 25% perforations, suspended from ceiling with porous absorbent material laid on top	0.40	0.60	0.80	0.80	0.90	0.80

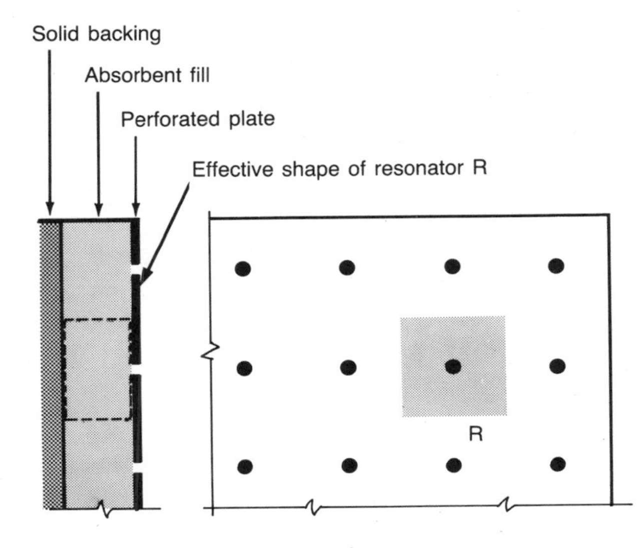

Fig. 11.3.2. Sound-absorbing panel formed by a sheet of perforated particle board and a backing of mineral wool.

Any cavity acts as a resonator; this is readily demonstrated if one blows across a closed tube or an empty bottle. The Helmholtz resonator used in concert-hall acoustics is a cavity with a narrow opening that absorbs sound of the particular frequency at which it resonates.

The same principle is used in this absorbing panel. About 10% of the outer panel is perforated, and the mineral wool, about 50 mm thick, behind each hole, and the solid wall behind, form a broadband Helmholtz resonator. The fact that these holes communicate with one another detracts only slightly from the sound-absorbing efficiency.

Sound-absorbing material fixed in patches, intermingled with reflecting material, is slightly more effective than the same area of absorbing material fixed all in one area. Sound waves arriving at the junction of the absorbing and the reflecting material are bent by diffraction toward the absorbing material (Fig. 12.2.2). As a result the edges of absorbing material are more effective than the middle part.

Perforated material backed by absorbing material is very effective for sound absorption, because it acts as a resonator (Fig. 11.3.2).

Sound absorption is measured in sabin. The unit was introduced by Wallace C. Sabine, the American pioneer of architectural acoustics, in

1911, and the name sabin was given to it in 1937 by the American Acoustical Society.

$$S = \alpha A \qquad (11.3)$$

where S is sound absorption in sabin;
α is the sound absorption coefficient of a material at a particular frequency (Table 11.3);
A is the surface area of that material in the room, in square meters. (The dimensions of the sabin are m^2 if A is measured in m^2, and ft^2 if A is measured in ft^2.)

Figure 11.3.3 shows a room with a sound source S. The listener L receives sound directly from S and also receives sound that is reflected several times by the various surfaces of the room. The sound is reduced by distance in accordance with the inverse square law, and it is reduced by absorption, whenever it strikes an absorbing surface. If the absorption coefficient of all surfaces is 0.8, the reflected sound will be damped very quickly, and the room will appear to be "dead." Even if the coefficient is only 0.5, the sound will be greatly reduced by absorption. A fully tiled room with a sound absorption coefficient of 0.01 is highly reverberant.

Absorbing materials have no effect on the direct sound. If the sound source is a typewriter, this does not matter, as a sufficient reduction in noise can be achieved through absorption by the ceiling and some of the walls. If the sound source is a machine that produces, say, 90 db at a distance of 3 m, it would be appropriate to place a partition of absorbing material between S and L.

In practice there are a number of limitations on the use of sound-absorbing materials, because the design of the room must also satisfy the requirements for the thermal [Chapters 5 and 6] and the luminous environment [Chapters 9 and 10].

There is no conflict between the sound-absorbing properties of a light-colored ceiling and any other environmental requirements.

The main object of curtains is to admit daylight during the day and ensure privacy at night; they also have a subsidiary thermal function, to reduce heat loss through the windows at night. Curtains drawn back from the glass windows have a lower sound absorption than the closed curtains. Excellent sound absorption at high frequencies is obtained with a carpeted floor, but this means that it cannot be used as a thermal store [Section 3.4]. Similar considerations may apply to the rear wall in passive solar designs [Section 5.15]. The more dominant environmental requirement determines the design, and this will vary from building to building. Sound absorption within the room simplifies the problem of sound insulation [Section 11.4], as there is less noise to be transmitted. The absorption of all the materials in the room is calculated. To this must be added the absorption due to furniture, the people in the room, and the air (Table 11.4).

We then consider how the absorption could be improved, for example, by making the ceiling more absorbent, by adding some curtains, or by putting a carpet on a timber floor. If the absorption of the room before the improvement is S_1 sabin and the absorption afterwards is S_2 sabin, the sound intensity of the diffuse sound is changed in the ratio S_2/S_1. This change is translated into decibels by calculation or by using Table 11.1.

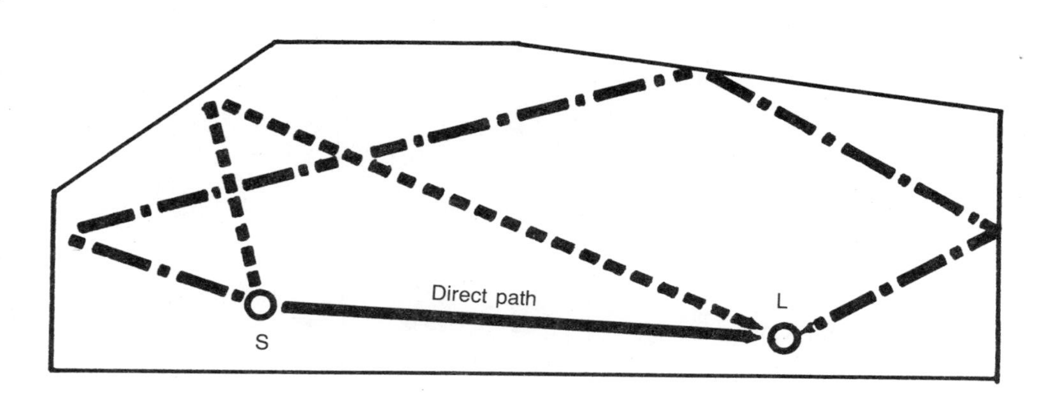

Fig. 11.3.3. A sound source S emits sound in all directions. Some of this sound reaches the listener L directly. The remainder is reflected by the walls, the floor, and the ceiling. If any of the room surfaces are made of acoustically absorbent material, some of the sound is absorbed depending on the frequency spectrum of the sound and the sound absorption coefficient of the material (Table 11.3). Furthermore, the sound pressure level falls off proportionally with distance from the sound source in accordance with the inverse square law of sound power.

Example 11.9 An office containing five typists is considered too noisy. Its absorption is 13 sabin at 125 Hz, 13 sabin at 250 Hz, 12 sabin at 500 Hz, 13 sabin at 1000 Hz, 15 sabin at 2000 Hz, and 15 sabin at 4000 Hz. To reduce the noise level the plasterboard ceiling, measuring 20 m^2, is replaced with one of perforated acoustic tiles. Determine the reduction in the noise level.

Solution The calculation must be performed for each frequency in turn, taking the absorption coefficients of the plasterboard, which is being replaced, and the acoustic tiles, which are replacing it, for that particular frequency. We will perform the calculation only for the middle frequency of 1000 Hz. If the coefficient of absorption of the plasterboard at that frequency is 0.1 and that of the acoustic tiles is 0.8, then the absorption is increased by

$$(0.8 - 0.1) \times 20 = 14 \text{ sabin}$$

Therefore the sound intensity is changed by a factor of

$$\frac{13 + 14}{13} = 2.08$$

From Table 11.1 the noise level at 1000 Hz is reduced by 3 db. The calculation should be repeated for the other frequencies, because the absorptive properties of the material and the noise level both vary with the frequency.

11.4 Transmission of Airborne Noise

Outside noises that disturb people inside buildings are mostly airborne. They include noise due to traffic, to airplanes, and to construction work on other buildings.

Noise generated inside a building may be airborne sound or impact sound. Impact noise is caused mainly by people walking on floors, and the measures needed to prevent its transmission to the story below are considered in Section 11.5.

The sound insulation against airborne noise, whether it comes from outside the building or from another room in the same building, is almost proportional to the mass of the wall, floor, or roof through which it has to pass; this is known as the mass law (Fig. 11.4.1). For the heavier materials, which are most useful for sound insulation, a doubling in thickness increases the sound insulation by approximately 5 db, that is, the proportionality of sound insulation to mass is not exact.

The requirements of a good thermal insulating material (Table 5.2) are almost exactly opposite those of a good sound-insulating material. The best thermal insulation is obtained from very light materials, such as mineral or glass wool, which have a low mass and provide poor sound insulation. Concrete is an excellent material for sound insulation but not for thermal insulation [Examples 5.2 and 5.3].

In practice the two requirements rarely conflict. Many airborne noises originate within the building; they are caused, for example, by a family member's tape recorder, by a party in the apartment on the story below, or by the typing pool in an office. Internal walls and floors do not require thermal insulation, so that the design needs to consider only the sound insulation.

Noise from traffic and airplanes may cause great irritation. If the building requires thermal inertia [Section 5.6] for its roof and outer walls, the sound insulation requirement can be met by the same massive material. The thermal and acoustic requirements conflict only for buildings near airports and major highways that are either heated only intermittently [Section 5.6] or are situated in the hot-humid tropics [Section 5.11].

There is a greater conflict between insulation against airborne sound and the widely felt need for natural ventilation in warm weather. There is no way of stopping the noise from a radio if the person who listens to the radio and the person who wishes to be undisturbed in the next room both have their windows open.

In a residential building this problem can be overcome to some extent by tact and good manners; but it is evidently not possible to stop noisy machines in an office or a factory when the

Table 11.4

Sound Absorption of Furniture, People, and Air (in metric sabin)

	Sound Absorption Coefficient at a Frequency of					
	125 Hz	250 Hz	500 Hz	1000 Hz	2000 Hz	4000 Hz
People seated on upholstered wooden chairs, per person	0.15	0.35	0.40	0.45	0.45	0.40
Upholstered wooden chairs, per chair	0.10	0.10	0.15	0.15	0.20	0.20
Air, per cubic meter	0	0	0	0.003	0.007	0.02

weather is warm, and the only solution is the air conditioning of noisy rooms so that the windows can be kept closed [see also Section 11.5].

The sound insulation of a solid wall or floor is easily determined from the mass law.

Example 11.10 *Determine the sound insulation of a solid reinforced concrete floor slab (a) 200 mm thick, (b) 300 mm thick, and (c) 400 mm thick.*

Solution *The mass of reinforced concrete depends on the type of aggregate used and on the amount of reinforcement. We will take it as 2500 kg/m³. Thus the three slabs have, respectively, a mass of 500 kg/m², 750 kg/m², and 1000 kg/m². From Fig. 11.4.1 the insulation value is (a) 51 db, (b) 54 db, and (c) 56 db.*

The sound insulation values of typical walls, floors, ceilings, and partitions are listed in textbooks (for example, Ref. 11.2, pp. 286–87, or Ref. 5.6, pp. 1284–94) and in the trade literature. The sound insulation varies with the frequency; for example, for a 200-mm reinforced concrete slab it ranges from 38 db at low frequencies to 60 db at high frequencies.

However, the problem of sound insulation is more complex. The wall or floor that provides the sound insulation is vibrated by the sound waves, so that it becomes itself a radiator of sound (Fig. 11.4.2), and this reduces its insulation value to some extent.

Furthermore, the direct sound path may not be the critical one if a wall has a high sound insulation but other parts of the building are less solidly constructed (Fig. 11.4.3). The floor, the ceiling, and all the walls of the sound source room are vibrated by the sound, and the sound is transmitted along the solid material, although its intensity is reduced by the length of the sound path. Thus a solid wall of high sound insulation between two rooms may not be effective if it is flanked by walls of low insulating value, or if there are doors and windows through which the sound can be transmitted more readily.

False ceilings are often a weak link in the sound insulation of multistory buildings. Partitions generally terminate at ceiling level to allow for the reticulation of the building services in the space above. If the ceiling has a low sound insulation value, noise is transmitted through that space from one room to the next, unless a barrier against airborne sound is placed around the ducts in the space between the false ceiling and the floor above [Section 11.7].

Evidently sound insulation becomes impossible if windows or doors are left open. Even when they are closed, there can be an appreciable leakage of sound through cracks. Doors and windows that have been made weathertight to reduce winter heating or summer cooling also provide good

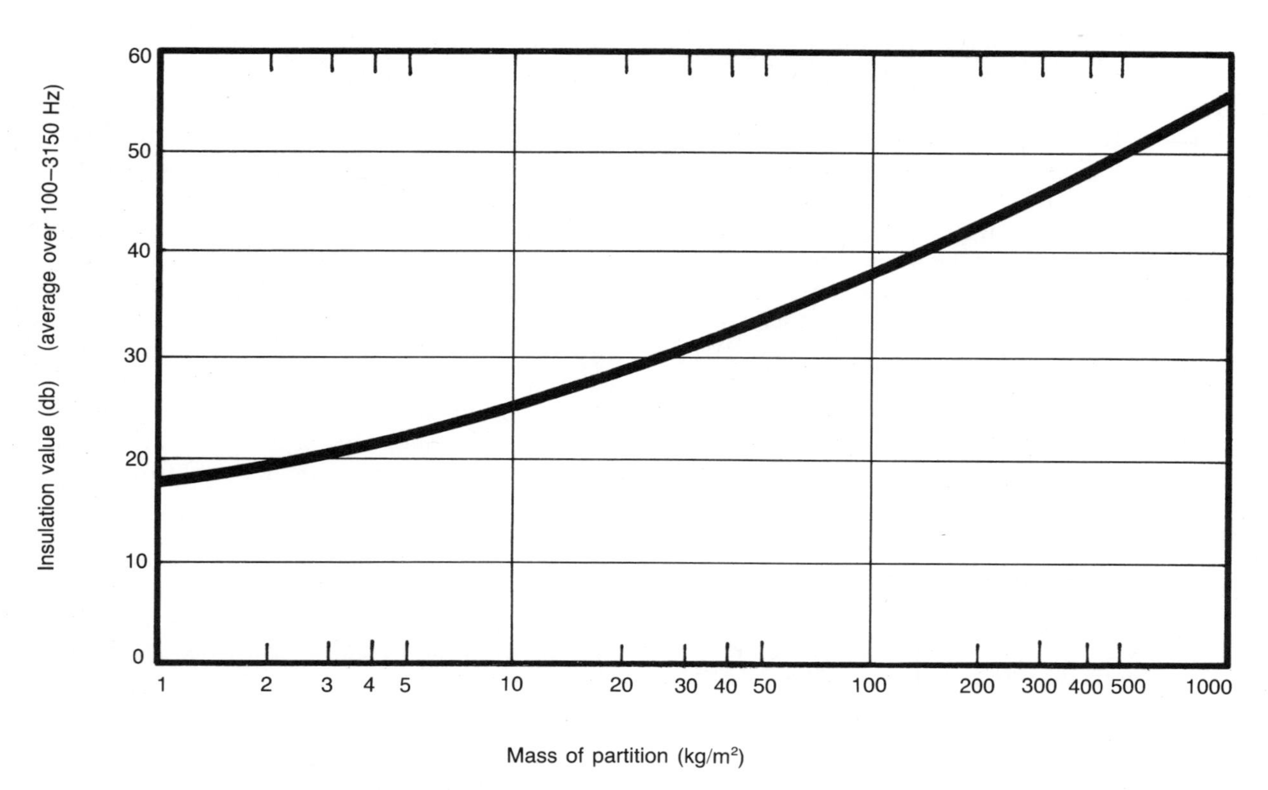

Fig. 11.4.1. The mass law of sound insulation. The experimental relation between sound attenuation, or reduction, and mass is an almost straight line. The decibel scale is logarithmic, and an increase in the sound insulation value by 6 db means that the sound has been reduced by half (see Table 11.1). In the range of 200 to 1000 kg/m² a doubling of mass produces a sound attenuation of approximately 5 db.

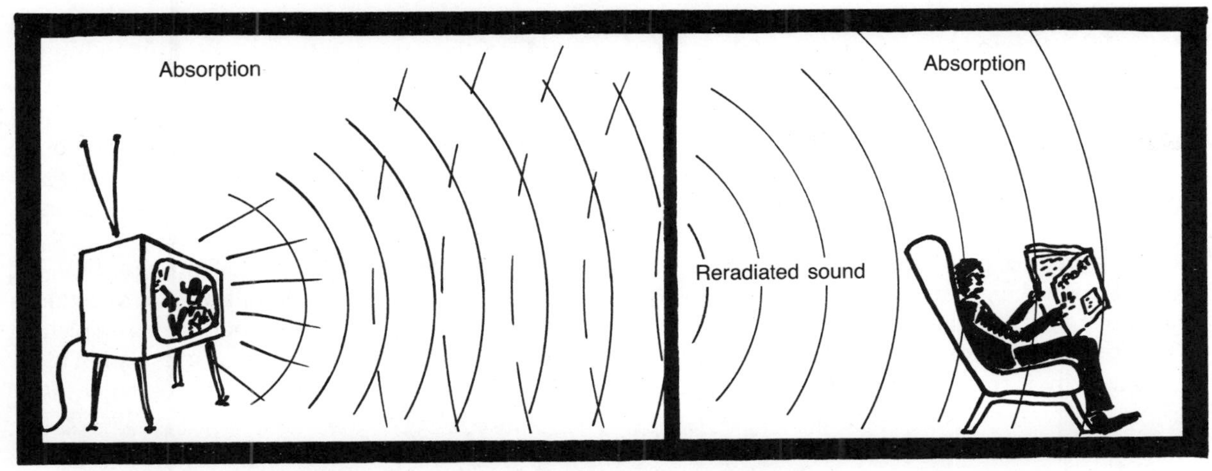

Fig. 11.4.2. The wall that provides the sound insulation between two rooms is vibrated by the sound waves, and thus becomes itself a source of sound. The wall reduces the noise transmitted, but it makes also a small addition to the (reduced) noise level. This addition depends on the reverberation time of the room [Section 12.3] and on the sound absorption of its surfaces and its contents.

insulation. A similar treatment may be needed for internal doors. Spaces between the door and its frame, cracks between a partition and the floor or ceiling, and holes for pipes or conduits that have not been sealed all reduce sound insulation, sometimes appreciably.

Air spaces increase sound insulation, particularly at the higher frequencies. Thus a plywood partition made of two sheets separated by an airspace provides several decibels more insulation than the same mass of material used in one solid sheet. For the same reason a considerable improvement in sound insulation is achieved by interposing a room or a passage (Fig. 11.4.4).

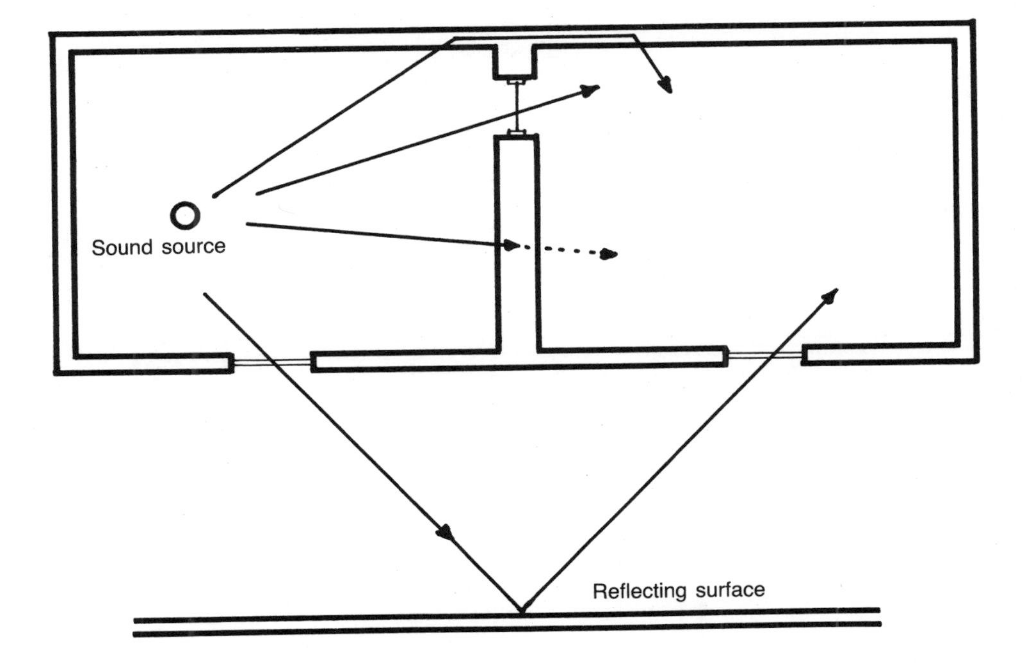

Fig. 11.4.3. If the sound insulation of a wall is particularly high, the sound path through it may not be critical, because the sound transmitted along other paths may produce a higher noise level in the adjacent room.

11.5 Impact Noise

Impact noise is transmitted directly to the structure or the fabric of the building. We noted (Fig. 11.4.2) that airborne sound causes the structure to vibrate, but the noise so generated is relatively small. With impact sound there is no air cushion; the noise is directly transmitted to the structure, and the noise intensity produced by the vibration of the structure in other rooms can be very high.

Impact noise is transmitted by the *continuous* solid parts of the structure and the fabric. Additional mass does not significantly reduce it. It is necessary either to interrupt the sound path or to cushion the impact.

The most common source of impact noise is footsteps on the floor above. Some machines attached to the floor or a wall feed impact noise into the structure, and closing of a valve or faucet causes "water hammer" that can travel long distances through plumbing.

The sound of footsteps can be suppressed at the source by a thick carpet with a resilient underlay, or it can be reduced by cork, rubber, or vinyl tiles. However, none of these materials significantly improves airborne insulation, and they may interfere with the thermal performance of the building [Sections 3.4 and 5.16].

Impact noise is effectively controlled by breaking the sound path, that is, by floating the floor on a blanket of resilient material, such as rubber or mineral wool. A suspended ceiling that has only a light connection to the floor above provides an additional break in the sound path, but it is not usually sufficient by itself. A sound-absorbing blanket in the space between the floor and the suspended ceiling is also helpful.

Typical construction details of floating floors that provide good impact sound insulation are given in some textbooks on acoustics and in most books on building construction (for example, Ref. 11.4, pp. 409–13, and Ref. 11.5, pp. 39–49).

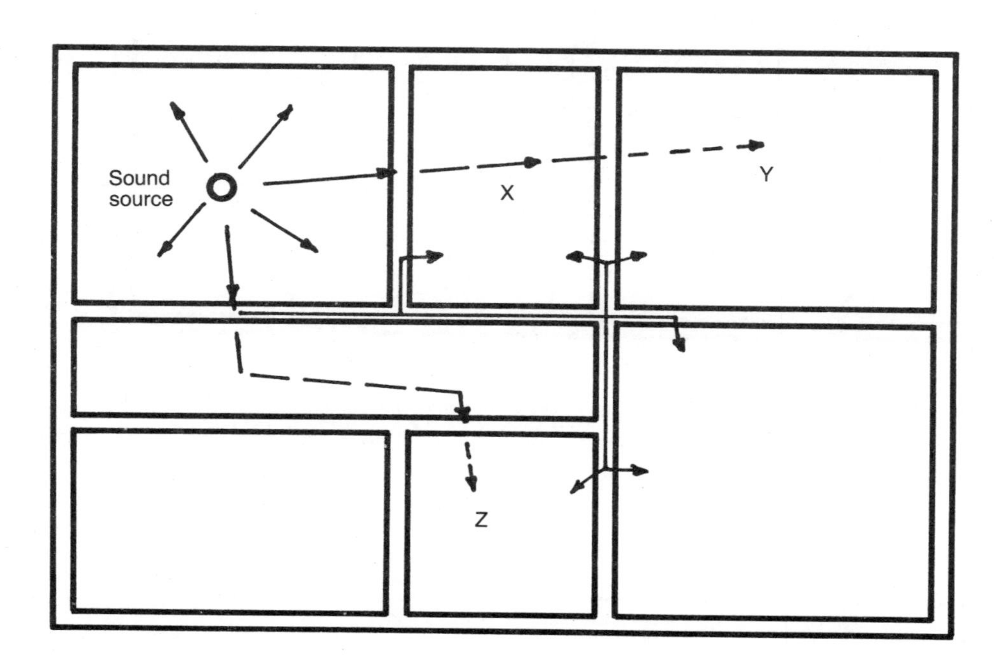

Fig. 11.4.4. The noise level can be reduced appreciably by interposing a room or a passage. The extent to which this can be done depends on the construction. If all the walls are of the same type, the noise level in room Y will be significantly less than in room X, and that in room Z will be even lower. A concert hall, for example, will be much less troubled by traffic noise if auxiliary rooms are placed between it and the outside walls.

11.6 Transmission of Noise in Ducts

Before the telephone was invented, speech was transmitted over quite long distances by speaking tubes. In that way the officer of the watch on a ship communicated with the engine room. Sound travels as easily in the ducts of air conditioning or ventilation systems.

Since the velocity of sound is 340 m/s, and the velocity of air is less than 20 m/s even in a high-speed duct, sound travels almost as fast against the direction of the air flow as with is, so that ducts for return air are as noisy as air supply ducts.

All noise is caused by vibration. Some noise is transmitted from the air blower, and this can be reduced by inserting a sound silencer between the fan and the duct.

The rest of the noise is caused by the air flow in the duct, particularly by constrictions or sharp edges that cause eddies; these are transformed into noise when they come into contact with the duct walls. The most effective method of reducing duct noise is to line the duct with a sound-absorbing material. The lining is required (a) to reduce the sound within the duct, and (b) to insulate the duct itself and prevent sound from it getting through the thin duct wall into the ceiling space.

Noise increases with the velocity of the air in the ducts; on the other hand, high-velocity ducts need less space. Bends in ducts, particularly if they are sharp, greatly reduce the transmission of low-frequency noise, but they also reduce the air flow.

The noise attenuation required in a duct depends on the use of the room. Some background noise is acceptable, but it must not be so high as to interfere with speech intelligibility.

11.7 Speech Privacy and Masking Noise

Air ducts are a major cause of "cross talk" between rooms. Since sound travels much faster than the air in the duct, it can enter through an *outlet* grille in one room and escape through the outlet grille in an adjacent room. Thus people in a waiting room may be able to hear a conversation between a medical practitioner and his patient, or a lawyer and his client. This can be prevented by placing sufficient absorption and sufficiently sharp bends in the duct to make the speech unintelligible.

The space between the false ceiling and the floor above is another potential source of leakage. Partitions generally terminate at the false ceiling. A barrier is needed in the space above the ceiling that fits around the ducts and provides an acoustic baffle in the ceiling space [Section 11.4].

As an additional precaution, a masking noise may be used to blur the intelligibility of speech at a distance and protect the privacy between offices. When several people are working in the same office, it reduces the disturbance caused by a conversation to the people farther away who are not a party to it.

Most people are more disturbed by complete silence than by a low level of neutral noise. Thus buildings with sealed windows and well-fitted doors can actually be too quiet. The masking noise, sometimes called a sound perfume, can be supplied by air ducts. Provided the intelligibility of speech is ensured where it is required, the sound silencing need not be complete. Music played quietly often provides a more expensive, and generally less acceptable, alternative.

11.8 Noise and Site Planning

The single-family detached house offers greater acoustic privacy than can be achieved in a town-house or an apartment building. It is not generally economically feasible to provide the standard of sound insulation in an apartment building that is naturally achieved in a detached house.

The same applies to industrial activities. If a part of a manufacturing process is particularly noisy, it may be simplest to place it in a separate building at some distance from other activities. This ensures at least that the noise does not get into the structural steelwork or the air conditioning ducts of the main building and emerge from it in unexpected places.

Noisy activities should, as far as possible, be separated from rooms that need to be quiet by placing corridors and "neutral" rooms between them (Fig. 11.4.4).

A single row of trees does not provide any sound insulation because of the interreflection between the trees; but multiple rows form a useful screen from the traffic noise of a major highway. Closely spaced trees on a strip 50 m wide reduce the noise by up to 6 db.

References

11.1 ANITA B. LAWRENCE: *Architectural Acoustics*. Elsevier, London, 1970. 219 pp.

11.2 P.H. PARKIN, H.R. HUMPHREYS and J.R. COWELL: *Acoustics, Noise, and Buildings*. Fourth Edition. Faber, London, 1979. 297 pp.

11.3 L.L. BERANEK (Ed.): *Noise Reduction*. McGraw-Hill, New York, 1960. 752 pp.

11.4 W.C. HUNTINGTON and R.E. MICKADEIT: *Building Construction*. Fifth Edition. Wiley, New York, 1981. 471 pp.

11.5 BUILDING RESEARCH STATION: *Principles of Modern Building*. Volume 2. H.M. Stationery Office, London, 1961. 189 pp.

Suggestions for Further Reading

References 11.1 and 11.2.

11.6 L.H. SCHAUDINISCHKY: *Sound, Man and Building*. Applied Science, London, 1976. 413 pp.

11.7 L.E. KINSLER and A.R. FREY: *Fundamentals of Acoustics*. Wiley, New York, 1962. 524 pp.

11.8 J.E. MOORE: *Design for Noise Reduction*. Architectural Press, London, 1966. 150 pp.

11.9 CLIFFORD R. BRAGSON: *Noise Pollution—The Uniquiet Crisis*. University of Pennsylvania Press, Philadelphia, 1970. 280 pp.

Chapter 12 Acoustics

The previous chapter dealt with the negative aspect of acoustics, the control of noise. This chapter examines briefly the acoustics of auditoria.

The history of theater design is reviewed, and the classical theory of geometric acoustics and echoes, which was known in the nineteenth century, is examined. To this W.C. Sabine added reverberation time in the year 1900. Since that time we have acquired many empirical data, but acoustic design remains to a large extent a matter of experience and judgment.

Acoustics in lecture rooms and halls for popular music by sound amplification is briefly discussed.

12.1 Auditoria Before the Twentieth Century

Vitruvius' description of Greek acoustics was quoted in Section 1.1. The Greek theater at Epidauros in southern Greece, dating from the fourth century B.C., and the Roman theater of Herodes Atticus in Athens (Fig. 1.1.1), built in the second century A.D., have both been reconditioned for present use, and both have remarkably good acoustics for open-air theaters (Ref. 12.1).

Few Greek theaters survive because many were altered by the Romans, but about forty Roman theaters remain in various countries ranging from France to Turkey and North Africa. The larger Greek theaters accommodated several thousand people, and some Roman theaters could seat tens of thousands. Nothing comparable in size and acoustic quality was built again until the late nineteenth century.

The sound produced by the actors was reinforced near its source by reflection from the *orchestra,* the paved platform that is visible in front of the stage in Fig. 1.1.1 [Section 1.1], and from the *skene,* the wall at the back of the stage, also visible in Fig. 1.1.1. These two reflectors were

Fig. 12.1.1. The Teatro Olimpico, designed by Andrea Palladio in Vicenza in 1580, followed the general arrangement of the Roman theater, although it was much smaller. The seats were steeply banked in elliptical rows. Palladio designed the theater without a roof; the present roof dates from the nineteenth century. Like the traditional Roman theater, it has a long and narrow stage backed by a proscenium with five entrances.

In the Teatro Olimpico, however, passages built as city streets radiate from the five entrances. For perspective these streets narrow and the floor rises as they recede from the stage front. Every spectator can see at least one of these passages. Actors at the rear of any one of the five streets can be heard throughout the theater.

so close to the sound source that the reflected sound appeared to arrive instantaneously with the direct sound. The seats were steeply banked so that every spectator had a direct unobstructed sound path from the stage, the *orchestra,* and the *skene*. If the theater was full, the audience acted as an absorbing surface, and there were no reflecting surfaces behind it. If the theater was partly empty, there could be some reflected sound from the rear seats, but the farthest spectator was then closer to the stage and could thus hear better.

Roman theaters fell into disuse during the Middle Ages. Medieval plays were frequently performed in the marketplace, and there are no reports of buildings especially designed for drama in the Middle Ages.

The theater made a comeback in the Renaissance, although audiences were much smaller. Andrea Palladio's Teatro Olimpico in Vicenza followed the general arrangement of a Roman theater (Fig. 12.1.1). It is the only theater surviving from the sixteenth century. In 1618 Giovanni Battista Aleotti built the Farnese Theater in Parma, which also survives. Here the Roman proscenium remained only as a decoration, and the action took place behind a large opening that could be closed by a curtain, so that changeable scenery painted on movable screens could be introduced. The theater had a roof, and the ceiling now replaced the Roman *skene* as the principal reflector of sound.

Instead of using a semicircular plan, as in Greek and Roman theaters, Aleotti's auditorium is U-shaped, probably following the arrangement of the temporary seating customary for the masques and court plays staged in banqueting halls in the Renaissance. By 1690, in Ludovici Burnacini's Imperial Theater in Vienna, which no longer exists, the U had become a horseshoe; the seats lining it were arranged in boxes rising to the ceiling, and the floor within the horseshoe was filled with further seats. There is no indication that any rules, empirical or otherwise, were used for the acoustical design. However, Baroque theaters had two great advantages over Roman theaters: they were much smaller, and they had a roof that acted as a reflector and restricted external noise. The good and relatively short sightlines ensured good audibility except for people in the rear of the upper rows, where visibility was also very poor.

This remained the standard layout for theaters and opera houses until the twentieth century. Some of the seats were placed on a gently sloping floor, below the level of the stage, the rest were arranged in multistory horseshoe galleries, as in the Teatro San Carlo in Naples, built in 1737 (Fig. 12.1.2), the Teatro alla Scala in Milan (1778), the Royal Opera House at Covent Garden in London (1858), the Staatsoper in Vienna (1869), the Théâtre National de l'Opéra in Paris (1875), and the Metropolitan Opera House in New York (1883).

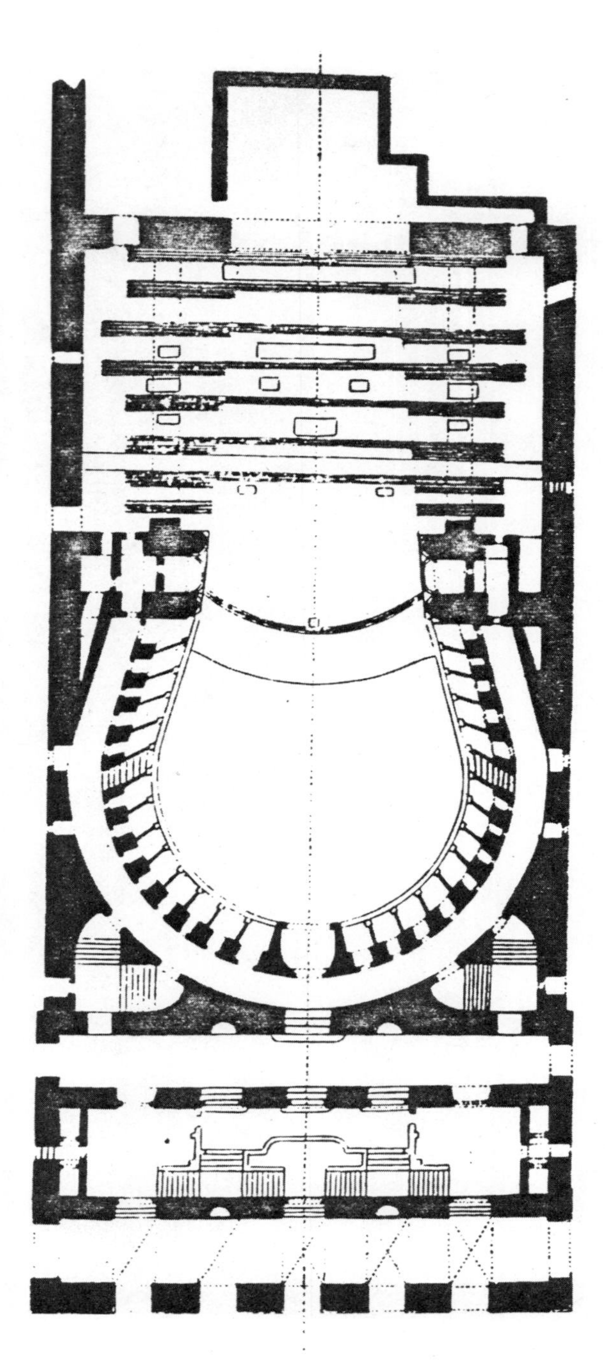

Fig. 12.1.2. Plan of the Teatro San Carlo in Naples, the most important opera house in the eighteenth century. The musicians are seated in a pit below the stage, separate from the singers, and the acoustical balance of the music produced by them now becomes a major problem.

Fig. 12.1.3. Auditorium of the Town Hall in Sydney, built in 1889. It has a seating capacity of 1950, and it is a typical municipal auditorium of the late nineteenth century: a rectangular room with a horizontal ceiling and a gallery on three sides. It has a reverberation time of 3.0 seconds when it is empty and of 2.1 seconds when every seat is taken. Since the completion of the Concert Hall of the Sydney Opera House in 1973 it is used only occasionally for music. (*Photograph by courtesy of the Town Clerk of the City of Sydney.*)

Although the shape of the nineteenth-century theater or opera house was not primarily due to its acoustical requirements, it worked well. Some of the best existing opera houses were built during the eighteenth and nineteenth centuries. However, there is a law of the survival of the fittest for buildings. Good auditoria are carefully looked after, and exactly rebuilt if they are destroyed by a war. Those less satisfactory are demolished or altered when the opportunity arises. By the middle of the nineteenth century the geometry of acoustics [Section 12.2] was well understood, as were the effects of reflecting and absorbing surfaces. It was entirely feasible to correct minor acoustic faults during redecoration.

This gradual adaptation also occurred in concert halls and in churches. The Gothic cathedrals provide superb auditoria for organ music today, but the modern organ developed only in the sixteenth and seventeenth centuries, long after the end of the Gothic era. The oldest surviving organ, a small instrument in Syon in Switzerland, was built in the late fourteenth century. It seems likely that the organ was designed to fit the acoustics of the existing cathedrals, and not vice versa.

Music before the nineteenth century tended to conform to the acoustical qualities of the auditoria, rather than vice versa. Gregorian chant sounds best in an auditorium, such as a medieval cathedral, with a long reverberation period, which may be as high as ten seconds [Section 12.3].

The Reformation took over old Catholic churches, many of which had long reverberation times, but in other newly built Lutheran churches the congregation occupied galleries as well as the main floor, and it is possible that this may have affected the music of Protestant Germany. Leo L. Beranek (Ref. 12.2, p. 46) estimated the reverberation time of the Thomaskirche in Leipzig, for whose congregation Johann Sebastian Bach composed most of his religious music in the seventeenth century, as 1.6 to 2 seconds.

The fifteenth century did not produce so marked a change in the style of music as the Renaissance did in architecture, painting, and sculpture. The change occurred in the seventeenth century with composers such as Vivaldi and Corelli in Italy and Bach and Handel in Germany. This is the period of Baroque architecture in Italy and in Germany. In its own time Baroque secular music was normally performed in small rooms with hard, reflecting surfaces, such as ballrooms, small theaters, or large living rooms. Beranek estimated that the reverberation period of these rooms would have been below 1.5 seconds if the room were filled with people. The music therefore had high definition and, because of the many nearby reflecting surfaces, it sounded intimate. Again, it may be argued that the composer adapted himself to the existing architecture, and not vice versa.

The ballrooms of the aristocracy provided the pattern for the eighteenth-century assembly rooms in which the upper middle classes gathered for social functions, the these became the first public concert halls.

In the mid-nineteenth century the new centers of population created by the Industrial Revolution erected public buildings befitting their new status. Thus St. George's Hall was built in Liverpool, England, as a concert hall in 1850–54, and the Free Trade Hall in Manchester, England, followed in 1853–54. Many of the new town halls included a large room that could be used for an auditorium (Fig. 12.1.3). As the century progressed and cities tried to outdo one another in their public buildings, the capacity of these concert halls increased, and so did their reverberation time.

Halls were also built specifically for concerts in the older centers of population between 1850 and 1914. Most were enlarged versions of the assembly room: a rectangular room with a gallery along three sides. They include some of the best concert halls in existence: the Grosser Musikvereinsaal in Vienna (1870), the Concertgebouw in Amsterdam (1888), and the Grosser Tonhallesaal in Zurich (1895).

Before the nineteenth century music was mostly composed to fit the place in which it was to be performed, rather than vice versa; but in the nineteenth century a composer might have his music played in a number of different concert halls and theaters which, even if they had the same general shape, had different reverberation times and reflective surfaces in different locations.

Prior to the nineteenth century the music performed consisted mainly of contemporary and recent works that required the same acoustic qualities. A concert program in the late nineteenth century might contain music that required widely different concert-hall acoustics, for example, a Mozart piano concerto and a Brahms symphony.

The nineteenth century was a great age for music. The existing instruments were perfected and some new ones invented (Ref. 12.3). Music played an important part in the entertainment of the middle classes, and playing the piano had by the turn of the century become as much part of a young lady's education as reading and writing. Famous orchestral music would be performed in the home as a *Klavierauszug,* a piano transcription, which fulfilled roughly the same function as the modern tape or phonograph record.

The architects of the late nineteenth century concert halls were therefore dealing with an informed public. A great deal was written on architectural acoustics, but apart from the geometry of sightlines and reflections [Section 12.2] it was mostly guesswork. Charles Garnier [Section 1.5] in the 1860s and 1870s found it of little help in the design of the Paris Opéra, one of the most prestigious and successful auditoria.

12.2 Reflection and Diffraction of Sound, and Echoes

Sound, like light, is reflected by plane and curved mirrors (Fig. 12.2.1), and both conform to the same laws. A mirror for light needs a highly polished surface because the wavelength of light is of the order of 0.001 mm. The wavelength of sound is much longer, so that a sound mirror can be formed by any correctly shaped surface with a low coefficient of sound absorption, such as plaster or concrete.

When an obstacle is placed in the way of light rays, it throws a shadow. Similarly, an acoustic screen throws an acoustic shadow (Fig. 12.2.2), but "diffracted" sound rays intrude at the edges. This bending of sound is due to its longer wavelength. The wavelength of light rays ranges from 0.000 4 mm to 0.000 75 mm; the wavelength of sounds generated for music ranges from 40 mm to 12 m. The wavelength of the deeper sounds may be much greater than the size of an obstacle, and diffraction is most marked for the low tones with a longer wavelength. Thus a column produces distortion of sound, rather than an absolute sound shadow. We noted in Section 11.3 that all sound inside a room is reflected to a greater or smaller extent by the room's surfaces, and the ear perceives this reverberant sound as part of the original sound. Even when the sound starts to decay, a reflection may blend into the original sound. However, the brain tends to perceive speech clearly reflected by a sound mirror and received more than 35 ms after the original sound as separate, that is, as an echo (Fig. 12.2.3).

The modern auditorium differs from the Ancient Greek theater by having a roof that reflects sound. It is dependent on reflection from that roof because its seats are sloped more gently, so that the direct sound toward the rear is absorbed to an increasing extent by the people in front.

We noted in Section 11.3 that different materials absorb sound to a different extent. Acoustic per-

Fig. 12.2.1. Focusing sound in a chamber with an ellipsoidal ceiling, from *Musurgia Universalis*, by Athanasius Kircher published in Rome in 1650.

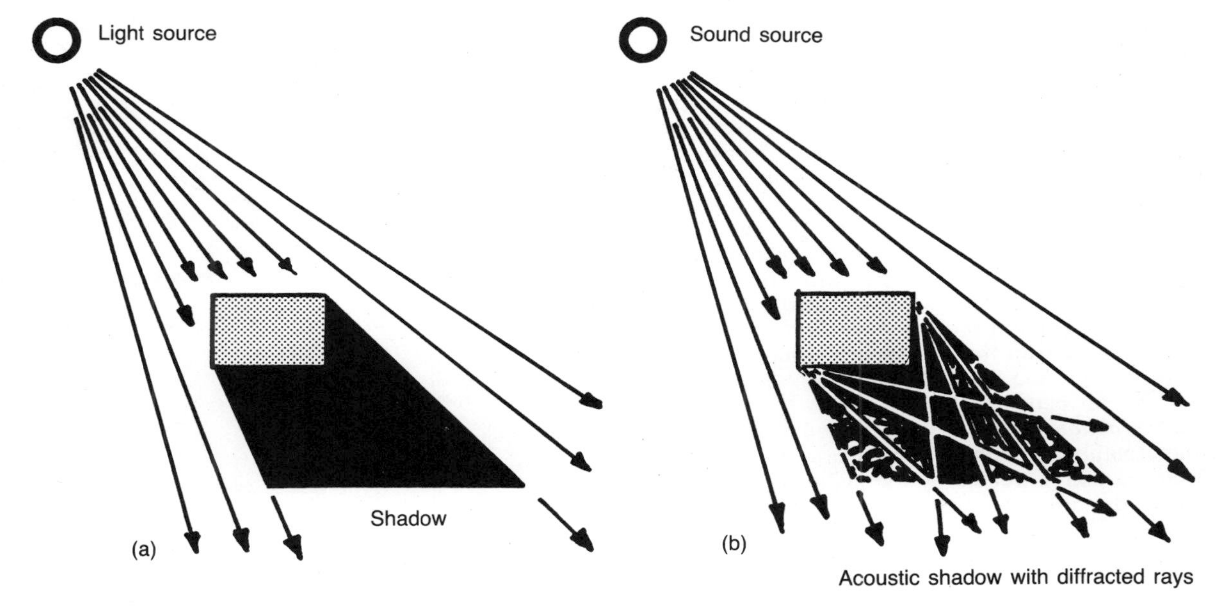

Fig. 12.2.2. If a screen is placed in front of a light source L, it throws a shadow (a). However, sound rays bend or "diffract" around the obstacle because the sound waves are much longer than the light waves (b). The long-wave low notes of sound are particularly diffracted. Thus a short screen or a column produces a distortion of sound rather than an absolute shadow.

formance can be greatly improved by reinforcing sound with reflecting surfaces in some parts of the auditorium, and absorbing it in other parts where reflection is undesirable. The design can be aided by drawing the sound-mirror reflection for the plan and for cross sections and longitudinal sections, but it should be borne in mind that this is a problem in three-dimensional geometry. Generally speaking, reflecting surfaces are needed near the sound source, and absorbing surfaces at the rear to protect people in the middle of the hall from echoes.

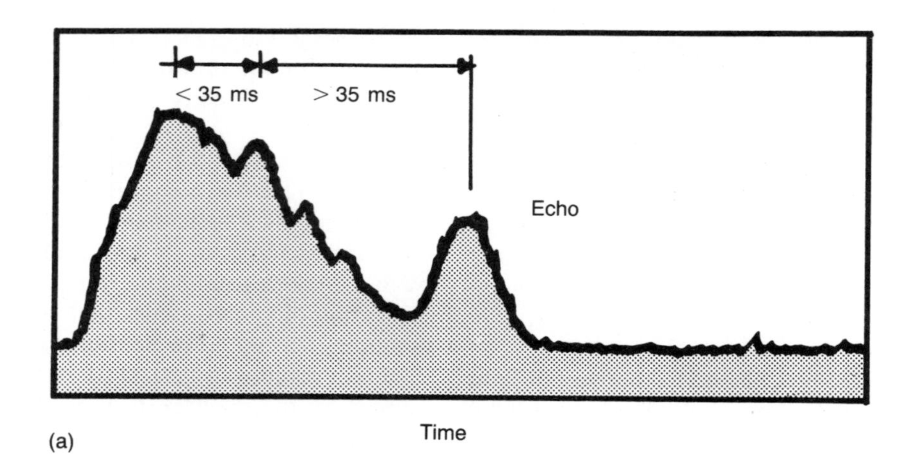

(a)

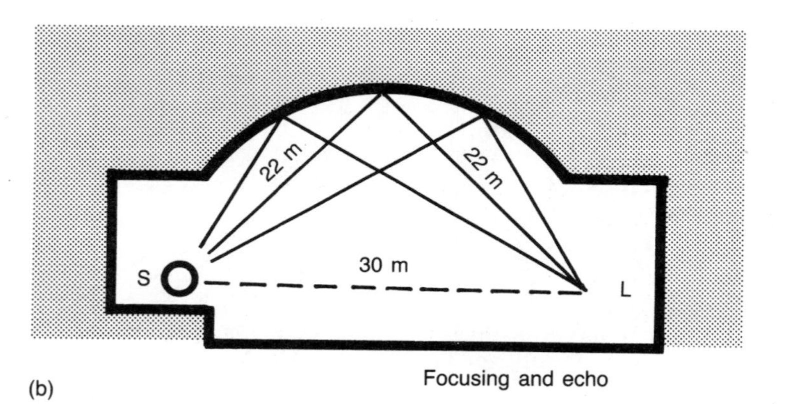

(b) Focusing and echo

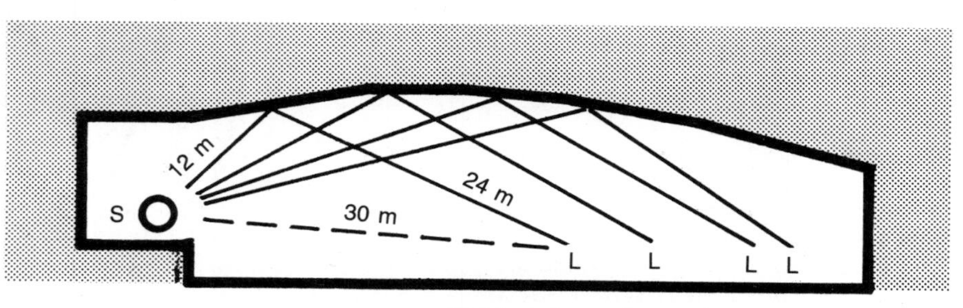

(c) Distributed sound, no echo

Fig. 12.2.3. (a) When a sound starts to decay, a reflected sound blends into the original sound and is perceived as part of the original sound. However, the brain tends to perceive speech clearly reflected by a sound mirror and received more than 35 milliseconds (ms) after the original sound as an echo, that is, as a separate sound. In the case of music, reflected sound blends with the original sound more easily; an echo would result only from a distinct reflection more than 70 ms after the original sound.
(b) While the velocity of light may be treated as infinite for all architectural problems, the much slower velocity of sound has an influence on acoustic design. In 35 ms sound travels [Section 11.2] $343 \times 0.035 = 12$ m, and in 70 ms it travels 24 m. Thus if the sound mirror produces a distinct reflected sound path that for speech is more than 12 m longer than the original sound path (24 m for music), the ear perceives an echo.
(c) A *gradual* change in the reflected sound and a sound path shorter than 12 m ensure that the reflected sound is treated by the auditory senses as part of the original sound.

12.3 Reverberation Time

The concept of reverberation time was developed in 1895 by Wallace Clement Sabine, professor of mathematics and natural philosophy at Harvard University, during work on acoustic improvements to a lecture room at the Fogg Museum at Harvard. He then employed it in the design of the Boston Symphony Hall, whose acoustic consultant he became in 1898. The hall was immediately recognized as an acoustic masterpiece. L.L. Beranek (Ref. 12.2) asked 23 of the world's most famous conductors in the 1950s to rate 54 of the world's best-known auditoria in order of preference. With one exception, all rated Boston Symphony Hall as America's best, and one of the three best in the world. However, the architectural style of the hall is conventional: It is a rectangular room with a horizontal, coffered ceiling, and it has galleries around three sides, in the nineteenth-century tradition.

Sabine defined reverberation time as the time required for a sound to decay 60 decibels, that is, to one millionth of its original sound intensity (Table 11.1). This is roughly the proportion between the sound produced by an opera singer, or an experienced public speaker without a microphone, and the background noise in a quiet auditorium.

Reverberation is determined from the graphic output of a sound level recorder, that is, a recording sound level meter [Section 11.2]. A starting pistol is a suitable sound source.

Corrections can be made to the hall by adding or removing reflective diffuse or absorbent material; some halls are designed so that an adjustment of this kind can be made easily.

Sabine determined the experimental relation between the volume and absorption of a room and its reverberation time at Harvard University. In metric units this is

$$t = \frac{0.16V}{S} \tag{12.1}$$

where t is the reverberation time in seconds;
V is the volume of the room in cubic meters;
S is the total absorption of the room in sabins, as calculated in Eq. (11.3), Section 11.3.

This is called Sabine's Law. There is general agreement on the validity of the law and its importance to acoustic design, but the desirable reverberation time for different types of use remains a matter of opinion.

Recording studios need a very low reverberation time, preferably less than 0.5 seconds; lecture rooms, conference rooms, and rooms for public debate also should have a low reverberation time, about 1.0 to 1.2 seconds; auditoria for music need a long time [Section 12.6]. For concert halls there is a conflict between the requirements for baroque and modern music on one hand, for which a reverberation time as low as 1.4 seconds might be suitable, and for choral works, organ music, and the large "romantic" symphonies of the late nineteenth century, which need 2 seconds or more. In opera houses there is a similar conflict between the requirements of an opera by Mozart or Donizetti and one by Wagner.

Example 12.1 *The calculation of the reverberation time of a realistic room is quite lengthy because of the large number of different materials and contents that need to be considered. The following example has been greatly simplified.*

A square room measures 20 m by 20 m. It has stepped seats and an average height of 5 m. The room has concrete walls and a concrete ceiling, and a concrete floor covered with 22-mm cork tiles. There are no openings, except four thick wooden doors. There are 500 upholstered wooden seats in the room. Calculate the reverberation time at 500 Hz, if 300 of the seats are occupied.

Solution *Neglecting the doors and the steps in the floor, we find that the absorption of the room at 500 Hz, from Eq. (11.3) and Tables 11.3 and 11.4, is*

Floor: $20\,m \times 20\,m \times 0.20$	*80 sabin*
Ceiling: $20\,m \times 20\,m \times 0.02$	*8 sabin*
Walls: $20\,m \times 5\,m \times 4 \times 0.02$	*8 sabin*
Air in room:	*0 sabin*
200 seats: 200×0.15	*30 sabin*
300 people in seats: 300×0.40	*120 sabin*
Total	*246 sabin*

The volume of the room is $20\ m \times 20\ m \times 5\ m = 2000\,m^3$.

From Eq. (12.1) the reverberation time is

$$t = \frac{0.16V}{S} = \frac{0.16 \times 2000}{246}$$
$$= 1.3 \text{ seconds at } 500 \text{ Hz}$$

12.4 Rooms for Speech

The principal requirement of a schoolroom, a university or adult education lecture room, a committee room, a conference chamber, a debating chamber, or a drama theater is the same: adequate intelligibility of speech. This requires sufficient loudness of the speaker's voice, the absence of echoes, a low noise level, and a low reverberation time. If the reverberation time is too low, the room might seem unpleasantly dead, but this does not interfere with intelligibility. Sound paths will be shortest, and therefore intelligibility highest, if the room is made as wide as it is deep.

(a)

(b)

The simplest problem is the lecture room (Fig. 12.4.1), because the speaker occupies a position that is more or less fixed. Direct sound is the most useful, and it is therefore important that everyone should be able to see the speaker comfortably. Sound that passes between or within grazing distance of the heads of an audience is reduced by absorption. It is best to raise each row of seats by 75 to 100 mm above the preceding row. If that is not possible, the lecturer should be placed on a platform. In large auditoria the falling off of direct sound intensity in accordance with the inverse square law must be compensated by reflected sound, but without causing echoes [Section 12.2].

People are generally more absorptive than empty seats (Table 11.4). In concert halls and important debating chambers it is economically feasible to use seats that have as high an absorption when empty, as the same seat occupied by a person. However, these seats are expensive and cannot be justified for a lecture room. It is advisable to design the room for an audience that fills only two thirds of the seats, and additionally consider the acoustics for a room half full and entirely full, to determine whether it falls within acceptable limits.

Fig. 12.4.1. (a) Inside view of a lecture theater at the University of Sydney, with a seating capacity of 250 persons. The theater is almost as wide as it is long, and each row of seats is raised 300 mm above the preceding row. The room has excellent acoustics.
(b) External view of the same lecture theater. "Form follows function," but that does not assist the esthetics of this building, although the banked seats create a pleasant space for a coffee bar below the rear seats. (*Photographs by Mr. Frank Morand.*)

A lecture room without electronic sound reinforcement should have a reverberation time of 1.0 to 1.2 seconds. If the reverberation time falls much below 0.8 seconds, the room may sound "dead," that is, it may be unpleasant to speak in it or to listen to the speaker. With a reverberation time above 1.5 seconds the audience may have difficulty understanding the speaker.

Having determined the amount of absorption required to achieve the desired results, we subtract the absorption due to the audience and then determine how to distribute the rest in the room. It is best to back the speaker with a reflective surface and to use the absorptive material at the back of the room and in front of any gallery. These may otherwise produce echoes, particularly if they are curved inwards. If more absorptive material is required, it is best to place it in the parts of the ceiling near the walls, so that the center of the ceiling remains reflective, and in the upper parts of the wall.

Important debating chambers (see Fig. 12.7.2) are now generally provided with electrical sound reinforcement, which simplifies the problem. Where this is not economically feasible, it is advisable to keep the chamber as small as possible by placing seats close together and by limiting public access. Each person should be able to see every other person who is permitted to speak to ensure a direct sound path. To give reflections, the ceiling should be kept as low as the dignity of the architecture will permit.

Whereas the speakers should all be able to hear one another, the public admitted to the

Fig. 12.4.2. York Theater at the Seymour Center for the Performing Arts, University of Sydney, with a seating capacity of 780. All the walls are of thick timber, and are backed by concrete. (*Photograph by Mr. Frank Morand.*)

chamber should be able to hear but not be heard. Therefore the public gallery should have a carpet and an absorbent ceiling and walls.

Theaters present a more difficult problem, particularly as sound reinforcement, which may be entirely acceptable for large lecture rooms and debating chambers, is rarely used in theaters. This is partly because it is difficult to provide sound reinforcement, even with radio microphones, when people move about and perhaps perform acrobatics; furthermore the concept of sound reinforcement is anathema to many actors and producers, who consider themselves heirs to a long tradition of performing without it.

The thrust stage, which has the audience on three sides (Fig. 12.4.2), and the theater-in-the-round avoid some of the acoustic problems of the conventional stage. However, as the human voice is directional, the people to the sides (and the back) will lose some of the sound, particularly at the higher frequencies that are important for intelligibility. It is desirable to have reflecting surfaces on all the walls to overcome this problem. Since these theaters are generally small, no member of the audience is far from the stage.

The conventional theater, whose stage is behind a proscenium arch, does not encounter that problem. However, the fly tower that contains the scenery may have a volume as great as the auditorium. When it is empty, it adds to the reverberation, but when it is full of scenery its absorption may cause the stage to sound dead. This may vary from one performance to the next.

The proscenium arch traditionally contains reflecting surfaces at the top and the sides, and the ceiling reflects sound to the rear seats. The auditorium should be carpeted to reduce audience noise, and this provides necessary absorption. As in lecture rooms, absorbents are needed at the rear of the auditorium and in front of the balcony, if any, to prevent echoes or blurred sound. Absorbent or discontinuous surfaces are also sometimes needed on the side walls near the stage to stop flutter echoes and increase diffusion of sound for the front seats. The boxes to the side of the stage served this purpose in the traditional nineteenth-century theater.

Drama theaters should not be too large, to ensure that no seat is too far from the stage. This is necessary for speech intelligibility; it is also possible because drama costs less to produce than opera.

An opera house needs a larger audience because of the high cost of production. Fortunately, few people expect to be able to hear every word of an unknown opera libretto. The desirable reverberation time ranges from 1.2 seconds for Mozart to 1.7 seconds for Wagner.

12.5 Electro-Acoustic Aids to Speech

Today only actors learn how to project their voices so that they can be heard by a large audience. A trained actor can speak distinctly with a voice about 9 db louder than the average person, without apparent effort. In the nineteenth century this skill was taught more widely, because there was no other way in which a politician or a preacher, or even a scientist, could address a large audience.

In designing an auditorium we should assume that people probably will not have the ability to address an audience larger than 200 persons, and thus an electro-acoustic system should be provided.

A room equipped with loudspeakers does not need the same careful design as an auditorium without them. Echoes are less of a problem, as loudspeakers can be placed close enough to each person. Reverberation time is less critical, and a certain amount of background noise is acceptable. It is not necessary, for acoustic reasons, to place the lecturer on a platform, or alternatively to provide ascending seats, since the loudspeakers can be placed well above the heads of the audience; however, for visual reasons it may still be desirable that everyone be able to see the lecturer.

The most effective microphone is one that hangs from a band around the lecturer's neck. This ensures that his mouth is always equidistant from the microphone, which travels with him if he moves around, for example to point to some part of a picture on a projection screen. If necessary this microphone can be connected to a radio transmitter so that no electrical cord is needed.

For an auditorium of moderate size it is sufficient to use a single loudspeaker array, or a pair of loudspeaker arrays on each side of the room. These should be near the ceiling, or about 6 m above the head of the lecturer if the room is tall. They can be operated at as high a level of sound as is acceptable to the lecturer and to people in the front row. The loudspeakers virtually replace the voice of the lecturer, but like his voice their sound is reduced in accordance with the inverse-square law [Section 11.2].

The alternative system, which can be used for auditoria of any size, employs a larger number of loudspeakers, each reaching only a few people and operating at a much lower level of sound. The electrical current travels through the loudspeaker cable about a million times faster than the sound travels through the air, and the sound from the nearest loudspeaker would reach the listener first, followed by sound from adjacent loudspeakers, and eventually the direct sound from the lecturer. This would not merely detract attention from the lecturer but could also cause echoes [Section 12.2]. The loudspeakers are therefore supplied with delayed signals so that the nearest loudspeaker broadcasts its sound about 20 milliseconds after the direct airborne sound from the lecturer arrives. The listener is therefore conscious of the sound coming from the lecturer, which is reinforced but not dominated by the much stronger sound from the loudspeaker. There are now reliable delay units that can provide any number of delayed signals at any required delay times.

In deciding whether to install loudspeakers one should bear in mind that a sound-reinforcing system functions satisfactorily only if it has an operator to adjust the controls, as the sound input varies from speaker to speaker.

12.6 Rooms for Music

The musical quality of a performance, given good performers, is to an appreciable extent determined by the acoustics of the room, and its appropriateness for that type of music.

L.L. Beranek defined the essential qualities (Ref. 12.2, pp. 34–43):

1. *Fullness of tone* depends mainly on the reverberation time of the room and on the ratio of the sound received directly from the musicians to that reflected by the room. The greater the reverberation time, the fuller the tone. The greater the ratio of the loudness of the reverberant sound to that of directly received sound, the fuller the tone. Bach's Organ Toccata in D Minor needs fullness of tone.
2. *Clarity* or *definition* requires the opposite qualities in a room. The listener will be able to hear all the individual sounds distinctly and clearly, provided that the musicians play them distinctly and clearly, if the direct sound and the sound reflected by overhead reflectors above the orchestra is louder than the reverberant sound from other surfaces, and if the reverberation time is low. A Mozart piano concerto needs clarity.
3. *Intimacy* is determined by the time difference between the arrival of the direct sound and the reflected sound. This is much lower in a small hall. Most chamber music needs intimacy.
4. *Timbre* is the quality of sound that distinguishes one instrument from another, and one voice from another. *Tone color* is the effect produced by a combination of timbres. If the absorbing surfaces of the hall absorb the low and middle frequencies to a greater extent than the high frequencies, the full orchestra may appear deficient in cellos, double basses, and bassoons. The reflecting surfaces may project the sound of certain instruments toward certain parts of the hall. For example, the cellos and double basses are usually seated on the right-hand side of the stage, and the sound from them, because of differential reflection, may be so strong on the left side of the hall that it destroys the balance of the orchestra, while it is missing in other parts. These are faults in the design of the hall that can usually be corrected.
5. *Ensemble* is the ability of the musicians to hear one another, and of the conductor to hear them in the correct balance. It is evidently difficult for a bassoon player to play his part properly if he can hear nothing but the trombones behind and the cellos to his left. In a concert hall it is generally possible to provide special reflectors to ensure ensemble, but this is more difficult in the orchestra pit of an opera house because of space limitations, and because of the added complication of having the singers at a higher level.
6. *Dynamic range* is the spread between the faintest pianissimo and the loudest fortissimo. It is limited at one end by the noise that penetrates into the auditorium from the outside (which ideally should be zero) and by the background noise generated by breathing and movement of the audience. At the other end of the range, a fortissimo should be loud but not so loud as to be painful. Verdi's *Requiem* or Walton's *Belshazzar's Feast* can sound painful in a hall that is too small and very "live."
7. *Liveness* and *warmth* are subjective impressions of reverberation. Liveness is related to the reverberation times at the middle and high frequencies above 500 Hz. Warmth is produced by reverberation at low frequencies. Both require high reverberation times, but a hall can be "live" while lacking in "warmth," and vice versa.

It is possible, although sometimes expensive in the center of a busy city, to design an auditorium that is free from intrusion by outside noise [Section 11.4]; to eliminate echoes [Section 12.2]; and to achieve the correct ensemble, tone, color, and dynamic range.

The other qualities are to some extent in conflict with one another. A room can be designed for a specified reverberation time if allowance is made for adjustment after completion; however, it is impossible to design a hall in which perfect performances can be given both to a Haydn string quartet and to Handel's *Messiah*.

This conflict can be resolved by using different rooms for chamber music, and for symphonic, choral, and organ music. Only a few players are needed for chamber music, so that it can be performed economically in a small auditorium that has the necessary intimacy and definition. Many existing multipurpose rooms are suitable for chamber music, or a suitable space may be available in the building for the main concert hall.

(a)

The main auditorium need then be designed only for symphonic and choral music. It may or may not be provided with an organ. A reverberation time ranging from 1.7 seconds to 2.2 seconds is suitable for a modern concert hall. The lower time gives more definition, and the higher time gives greater fullness of tone. If the authority operating the hall is prepared to accept a seating capacity of 1500, a reverberation time of 2.2 seconds can be achieved with comfortable seating. This is an acceptable concert audience in many central European cities where music is heavily subsidized.

In America and in Australia halls are designed for larger audiences because subsidies are lower and more tickets must be sold (Fig. 12.6.1). When the size of the hall is increased, echoes become a problem, and when absorbent materials are applied to eliminate echoes, it becomes difficult to reach the required reverberation time. One possible solution is the closer spacing of seats, which is accepted in some of the famous older halls but is unpopular in new auditoria; another is the use of ''clouds'' *suspended below* the ceiling that act as reflecting surfaces and reduce the length of the sound paths, and thus the risk of echoes.

Fig. 12.6.1. (a) Interior of the Concert Hall in the Sydney Opera House, which has 2690 seats and a reverberation time of 2.1 seconds. The shape was partly determined by the constraints imposed by the concrete exterior, which was built before the interior had been designed.
(b) Exterior view of the Concert Hall of the Sydney Opera House. Form does not follow function. (*Photographs by courtesy of the Government Architect of New South Wales.*)
(c) Relation of Concert Hall to roof structure.

(b)

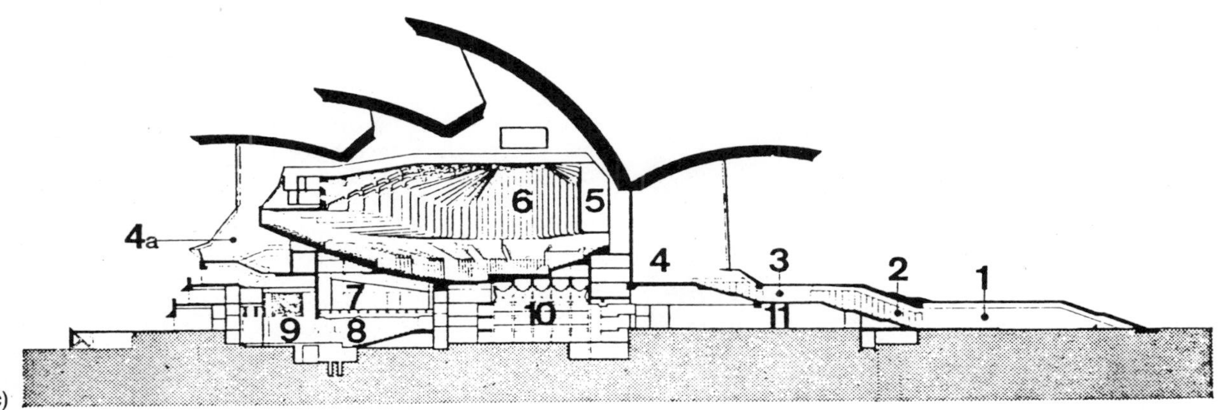

(c)

The current thinking of acoustic specialists on the design of auditoria may be found in Refs. 11.2, 12.2, 12.4, and 12.5. References 12.6, 12.7, and 12.8 are older books by distinguished acoustic consultants.

12.7 Model Analysis and Testing of Auditoria

Because of the complexity of the design of large concert halls and opera houses it is not possible to predict their acoustical performance at low and middle frequencies with any accuracy from theory. Models can be helpful as an aid to design. The laws of model analysis (Refs. 12.9 and 12.10) require that the model be exactly scaled, that is, that every part of the model be reduced in the same ratio. The time must be changed in the same ratio, and scale adjustments must be made to the absorbing materials.

If the size of the model is one tenth that of the auditorium, we have quite a large model, which is expensive to build and requires an appreciable space for testing. Anything smaller, however, would create problems with the time scale (Figs. 12.7.1 and 12.7.2).

Fig. 12.7.1. Acoustic model of the Concert Hall of the Sydney Opera House to a scale of 1:10. The model is about 7 m long. It was tested for "early decay time" and "inversion index," two concert hall criteria developed by V.L. Jordan, the acoustic consultant, using an electric spark as a sound source and microphones connected to a tape recorder (Ref. 12.4). (*Photograph by courtesy of the Government Architect of New South Wales.*)

Fig. 12.7.2. 1:10 scale model of the Chamber for the House of Representatives of the new Parliament House in Canberra, Australia, set up for testing. The building is scheduled for completion in 1988. *(Photograph by courtesy of the acoustic consultants, Louis Challis and Associates.)*

Even a model as large as this has several limitations. In theory it should be possible to record music on the speeded-up time scale, play it in the model, record the result, slow down the time scale, and find out how the auditorium sounds. In practice it is difficult to get unanimity of opinion from listeners, so that only simple sounds were used in the model of the larger auditorium in the Sydney Opera House (Fig. 12.7.1), one of the most elaborate ever built for a concert hall.

When an important auditorium is completed, it is always tested for reverberation time and for echoes. Test concerts are more difficult to evaluate, because the response is subjective. Each person can be given a questionnaire and asked to give his or her opinion on such matters as fullness of tone and definition, and also to observe any echoes or other disturbing aspects. It may be better to rely on two or three dozen trained observers, such as musicians, musical critics, and acousticians who are placed in various parts of the hall and who change seats from time to time. Adjustments are then made to the auditorium. If many faults are found, these can appreciably add to the cost.

References

12.1 F. CANAC: *L'Acoustique des Théâtres Antiques*. Centre National de la Recherche Scientifique, Paris, 1967. 181 pp.

12.2 L.L. BERANEK: *Music, Acoustics, and Architecture*. Wiley, New York, 1962. 586 pp.

12.3 JOHN BACKUS: *The Acoustical Foundations of Music*. John Murray, London, 1970. 312 pp.

12.4 VILHELM LASSEN JORDAN: *Acoustical Design of Concert Halls and Theaters*. Applied Science, London, 1980. 223 pp.

12.5 R. MACKENZIE (Ed.): *Auditorium Acoustics*. Applied Science, London, 1975. 231 pp.

12.6 FRITZ INGERSLEV: *Acoustics in Modern Building Practice*. Architectural Press, London, 1952. 290 pp.

12.7 VERN O. KNUDSEN and CYRIL M. HARRIS: *Acoustical Designing in Architecture*. Wiley, New York, 1960. 457 pp.

12.8 HOPE BAGENAL and ALEX WOOD: *Planning for Good Acoustics*. Methuen, London, 1931. 415 pp.

12.9 H.J. COWAN, J.S. GERO, G.D. DING, and R.W. MUNCEY: *Models in Architecture*. Elsevier, London, 1968. pp. 154–59.

12.10 B. DAY: Acoustic scale modelling materials, in Ref. 12.5, pp. 87–99.

Suggestions for Further Reading

References 11.2, 12.2, 12.3, and 12.4.

Chapter 13 People, Space, and Communications

Anthropometry is used to determine the dimensions of buildings and of furniture that fit the great majority of the adult population. Simple precautions that help to avoid accidents in the home are listed. Wheelchair access can be provided in most new buildings at relatively small cost if it is considered part of the design process.

Corridors, stairs, and escalators should be sufficiently wide to ensure that pedestrian traffic flows smoothly, even in the event of panic. The assessment of the traffic capacity of elevators is more complex and is determined from empirical data.

The equipment for elevators, communications, and security is briefly described.

The final sections discuss the arrangement of the various building services horizontally and vertically.

13.1 Measurement and the Human Body

The common measurements of length from the most ancient times until the nineteenth century were based on the size of the human body, notably the foot, about 300 mm, and the cubit, equal to the length of the forearm, about 500 mm. However, these were not fixed measures, except briefly during the Roman Empire. The length of the foot ranged from 295 mm to 350 mm, a variation of 18%.

During the Middle Ages it was not uncommon for foot measures of different length to be used in the same cathedral (Ref. 13.1). By the beginning of the nineteenth century the foot had become nationally standardized, but it differed slightly from country to country.

An international standard had been discussed since the seventeenth century, and in 1793 the new republican government of France invented the meter, designed as one ten-millionth of the distance from the north pole to the equator of the earth's meridian passing through Paris. It has since been redefined in terms of the wavelength of light at a particular frequency. The Napoleonic conquest spread the metric system to most of Europe.

The metric system made calculation much easier because it used decimal fractions, whereas the inch is traditionally divided into halves, quarters etc., up to sixty-fourths. This is a greater obstacle to computation than the fact that there are 12 inches to the foot.

It could be argued that no dimension of the human body has any simple metric equivalent. However, we cannot perform any calculations with a measurement such as 5 ft $7\frac{1}{2}$ in. until we have converted it to 5.625 ft or 67.5 in. A decimal system is a necessary prerequisite for any statistical analysis.

13.2 Anthropometry and Its Effect on the Size of Spaces

The size of the human body varies with time and place. Many of the surviving suits of medieval armor today fit only relatively small people, although one would assume that most knights were tall for their time. People born and brought up in northern Europe, America, and Australia tend to be a little taller than people from southern Europe and East Asia. However, in Japan there has been a marked increase in the height of people during the last 25 years. Since most applications of anthropometric measurements to architecture were made in Europe or America, including the data in Table 13.1 (Ref. 13.2, p. 3), they may need adjustment when they are applied to a population different from that measured.

These figures show that 2.0 m is an adequate height for most doors and that 0.8 is an adequate width if only one person at a time is required to pass through it. 1.2 m is needed for doors and corridors in which people need to pass one another, as they expect to be able to do so without touching. For continuous traffic, 1.6 m is needed for corridors if doors open into the rooms, and 2.4 m if the doors on both sides open into the corridor, so that a person can pass unimpeded between doors open on both sides.

The design of furniture and shelves is more appropriately determined by experiments with typical persons (Figs. 13.2.1 and 13.2.2).

Drawings showing the appropriate sizes for doors and windows; for bathtubs, toilets, beds, chairs, and tables; for corridors, bathrooms, kitchens and bedrooms, and so on, have been available since the late nineteenth century, and they are an essential aid to architectural design (for example, Refs. 13.4, 13.5, and 13.6).

Table 13.1

Principal Dimensions of the Human Body in Millimeters (Without Shoes)

	Men			Women		
Dimension	97½% are more than	Average	97½% are less than	97½% are more than	Average	97½% are less than
Overall height	1580	1720	1860	1470	1610	1750
Height of eyes, standing	1470	1610	1750	1360	1500	1640
Shoulder height, standing	1290	1420	1550	1180	1310	1440
Breadth of shoulders	400	450	500	360	410	460
Span of arms	1550	1750	1950	1350	1550	1750
Elbows above floor, standing	980	1080	1170	890	990	1090
Elbow height above seat	190	240	290	205	235	270

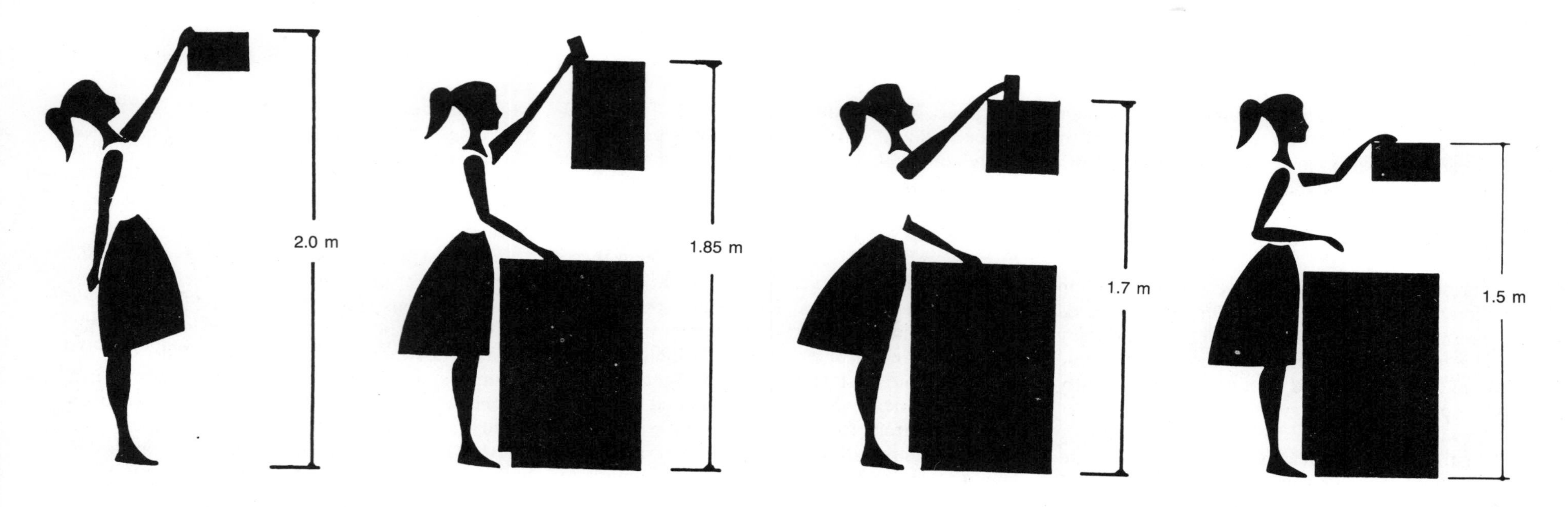

Fig. 13.2.1. Appropriate height for kitchen fittings for a woman of average height, that is, 1.61 m (Ref. 13.2, p. 14b).

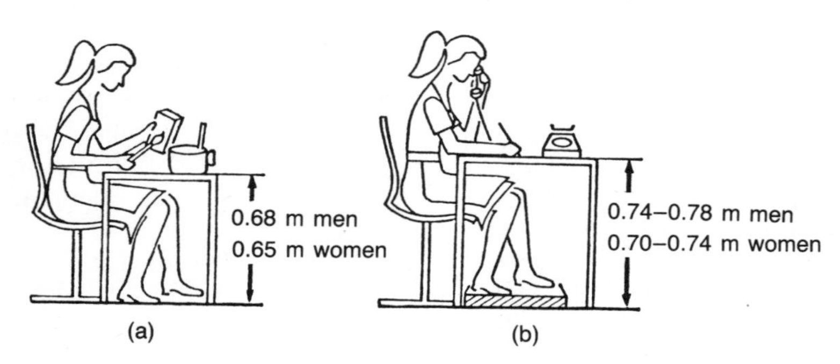

Fig. 13.2.2. Appropriate height for a desk (a) for a person performing light manual work, such as typing or assembling electronic equipment, and (b) for a person reading or writing (Ref. 13.3, p. 74).

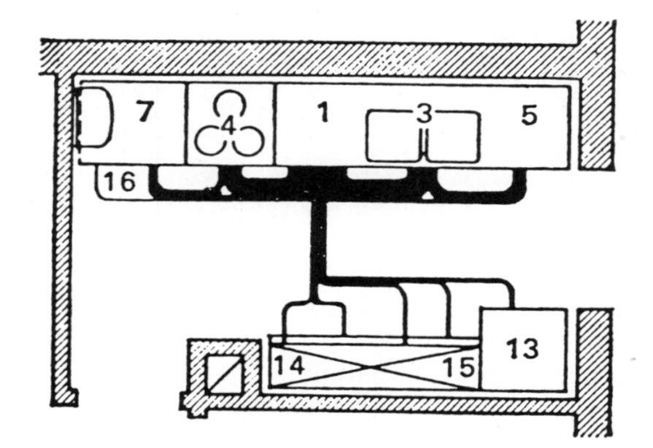

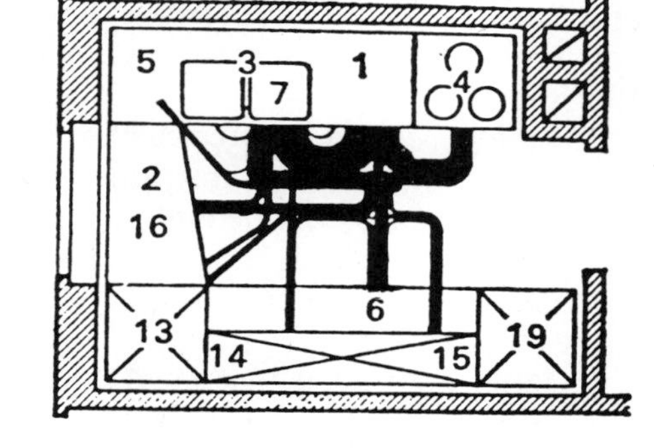

Fig. 13.2.3. Comparative study of the movements necessary in a kitchen with working surfaces only on opposite walls, and in a similar kitchen with working surfaces on three walls. Each movement is shown by a thin line, so that the thicker the line the more numerous the movements.

The left-hand kitchen occupies a floor space of 7.0 m^2, and requires 471 operations and 329 m paths; the right-hand kitchen of 6.2 m^2 requires 362 operations and 311 m of paths for the same tasks.

1, 2, 5, 6, and 7: surfaces where work is carried on or where articles can be deposited; 3: kitchen sink; 4: cooker; 13, 14, 15, 16, and 19: cupboards and refrigerator (from Ref. 13.8).

During the Second World War a number of investigations were undertaken to design tanks and airplanes that could be operated with the least physical exertion. These studies were then extended to factories, to office buildings, and finally to residential buildings. Several books have been written on investigations of this type, which are known in America as human factors engineering (for example, Ref. 13.7) and in Europe and Australia as ergonomics (for example, Ref. 13.3).

Ergonomic studies can be used not merely for the design of shelves, tables, chairs, and other furniture but also for studying the layout of rooms (Fig. 13.2.3).

13.3 Accidents in the Home

Accidents in the home are among the most common causes of injury and of death. They do not receive the same publicity as other causes of injury or death because they are not as newsworthy. Generally nobody is guilty of a crime or a misdemeanor, and nobody can be blamed except the person hurt or a relative; in a traffic accident another person can often be blamed, and in an accident at work the employer bears at least partial responsibility. Nor does it seem likely that the situation would be greatly improved by research, as might be in the case of deaths from cancer.

The causes are generally quite simple, and in many cases the accident is avoidable. Statistics collected in various countries (Ref. 13.3, pp. 290–93) show that in all countries falls are the major cause of injuries and fatal accidents in the home. In Canada and the United States fires are the second most common cause, but elsewhere they take third place. Fire prevention was discussed in Chapter 7. Poisons are the most common cause of accidents that befall children. This is not a problem that can be solved by architectural design, nor can death from firearms, which is a common problem in the United States.

Electrocution is a relatively minor cause of death; evidently safety outlets that can be switched on only when a fitting is in the outlet are desirable where there are young children.

Another relatively minor cause is scalding from hot water, which can be prevented by keeping the hot water temperature sufficiently low.

A further minor cause is drowning in swimming pools; this can be reduced by a childproof fence between the house and the pool. Drowning of children while they are actually in a swimming pool or a bath can only be avoided by better supervision.

As already mentioned, both old and young are particularly liable to be hurt or killed by falls in the home. This cannot be entirely prevented without placing the old and infirm into nursing homes before it is otherwise necessary, but a number of precautions can be taken:

1. Steps are better than slopes if nobody requires wheelchair access; if wheelchair access is needed [see Section 13.4], it is desirable to have both a slope and steps if room permits. If a slope is needed, it should have a nonslip surface and a rail to hold onto.
2. One or two isolated steps should be clearly marked.
3. Outside steps should be lit with at least 75 lux [Section 8.2].
4. All steps should have the same rise and tread, but the tread should be greater than the rise.
5. Steps without vertical backs are dangerous to small children.
6. All stairs, except those of one or two steps, should have a handrail that affords proper support and that is structurally sound if a person leans heavily on it.
7. Where staircases in apartment buildings have lights fitted with time switches, they should allow enough time for old people who can walk stairs only slowly.
8. All light switches on stairs, in hotel rooms, and in other places that may be unfamiliar should be lit by luminous paint or a pilot light while the light is in the "off" position.
9. Steps should not be placed immediately in front of or behind doors. If this is unavoidable, the far side of the door should be marked appropriately.
10. The opening arcs of two doors should not intersect, so that the two doors cannot collide if opened simultaneously by different persons.
11. A door with a large sheet of glass should have a line or other mark to show that it is not an opening.
12. A window that extends down to the floor should have a protective grill.
13. A balcony needs a railing high enough to make it impossible for small children to climb over it. If it consists of parallel bars, the distance between them should not be so large that a child can place its head between them.
14. Floor surfaces should not become slippery when wet.
15. Baths and showers used by elderly and incapacitated people must have grip handles. They are desirable for all baths and showers.

13.4 Wheelchair Access

The provision of wheelchair access to public buildings is mainly a development of the last twenty years. In some places building regulations require wheelchair access for all new buildings that are open to the general public, such as offices, stores, and places of entertainment (Fig. 13.4.1). In the United States it is a precondition for federal funding. Even where there is no obligation to provide wheelchair access, many clients now include it in the brief to their architects, either because they believe it is the right thing to do or because they wish to consider public opinion.

Wheelchair access means that a person in a wheelchair can enter a building unaided and move freely around the publicly accessible areas. The main requirement is continuity: a single step can become an obstacle insuperable without aid. An accessible building must also contain, or be near, a toilet accessible by wheelchair.

In theory it is only necessary to know the measurements of people [Section 13.2] and of wheelchairs (Fig. 13.4.2). In practice it is easy for a designer who has never used a wheelchair to make mistakes. Slopes that seem gentle to walking people may be too steep for wheelchairs. Stiff doors at the end of a ramp can become insuperable obstacles. Door handles and light switches are often too high. The requirements

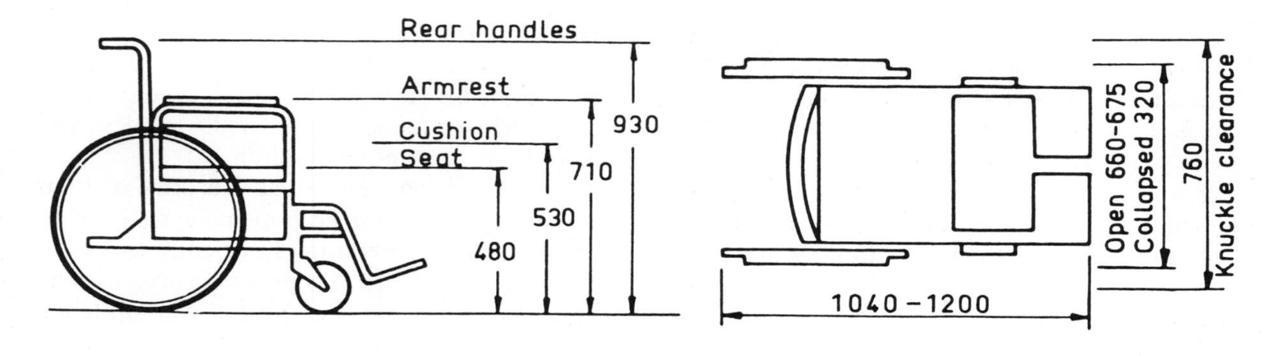

Fig. 13.4.2. Wheelchair dimensions in millimeters.

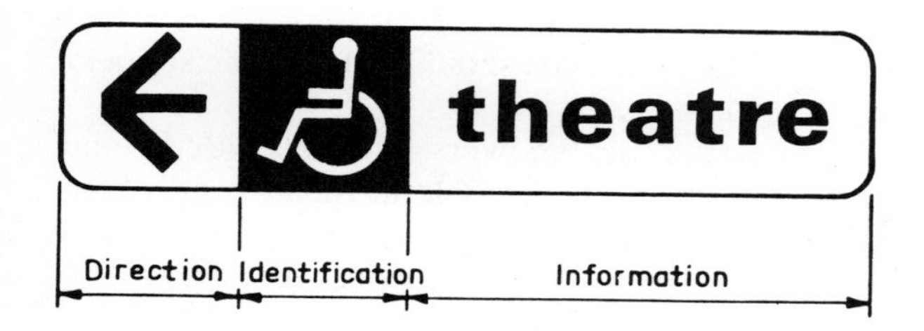

Fig. 13.4.1. Sign indicating that a theater has wheelchair access to the required standard.

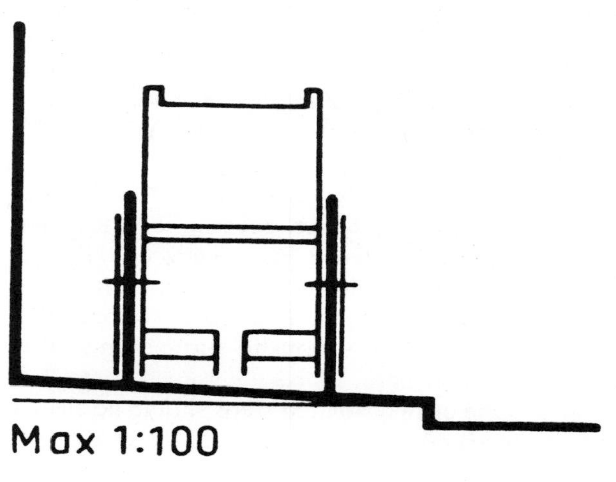

Fig. 13.4.3. Wheelchairs are not safe on walkways or ramps banked more than 1:100. (*Reproduced from Ref. 13.11.*)

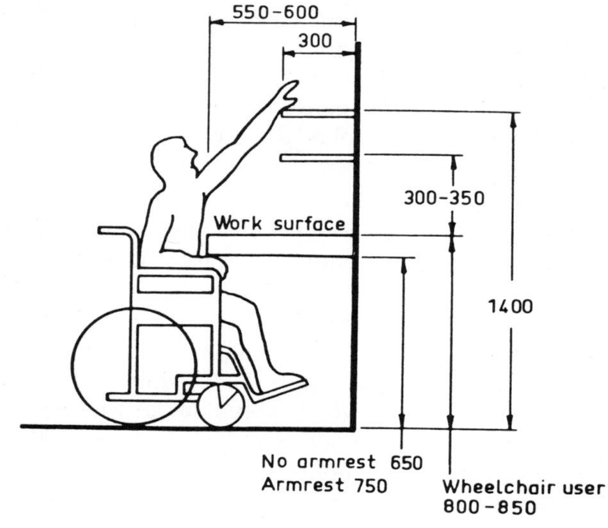

Fig. 13.4.4. Dimensions in millimeters for work surfaces and shelves for persons in wheelchairs. (*Reproduced from Ref. 13.11.*)

are set out in standard specifications (Refs. 13.9, 13.10, and 13.11) and explained in textbooks (Ref. 13.12).

If a person in a wheelchair is to move through a building unaided, there must be no steps. Differences in level are connected with ramps whose maximum slope is 1:12, except for short curb ramps, which may have slope 1:6. If the slope of a ramp exceeds 1:20, there should be handrails on both sides of the ramp. A walkway or ramp should not slope sideways by more than 1:100 (Fig. 13.4.3).

A corridor 0.8 m wide can be negotiated in a wheelchair provided that the corner is chamfered for 300 mm. However, it is better to have a corridor at least 0.95 m wide, in which case the chamfered corner is not required.

If a wheelchair user is to open a door by himself, it must be openable with one hand. It should have a horizontal handle, not a door knob, no higher than 1.2 m from the floor. If it is self-closing, the mechanism must not be too strong. Automatically opening doors operated by photovoltaic cells or by pressure plates are desirable.

Light switches and window controls must be no higher than 1.2 m if they are to be wheelchair-accessible.

The dimensions of work surfaces and toilets are shown in Figs. 13.4.4 and 13.4.5.

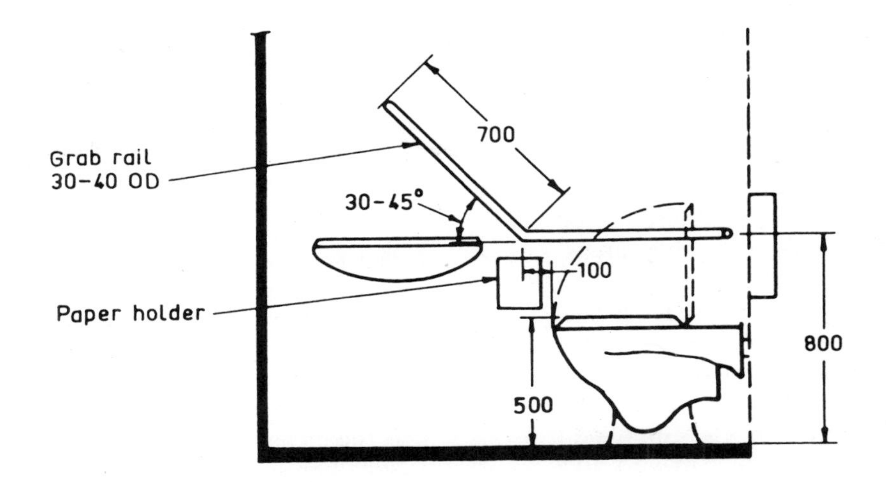

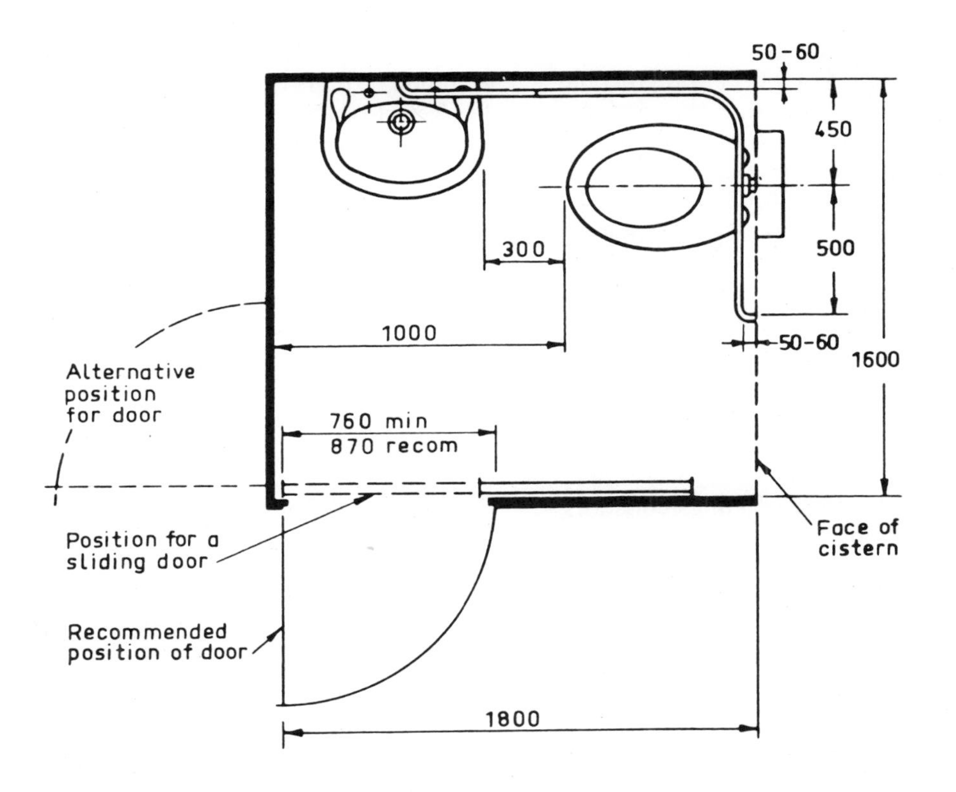

Fig. 13.4.5. Dimensions in millimeters for a wheelchair toilet. The correct positioning of the grab rail is important. It is best to place a wheelchair toilet separate from ordinary toilets for men and women and to make it accessible to both sexes. By that means a person of the opposite sex, for example a husband, wife, or mother, can render assistance if needed. (*Reproduced from Ref. 13.11.*)

All but the smallest elevators meet the requirements for minimum door and platform size. However, elevators are the weak link in wheelchair accessibility at present. We noted in Section 7.7 that it is not safe to use elevators in the case of a fire. Buildings are evacuated via the fire-resistant stairs. In tall buildings, where the occupants cannot reach the ground, "refuge areas" are provided that can be reached by a "safeguarded travel route" that may or may not include stairs.

The subject was discussed at a symposium in 1979 (Ref. 13.13), when it was recommended that buildings should have fire-safe elevators, or if that was not possible, they should have refuge areas on each story. It remains to be seen whether this expensive recommendation is implemented. In the meantime persons confined to wheelchairs are dependent on someone else to carry them out of the building in case of a fire. Many people in wheelchairs do not realize that the comparatively easy access and egress to which they are becoming accustomed in some buildings ceases in case of a fire, nor do most of the other persons in the building know that there is at present no provision for getting people in wheelchairs out of a burning building, and that it is their responsibility to carry them.

13.5 Horizontal Movement

We noted in Section 13.2 that 0.8 m is a sufficient width for a single person walking along a corridor and that 1.2 m suffices for two people to pass one another. In some buildings corridors and staircases are from time to time filled to capacity. This happens in theaters and concert halls at the end of the performance, in factory canteens at the end of the lunch hour, and in universities when classes change over. Most important, it may happen in any building, if it has to be evacuated because of a fire or a bomb threat; under those circumstances it is essential that the width of the corridor or stair should not merely *be* adequate but should *appear to be* adequate to avoid panic.

The flow rate of a crowd of people is given by

$$f = v\rho b \tag{13.1}$$

where f is the flow rate in persons per second (p/s);
v is the mean speed of the persons in meters per second (m/s);
ρ is the mean density of the crowd in persons per square meter (p/m^2);
b is the width of the corridor or stair in meters (m).

A number of investigators have studied the data (Ref. 13.14, p. 93). Pedestrian traffic flows freely for crowd densities of 0.25 p/m^2 or less. People have to start exercising care not to collide with persons going in the *same* direction when the density increases above 0.5 p/m^2, and their speed of walking is reduced. However, crowd densities in excess of 4 p/m^2 are possible. The walking speed depends partly on the crowd density, partly on the age and sex of the people in the crowd, and partly on their objective, that is, whether they are in a hurry or at leisure. P.R. Tregenza (Ref. 13.14, p. 97), after studying investigations of pedestrian traffic in various countries, suggests 0.3 p/m^2 as the upper density for free pedestrian traffic flow, and 1.4 p/m^2 for full design capacity, when people walk at less than natural speed and are aware of uncomfortable crowding (Table 13.2).

Very high flow rates of 300 persons per minute per meter width can be achieved if a crowd pushes through an exit door, but 60 p/m.m is a more likely figure under normal conditions. Revolving doors can slow pedestrian movement, as they allow only about 30 persons to pass per minute;

Table 13.2

Capacity of Pedestrian Passages for Free Flow of Traffic (0.3 p/m^2) and for Full Design Capacity (1.4 p/m^2)

Type of Person in Crowd	Walking Speed (meters per second)		Pedestrian Capacity (persons per minute per meter width of passage)	
	0.3 p/m^2	1.4 p/m^2	0.3 p/m^2	1.4 p/m^2
Commuters	1.5	1.0	27	84
Individual shoppers	1.3	0.8	23	67
Tourists, families shopping with their children	1.0	0.6	18	50
Schoolchildren (speed increases with age)	1.1–1.8	0.7–1.1	18–32	59–92

they should be made collapsible to make it possible to speed up flow in an emergency.

13.6 Stairs and Escalators

People move up a staircase more slowly than along a level corridor. Speed down a staircase is usually faster, but slower than movement along a level corridor. However, when people are crowded together, movement downstairs may be slower than upstairs. For fire escape stairs a reasonable design density is 60 persons per minute per meter width. An orderly group may attain 80 p/m.m, but under panic conditions this may be reduced to 40 p/m.m.

Building regulations for the evacuation of buildings in case of fire vary appreciably. They generally specify the number of persons per square meter that should be assumed for a particular type of occupancy, and the maximum distance of fire exits from any one place [Section 7.7]. Some codes also specify the exact dimensions of the stairs and passages.

Escalators perform essentially the same function as ordinary staircases. At low speeds their capacity is determined by the available standing room on the steps. As the speed increases, the number of empty steps becomes greater. The capacity increases at first, but at higher speeds it declines because people are deterred by the apparent speed and step with caution. The maximum capacity of escalators 1.2 m wide between balustrades in London Underground stations was observed to occur at a speed of 0.6 m/s, with 2.0 persons on each step, and this yielded a capacity of 180 persons per minute, more than double the capacity of a similar staircase (Ref. 13.14, p. 61). Similar observations have been made in North America (Ref. 13.15, p. 134).

Up escalators always consume power, but a heavily loaded "down" escalator actually regenerates electricity (Ref. 13.15, p. 137).

13.7 Elevators

The assessment of the traffic capacity of elevators is more complicated because the flow of people is not continuous. The theory of traffic flow can be derived from first principles (Ref. 13.14, pp. 15–62), but it is simpler to use empirical data. In some European countries elevators are often "made to measure," but in North America and Australia a limited number of standard car sizes and speeds are employed. Their characteristics are given in the trade literature and in some handbooks (for example, Refs. 5.6 and 13.4).

The elevators of American office buildings are traditionally designed for the morning peak traffic requirements. They should satisfy three criteria during that period (Ref. 13.15, p. 139):

1. The longest time taken for any passenger to reach his destination from the time he enters the elevator should not exceed $2\frac{1}{2}$ minutes in a single-purpose building, or 3 minutes in a building with diversified occupancy.
2. The maximum waiting period between elevator cars should not exceed 25 seconds for a single-purpose building, or 30 seconds for a diversified building.
3. The elevators should be able to handle all the people who arrive. This is measured by the 5-minute handling capacity, that is, the number of people who can be transported within 5 minutes by all the elevators. This should be at least 15% of the entire working population of a single-purpose building, and 11% of a diversified building.

In some buildings working hours have been staggered to an extent that the critical period occurs not in the morning but at some other time; in that case slightly different criteria apply (Ref. 13.15, p. 139–41).

The following simple example illustrates the three criteria.

Example 13.1 *A building containing diversified offices has 14 rentable stories above the lobby. Each floor measures 1100 m². The height of each story from floor level to floor level is 3.5 m.*

The elevator cars have a platform capacity of 1365 kilograms (3000 pounds), they operate at a speed of 2.5 meters per second (500 feet per minute), and they have a capacity of 16 passengers per peak period trip.

Solution *An average diversified office building has a population factor ranging from 11 m² to 14 m² per person (Ref. 13.15, p. 138). This means that the office layout is such that each person occupies, on the average, between 11 m² and 14 m². We will take it as 12 m²/p.*

Therefore the population of the entire building is

$$\frac{14 \text{ floors} \times 1100 \text{ m}^2}{12 \text{ m}^2/\text{p}} = 1283 \text{ persons}$$

The 5-minute handling capacity must be at least 11% of this population (criterion 3), that is, 0.11 × 1283 = 141 persons.

The height of the building above the lobby is

$$14 \text{ stories} \times 3.5 \text{ m between floors} = 49 \text{ m}$$

The round-trip time for a car to proceed to the top story, unload all passengers, and return to the lobby is obtained from a table or chart. A separate curve or column is needed for each type of car and for each elevator speed. The information is given in the trade literature and in a number of books, for example, Ref. 13.4, p. 569. The round-trip time for an elevator with stops at each of 14 floors is 148 seconds = 2 minutes 28 seconds. This satisfies criterion 1.

To satisfy criterion 3, we must transport 141

persons in 5 minutes, which is 300 seconds. Each car holds 16 persons, so that we need to make

$$\frac{141}{16} = 8.81 \text{ journeys}$$

The round trip takes 148 seconds, so that we can make

$$\frac{300}{148} = 2.03 \text{ round trips}$$

in five minutes. Therefore we need a bank of

$$\frac{8.81}{2.03} = 4.34, \text{ or 5 elevator cars}$$

Fig. 13.7.1. Elevator hoisting gear, built about 1876. The vertical steam engine is on the right, and its crankshaft is above. The cable drum on the left operates another winding drum on the top floor that hoists the elevator car. It works somewhat like a vertical cog railroad. (*By courtesy of the Otis Elevator Company.*)

Since we have 5 cars, and each comes every 148 seconds, the average interval between cars is

$$\frac{148}{5} = 29.6 \text{ seconds}$$

which just satisfies criterion 2. We do not wish to achieve a shorter time; otherwise, we waste elevator capacity.

We must now decide whether five cars fit into the layout of the building, or whether we would get a more satisfactory layout using elevators of a different size, or operating at different speeds. Higher speeds consume more energy. They are particularly worthwhile for express elevators that travel, say, 20 stories without stopping; for elevators that are liable to stop at every floor the round-trip time consists mostly of the time required to stop and start the elevator, and for the doors to open and close, plus the waiting time while people enter and leave the car. Smaller cars reduce the waiting time, partly because there are more cars for a given number of passengers, and partly because there are fewer passengers per car, so that it makes fewer stops per round trip. On the other hand, smaller cars cost more and require more space for a given passenger capacity.

The first passenger elevator was built in 1857. It was operated by a steam engine through gears and a drum (Fig. 13.7.1). The first electric elevator was installed in 1889.

For the slower speeds geared electric traction motors are used (Fig. 13.7.2). For the higher speeds gearless traction motors are more economical; they have fewer moving parts and thus generate less noise and require less maintenance. The driving sheave and the brake wheel are connected directly to the motor shaft (Fig. 13.7.3).

Since it is not possible to build electric motors for very slow speeds, gearless motors can only be used for the higher elevator speeds, that is, more than 2 m/s or 400 ft/min. Geared motors can use either alternating or direct current, while gearless motors must use direct current. The direct current is produced by a motor-generator set, or more recently and more economically by silicon-controlled rectifiers. All electrically driven elevators feed electricity back into the system when they brake, when they descend fully loaded, and when they go up empty, because their own weight is then less than the counterweight.

The maximum elevator speed employed in America at present is 9 m/s (1800 ft/min) and in Australia 7 m/s (1400 ft/min). Japan has the world's fastest elevators (10 m/s, or 2000 ft/min).

The motor for driving an elevator and the control gear are normally placed in a room above the elevator, so that the elevator is pulled up by a cable (Fig. 13.7.4). If the elevator operates up to the top story of the building, it requires a penthouse on the roof, unless a special story is provided for the building services. It is possible to operate the elevator from the bottom, by passing the cables over pulleys; however, this involves extra cable, which takes up space and causes friction loss and noise. For buildings only a few stories in height there is some advantage in the use of a hydraulic elevator (Fig. 13.7.5). This is pushed up by oil pressure. There is no need for a counterweight, and the elevator is lowered by gravity, restrained by the oil pressure. This type of elevator costs less for low-rise installations, and it does not require a penthouse above; its maximum speed is at present 0.75 m/s (150 ft/min).

More detailed information of elevator equipment may be found in specialist books (Refs. 13.16 and 13.17).

Fig. 13.7.2. Geared traction motor for elevator cars, suitable for speeds up to 2 m/s (400 ft/min). The casing is partly cut away to show the worm and gear that reduces the speed of the motor. This machine uses a direct-current motor whose speed can be varied more smoothly, but an alternating-current motor can be used for speeds up to 1 m/s (200 ft/min). (*By courtesy of the Otis Elevator Company.*)

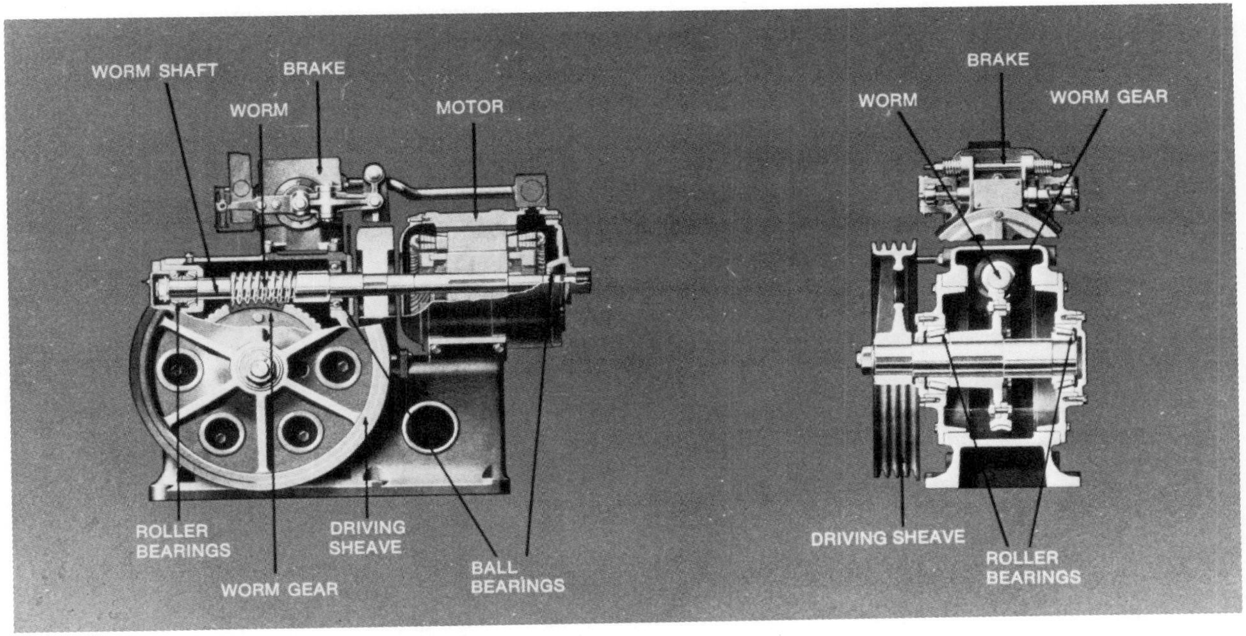

Fig. 13.7.3. Gearless direct-current traction motor for high-speed elevator, with casing cut away. (*By courtesy of the Otis Elevator Company.*)

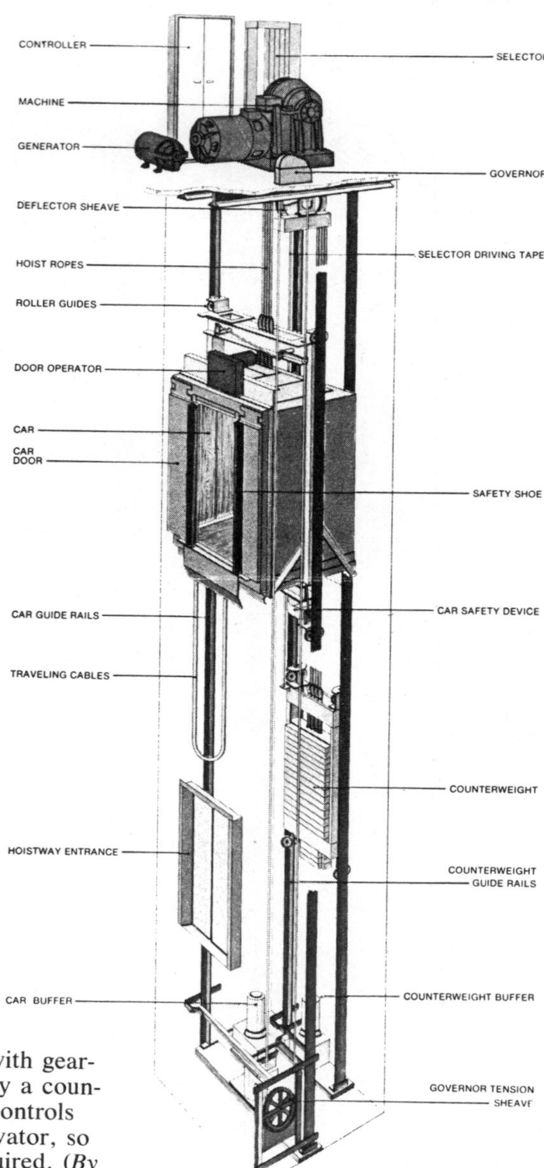

Fig. 13.7.4. High-speed electric elevator with gearless traction motor. The car is balanced by a counterweight. The hoisting machine and the controls are above the top story served by the elevator, so that a penthouse or a service story is required. (*By courtesy of the Otis Elevator Company.*)

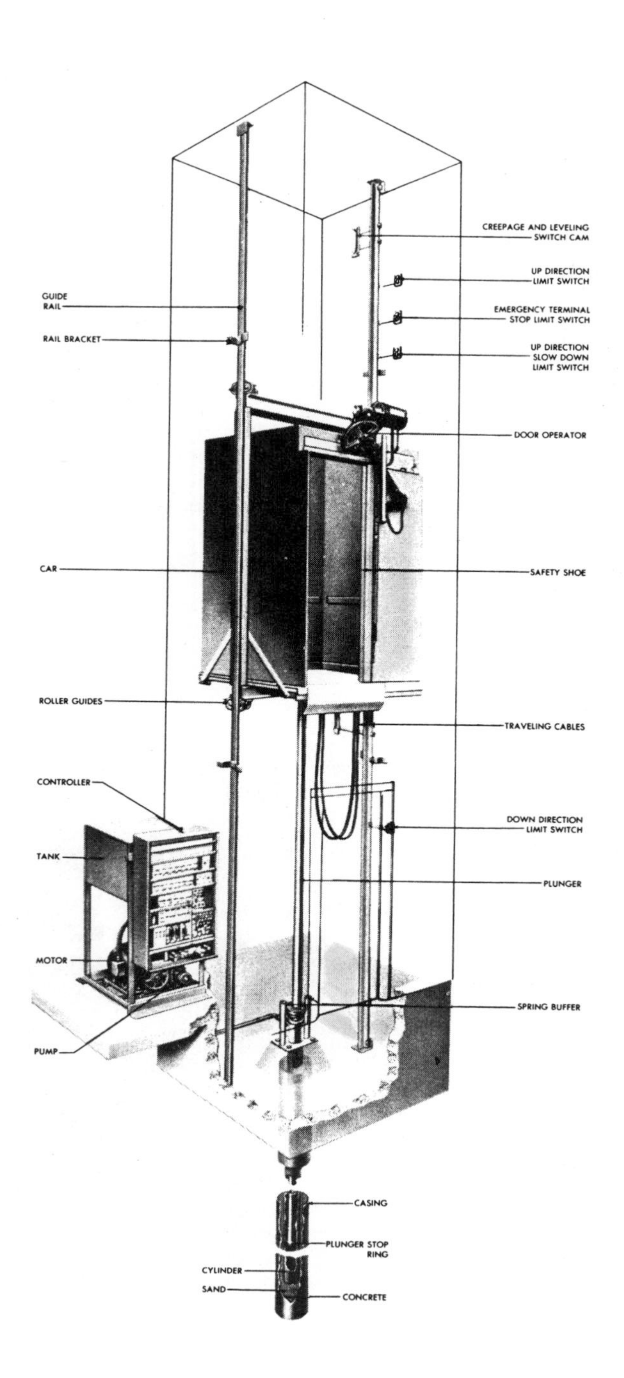

13.8 Elevators for Tall Buildings

In buildings more than 15 stories high it is advisable to zone elevators. For example, instead of allowing all elevators to go to the top and stop at any story (Fig. 13.8.1.a), some elevators go only as far as, say, the tenth story, whence they return to the ground, while the remainder go express to the tenth story and then stop at all remaining stories, if required (Fig. 13.8.1.b). The number of stories in each zone depends on the traffic generated. Sometimes the lower stories have occupancies that require frequent access while the higher stories are visited by fewer people. The top story may require its own express elevator because it contains a club, a restaurant, or a lookout.

The space taken up by the elevators has always been a limiting factor for the economic height of buildings. In the late nineteenth century elevator speeds were less than 0.5 m/s (100 ft/min), and accordingly more elevators were required to provide adequate service than would now be necessary. The increase in the height of buildings after the year 1900 was to a large extent due to the increased elevator speeds that followed electrification. The building regulations that required setbacks for tall buildings, whatever disadvantages they may have had for urban planning, made the design of the elevators easier (Fig. 13.8.2).

Fig. 13.7.5. (Opposite) Hydraulic elevator suitable for low-rise buildings only. The car is pushed up by a piston and descends by gravity, restrained by oil pressure. The oil pressure is generated by an electrically driven pump. No penthouse is required, but the hydraulic cylinder requires a hole underneath. The maximum speed is 0.75 m/s (150 ft/min). There is no regeneration of electricity. (By courtesy of the Otis Elevator Company.)

The sky lobby concept, introduced in the 1960s, is useful for buildings over 40 stories. It utilizes high-speed shuttle cars to transport people nonstop from street level to a "sky lobby," which serves the upper floors in precisely the same way as the lobby at street level (shown schematically in Fig. 13.8.1.c).

In the World Trade Center in New York and the Sears Tower in Chicago, both with more than 100 stories, there are two sky lobbies, and in each of the three sections the elevators are in several zones. The elevators in the upper two sections use the same shafts as the elevators in the lower section, so that an appreciable amount of space is saved. However, the extra space needed for the sky lobbies and the express elevators that serve them must be offset against that.

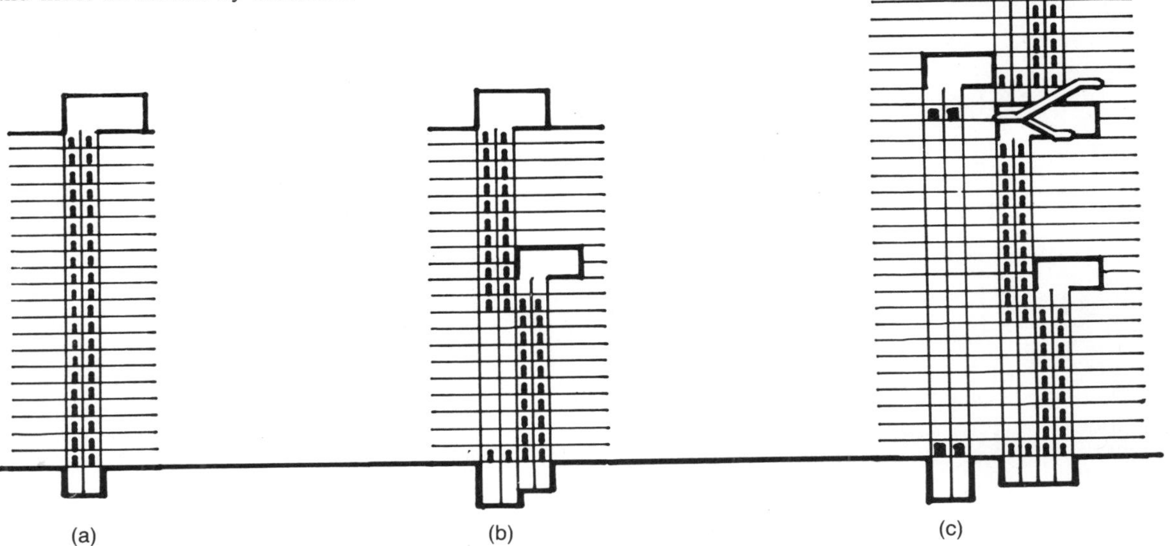

Fig. 13.8.1. Diagrammatic layout of elevator system.
(a) Single-zone system. Every car can go to the top story and stop at any story.
(b) Two-zone system. In the lower zone all cars can go to the tenth story and stop at any story. In the upper zone all cars go nonstop to the tenth story, and they can then go to the top story and stop at any story. This saves about half the space needed for the elevator shafts in the upper floors, but the plant room occupies some of that space.
(c) Sky lobby system. An express elevator goes nonstop to the sky lobby, which requires almost as much space as the lobby at street level. The elevator systems rising from the street lobby and the sky lobby are conventional two-zone systems. Because the elevators for the upper half of the building operate in the same shafts as those for the lower half, some stories near the sky lobby are not accessible and must be served by escalator.

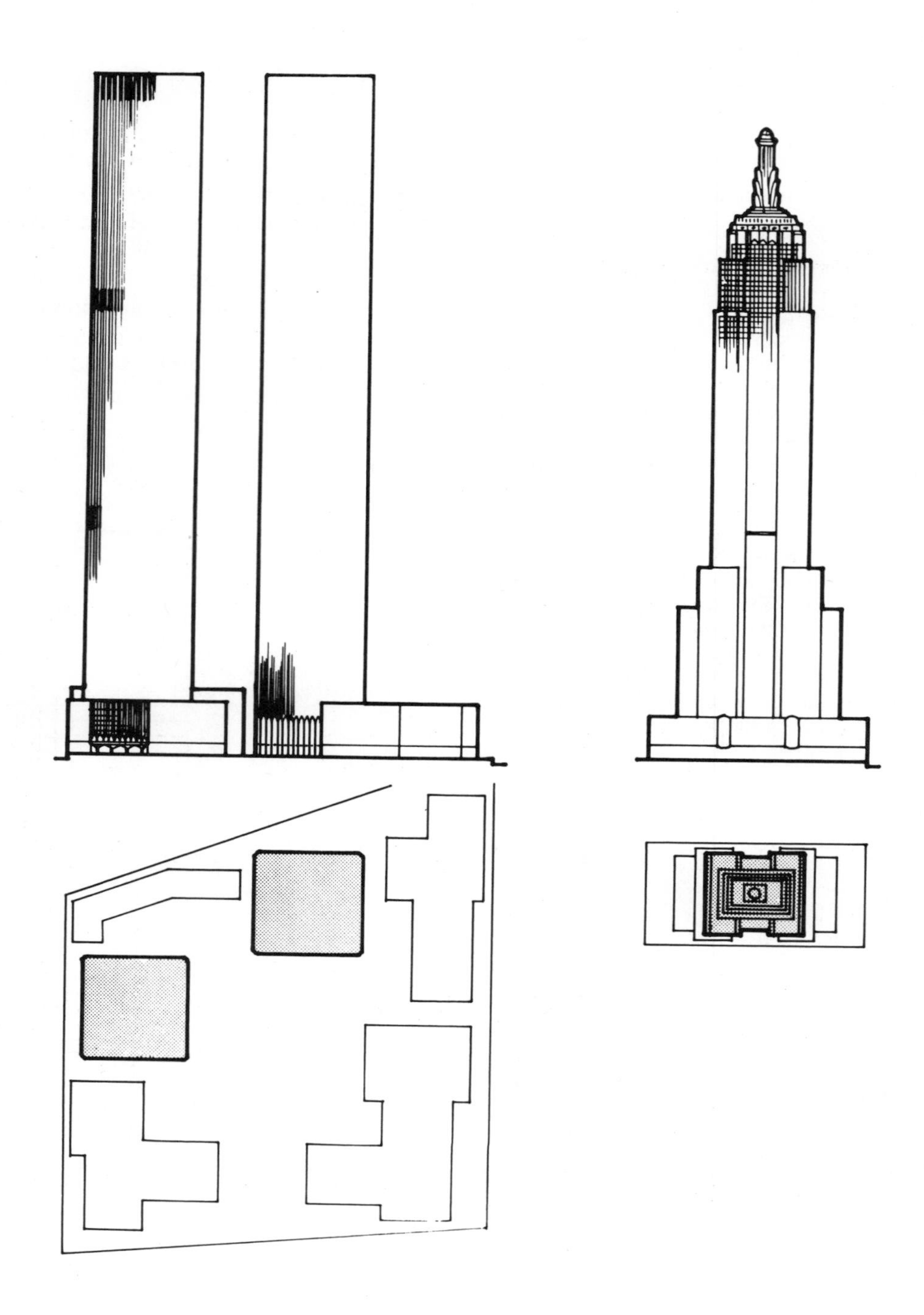

The double-deck elevator (Fig. 13.8.3) is another method for making better use of the space occupied by the elevator shafts. However, it does not transport twice as many passengers, because whenever the upper deck stops at a floor, the lower deck stops at the floor below, whether anybody wishes to get on or off the elevator at that level or not; the same applies in reverse. Thus a double-deck elevator transports only about 50% more people than a single-deck elevator of the same size operating in the same shaft.

Fig. 13.8.2. Comparison of the World Trade Center (left), completed in 1972, and the Empire State Building (right), completed in 1931, both in New York.

The Empire State Building occupies the entire site. It has 58 local and express elevators, all starting from street level. It relies on the spread of the lower floors of the building to provide enough usable space around the elevators.

The World Trade Center occupies only a part of its site. Each tower has 12 express shuttle cars to each of two sky lobbies. From each sky lobby 24 elevators start, arranged in 4 zones. In addition there are 2 express cars serving the observation deck. Thus each tower has 50 elevator cars at street level.

Fig. 13.8.3. (Opposite) Double-deck elevator. Wherever the upper car goes, the lower car must go the floor below. (*By courtesy of the Otis Elevator Company.*)

13.9 Communication Channels and Security

Communications have become much simpler during the last few decades through advances in telephone equipment. Most institutions, large offices, and factories use a PABX (private automatic branch exchange), which allows the installation of a large number of telephones at a reasonable cost. Internal calls and outgoing calls are made automatically by dial or press buttons. An operator receives and directs incoming calls, although many systems are equipped so that extension telephones can receive incoming calls automatically.

There is still a need for the transport of documents, although this may be rendered obsolete by closed-circuit television in the foreseeable future. These can be conveyed by dumbwaiters, which are small freight elevators; by horizontal conveyor belts; or by pneumatic tubes, which are much faster than either of the others; the papers are placed in a circular container and propelled along the tube either by a positive pressure or by a vacuum.

Fire alarms were discussed in Section 7.4. Their mode of operation is generally given the greatest possible publicity so that people should know what to do in case of a fire. Burglar alarms and other security precautions are kept secret as far as possible, and their design is usually in the hands of a security consultant. The known methods for triggering burglar alarms (Ref. 13.18, pp. 86–106) are:

1. The breaking of an electrical circuit;
2. The interruption of a light or ultraviolet beam;
3. The detection of infrared body radiation;
4. The detection of sound or vibration;
5. The detection of motion;
6. The variation in an electrical or magnetic field;
7. Observation with closed circuit television.

The recording of pictures with cameras or closed-circuit television does not detect a burglar, but it makes it easier to catch him.

Many large buildings have loudspeakers installed in each room, and some are required to do so by building regulations. These make it possible to direct people to refuge areas in case of a fire or another emergency, and they can be used to communicate with security guards. In large buildings fire and burglar alarms are centrally controlled for 24 hours a day.

13.10 Horizontal Transportation Between Tall Buildings

It is unlikely that the height of buildings is or ever has been limited by structural considerations. We know that Ancient Rome already had apartment buildings more than 70 Roman feet (21 meters or 68 American feet) high, because an edict of the Emperor Nero limited their height to 70 feet after a disastrous fire.

The height limitation set by staircases was removed by the development of the passenger elevator [Section 13.7], but before the end of the nineteenth century it was found that the space taken up by the elevator shafts set an economic limit to the height of buildings. As elevator speeds increased, this limit became higher.

The invention of the sky lobby and the double-deck elevator [Section 13.8] again raised the possible height of tall buildings. In most of them the internal vertical transportation system can now handle people more efficiently than the horizontal transportation system that brings them to and from the buildings. It is the latter that sets the present limit to the size of tall buildings.

Le Corbusier was an early supporter of the American skyscraper, but he noted that they were spaced much too closely. In 1924 he wrote in *The City of Tomorrow* (Ref. 13.19, p. 181–82):

> It is 9 A.M.
>
> From its four vomitories, each 250 yards wide, the station disgorges travellers from the suburbs. The trains, running in one direction only, follow one another at 1 minute intervals. The station square is so enormous that everybody can make straight to his work without crowding or difficulty.
>
> Underground, the tube taps the suburban lines at various points and discharges into the basement of the skyscrapers, which gradually fill up. Every skyscraper is a tube station.

In fact, most tall buildings, except for a few prestigious structures, are located on existing streets, usually in an old part of the city whose communication system was designed for a much smaller town. In the City of London some of the main roads still follow their alignment in the days of the Roman occupation, and their width is possibly no greater than it was in Roman times. High land values would make the realignments or widening of these roads excessively costly. In Sydney, where the roads in the central business district date only from the nineteenth century, the problem is precisely the same because the streets were laid out for a small town with little traffic. Even in cities like Chicago or Melbourne, where the streets are wider, there is insufficient parking and an inadequate public transport system. In Los Angeles, which has good freeways and fairly adequate parking, there are fewer very tall buildings. Contrary to Le Corbusier's *City of Tomorrow,* tall buildings are associated in practice with traffic congestion made even worse by the large number of people they attract.

Government departments and business firms seek office space in tall buildings because they wish their staff to be near one another. This need to concentrate a large number of people in a small area could disappear as a result of technical innovation. It may become cheaper to obtain a computer printout from a central data bank than to carry a file from one office to another. It may become simpler to have a business conference through television consoles than to congregate in one room. If business and government became dispersed as a result, the need for tall office buildings would vanish, and the problem of horizontal communication between them would vanish.

In the meantime this is an unsolved problem. A number of suggestions have been made for rapid transit vehicles, which are virtually horizontal elevators (Fig. 13.10.1), to connect major city buildings, but few have been realized.

The best solution at the present time is to walk from one tall building to the others. This is perhaps desirable for the health of people who spend too much time sitting down, but it is hardly in keeping with the advanced technology used in the design of the buildings.

Fig. 13.10.1. Personal rapid transit vehicle to connect buildings. (*By courtesy of the Otis Elevator Company.*)

Pedestrian circulation and access to subway stations would be made more agreeable if the passages were placed underground in a pleasant environment protected from inclement weather. This has been largely achieved in Montreal, perhaps in response to the very cold weather in winter.

13.11 Arrangement of Services, Conduits, and Ducts

When electricity and the telephone were first introduced into existing buildings, the wires were often located on the surface of the walls and ceilings; but when new buildings are designed to incorporate these services, the architect prefers to conceal the wiring.

In a building with fairly small rooms and a predictable location of furniture, conduits can be built in and seldom need relocation. The pipes for water supply and sewerage are larger than electrical or telephone conduits, but the places where they are needed are also predictable. In cheaper buildings these pipes may be located on the outside walls, where frost does not occur, but in more prestigious buildings it is usual to conceal them from view.

However, a large commercial building with service systems permanently built in is not sufficiently adaptable. Flexibility of arrangement is needed because:

— office plans are frequently rearranged;
— the type of telecommunication equipment, and the wiring for it, are rapidly changing;
— many work stations require electricity or data channels for machines that may not have been invented when the building was designed;
— light fittings, air conditioning outlets, fire alarms, and fire sprinklers may need to be relocated when a partition is moved.

Many buildings that are structurally sound have been torn down and replaced because their services systems and decor became outdated. In some cities, planning regulations no longer permit a building as large as the one already on a site. This causes building owners to prefer buildings that can be adapted and modernized without demolition.

Luminaires, air conditioning, fire sprinklers, and alarms are located on the ceiling. A false ceiling allows the beams and joists of the floor structure, as well as these services, to be located at will and to be only roughly finished, while providing a neat and smooth appearance. The depth of the false ceiling must allow for the worst condition of ducts and beams crossing each other.

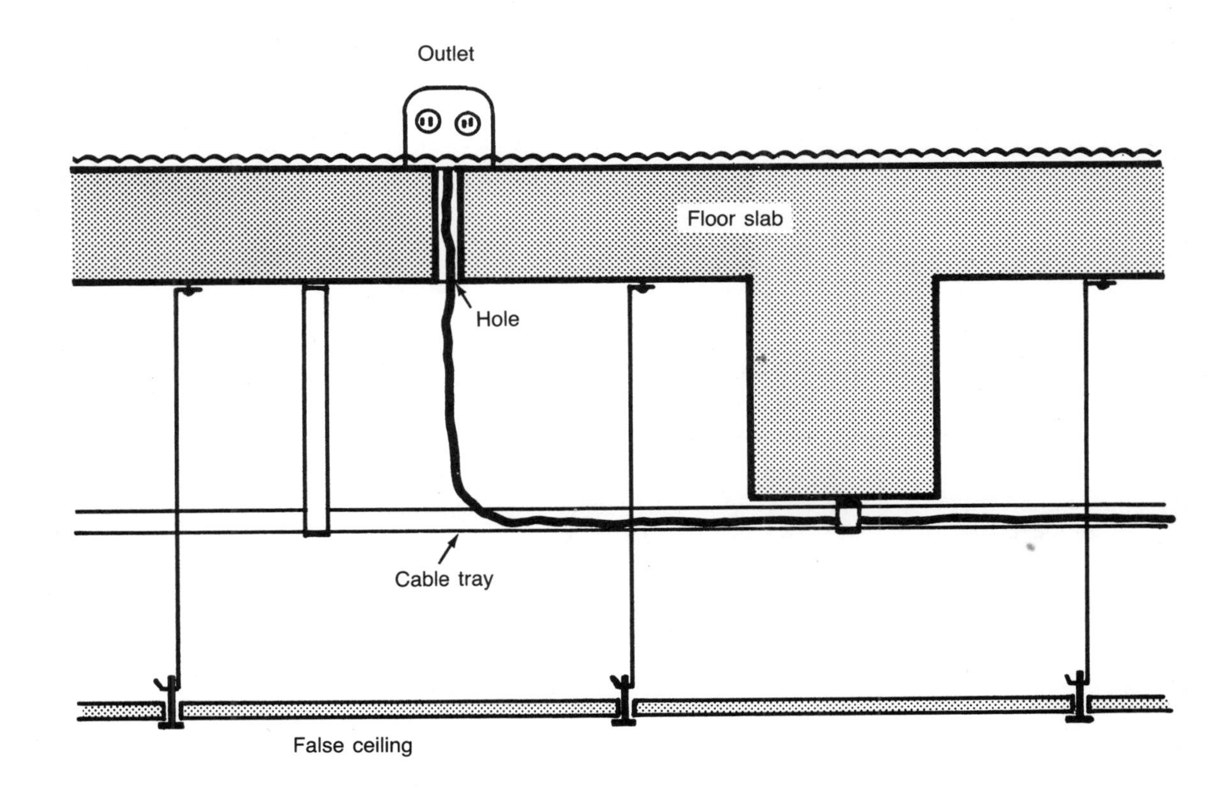

Fig. 13.11.1. Cables for power and telephone can be carried in cable trays in the false ceiling space below. Installation of a new outlet necessitates working on two stories of the building. The holes must be sealed for fire resistance of the floor.

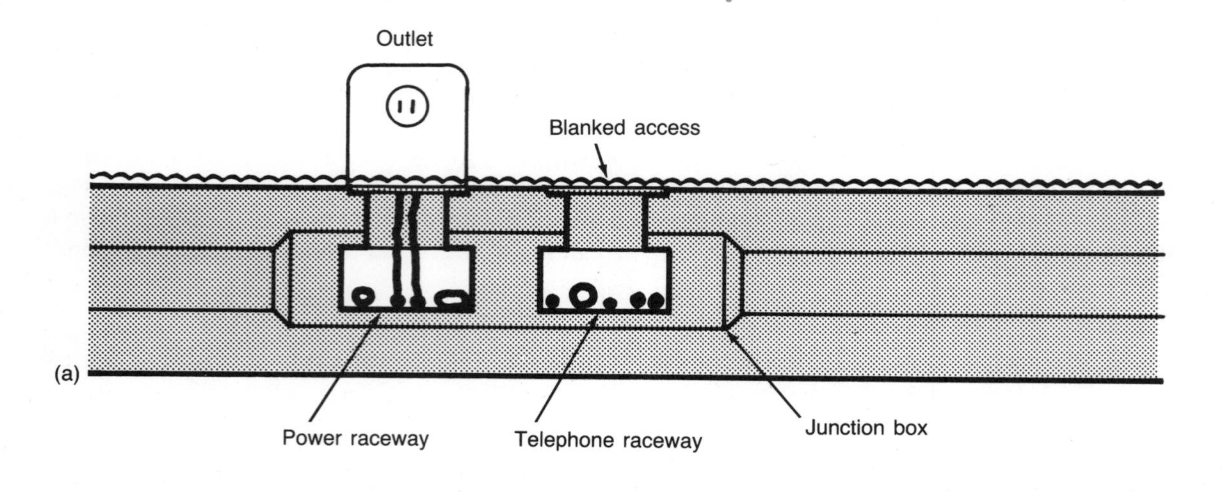

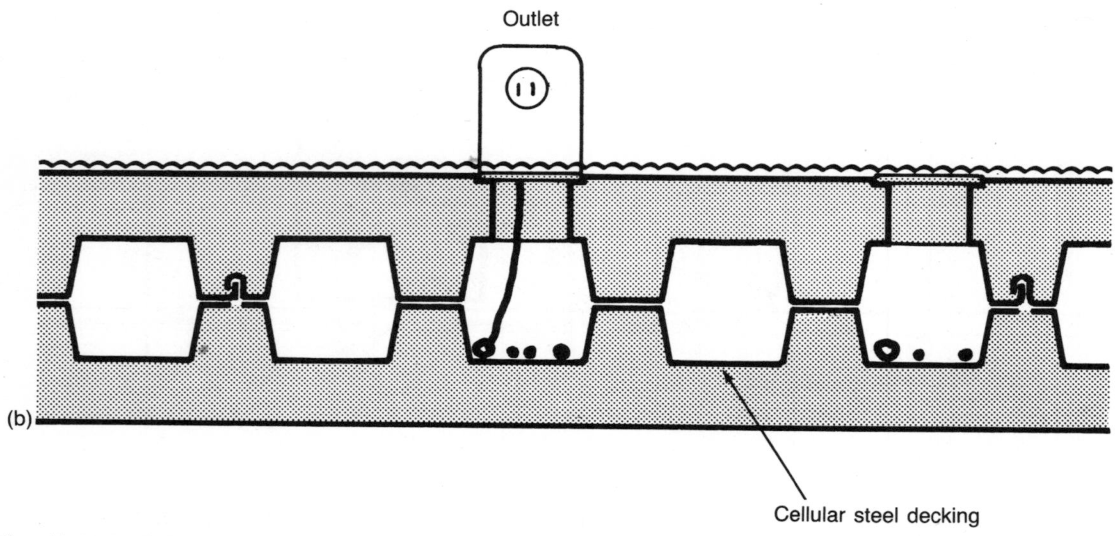

Fig. 13.11.2. Cables for telephone and power can be carried in raceways within the slab.
(a) In a structural concrete slab metal raceways are cast in near the middle of the slab. Access openings are left at predetermined locations and closed off unless needed.
(b) In a steel deck floor the cells of the deck can be used as the raceways.

The sizing of air conditioning ducts is discussed in Section 6.8. Light fittings, unless they are arranged in continuous strips, can usually be located to avoid ducts.

The electrical wiring to lights is easily accommodated in a false ceiling, but wiring for electrical power and telephones present greater difficulties. There are four possible locations: in the ceiling space below, within the floor slab, on the floor surface, and in the ceiling space above.

It is relatively easy to locate wires in the false ceiling space below. The floor slab can be designed with a grid of holes that accommodate the wiring as required, or a hole can be drilled whenever it is needed; all holes must be sealed for fire resistance. However, any reorganization of the floor above causes disturbance to those using the floor below (Fig. 13.11.1).

Raceway systems built into the floor slab are therefore preferable. In a steel-framed building the raceways can be formed in the structural-steel deck, covered by a concrete or gypsum topping slab. In a concrete-framed building the raceways are located either within the structural slab or on top of it and are embedded in a lightweight topping about 80 mm thick. In each case access openings are provided on a regular grid. The raceways are connected through headers to the main electrical and telephone risers (Fig. 13.11.2).

Some computers require a large number of cables between the various components of the system that cannot be accommodated in conventional ducts. An elevated floor, about 300 mm above the structural floor, consisting entirely of removable panels supported on small struts, is useful in offices with a large number of data processing machines, because the cables used to interconnect the machines can be installed without removing the plugs from their ends (Fig. 13.11.3). Care is needed to accommodate the steps or ramps

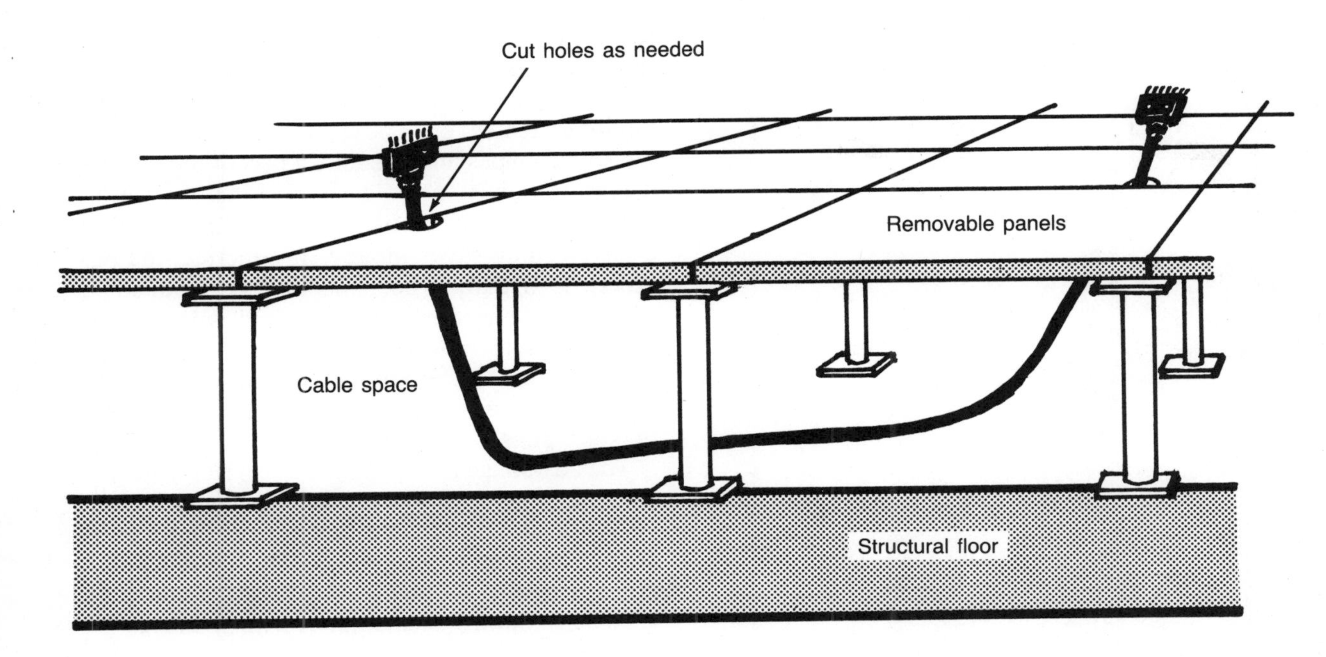

Fig. 13.11.3. An elevated "computer floor" system enables large cables, and cables with plugs on their ends, to be installed and moved at will. The removable floor panels are not part of the fire-resistant construction; they can be made of plywood, and the holes made through them do not need to be sealed.

that allow for the change in floor level when only a part of the story has "computer flooring."

Factories commonly use cable trays near the ceiling to hold the cables for the machines, which are thus protected from the mechanical damage and spillage that can be expected at floor level. This system is rarely used in commercial buildings, because the appearance of many cables dropping from the ceiling to the desks for the office machines is unattractive. However, it is possible to use this system if each work station is equipped with task lighting in the form of a luminaire supported from the desk. The cables can then be incorporated in an extended post that reaches almost to the ceiling (Fig. 13.11.4). The cable trays are usually enclosed in a false ceiling, but they may be exposed.

The false ceiling, commonly used in the commercial buildings of the last half century, hides the structure and services behind a smooth, uni-

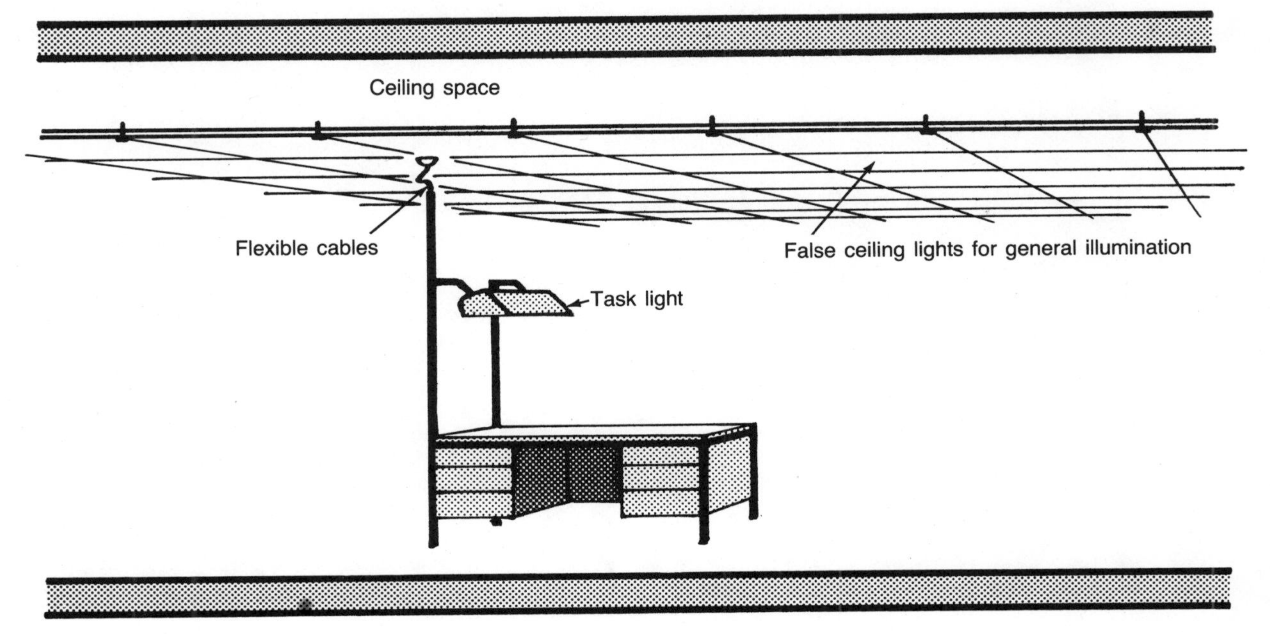

Fig. 13.11.4. Cables for power and telephone can be carried in the false ceiling space above, but some means is needed for carrying the cables down to the desks. If localized "task lighting" is used to supplement the general illumination, it is possible to design each desk with a column that serves this purpose.

form surface. In contrast the exterior form of the same buildings "followed function" and expressed the structure and the purpose of the building. However, a few buildings have made the service pipes and conduits a part of the interior decoration.

It is sometimes argued that the elimination of a false ceiling saves money. In practice, the opposite is usually true. The ductwork must be designed, fabricated, and installed with greater care if it is to remain on show, and the task of carefully painting around all the surfaces is more expensive than painting a flat ceiling or the prefinished panels of a false ceiling.

The advantage of exposing the services is mainly esthetic. The ceiling is the largest single surface visible inside a building, and it may be uninteresting if completely smooth. Exposed ducts and pipes are usually painted in bright colors and used as sculptural elements. The most successful examples have fairly simple ductwork. A tall gymnasium or transport terminal, with one or two large air ducts across the ceiling, is easier to treat in this way than an office building with low ceilings and a proliferation of ducts. Sometimes ducts are exposed both inside and outside a building. An extreme example is the Pompidou Center in Paris, in which all the ducts, pipes, elevators, and the structure are exposed to view (Fig. 13.11.5).

Fig. 13.11.5. The Pompidou Center, Paris. This exhibition building has most of the ductwork, as well as the elevators and escalators, exposed on the outside of the building. The air ducts appear larger than usual because they are covered with insulation, and then with a waterproof covering to protect the insulation from the weather. The painting and maintenance of such a building would be costly because so many separate elements are exposed to the weather.

13.12 Arrangement of the Vertical Building Services

The vertical communications in a multistory building are essential to its operation and are an important element in the planning. The elevator shafts extend the full height, and their location dictates the main circulation of people on each level. The location of fire stairs also influences the layout of corridors; but unlike elevators, stairs can be offset from one story to another. It is often necessary to locate the first-floor stairs in a different place from those on upper levels, because of access from the street. When this is done, there must be a continuous path from the upper floors to the street exit, within fire-rated passage and stairs (Fig. 13.12.1).

The elevators and stairs form a substantial vertical element, which can be used for structural purposes. Toilets and utility rooms, air ducts, plumbing and electrical ducts are often incorporated into the core so that the remainder of the floor plan is free from vertical ducts.

All vertical ducts that serve typical stories must be located to give easy access to the horizontal branches from that story. A duct hidden behind elevator shafts or stairs is of no use on a typical floor, although it may be employed to take pipes

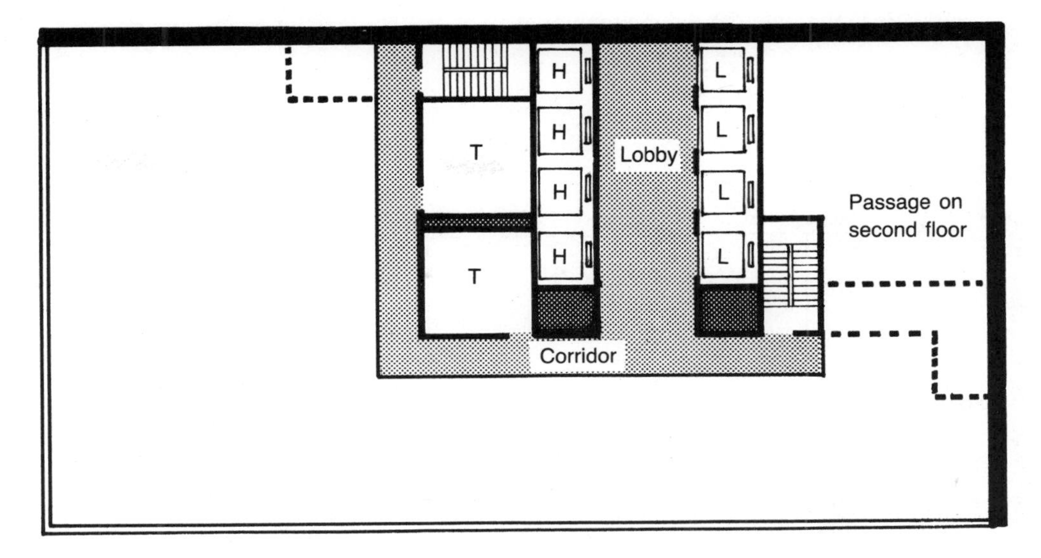

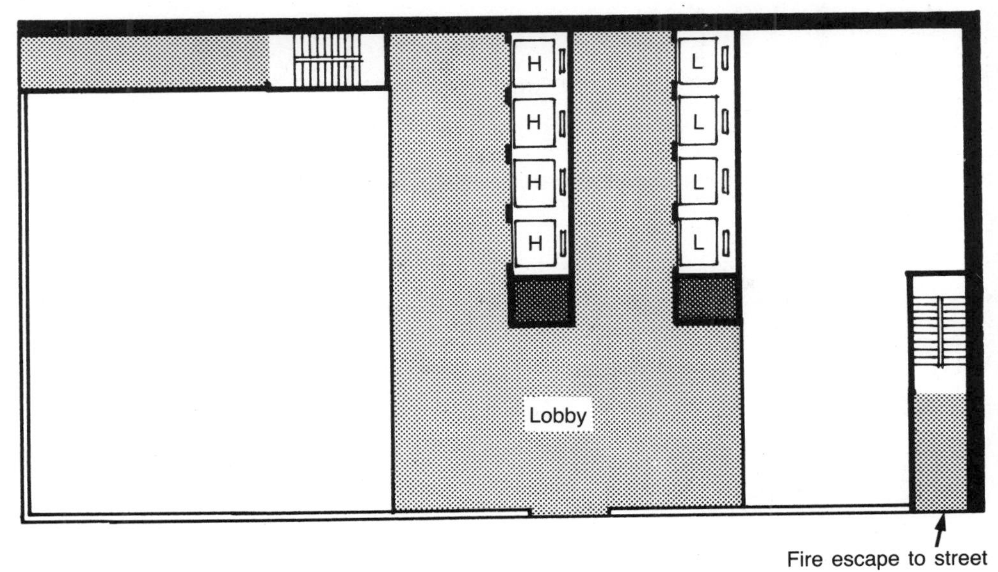

Fig. 13.12.1. The location of the elevator shafts must be the same on each floor. However, a location that is convenient on the upper floors may not be ideal for access from the street at the first floor. The location of the fire stairs must comply with the city fire code, but it should also avoid unnecessarily long corridors on the upper floors. Sometimes the location of the stairs is changed at second floor level, to allow them to discharge directly to the street and to give an uncluttered plan at street level.

directly from the basement to the roof, provided there is sufficient access for installation and maintenance (Fig. 13.12.2).

A single vertical duct supplying air to the whole of a story may require a large horizontal branch and a great length of distribution ductwork, to reach all parts of the floor. A simpler layout is often achieved with two or more supply ducts. If a separate air conditioning zone is used on each facade, it can be supplied by separate vertical ducts in the core, but it usually requires more ductwork. If the perimeter zones are supplied from vertical ducts built into the column enclosures on each facade, there is no need for horizontal ducts over one another in the ceiling (Fig. 13.12.3).

In hotels it is desirable to locate the units back to back, so that two bathrooms can share the same vertical plumbing duct (Fig. 13.12.4). For buildings containing small apartments it is worthwhile to locate two sets of bathrooms, kitchens, and laundries around a single vertical plumbing duct, if it can be done without spoiling the plan of the units. However, in larger apartments the plan may be improved by providing a second vertical plumbing duct, and this usually produces a simpler and more useful layout.

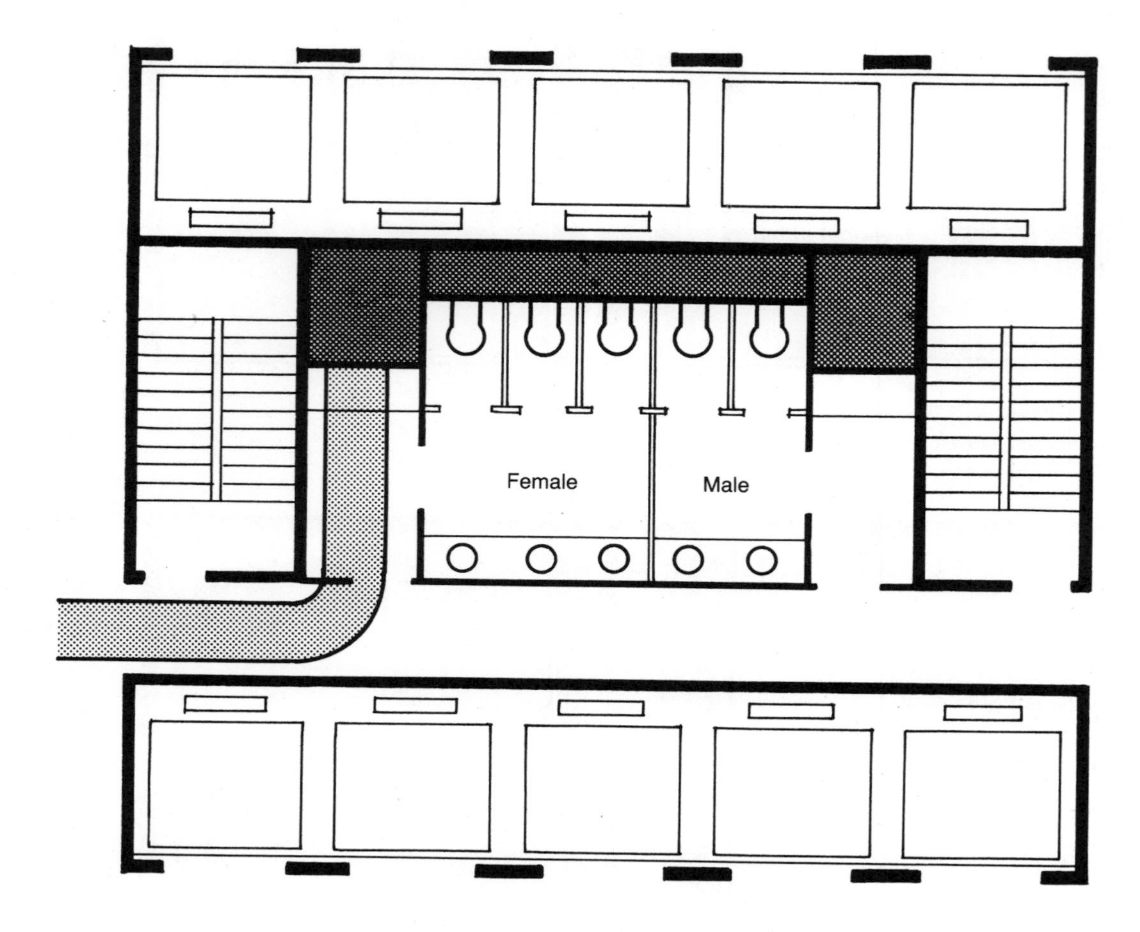

Fig. 13.12.2. The service core can form a compact, stiff structural element. Toilet rooms are often located within the core. In this diagram, all the toilets have direct access to a plumbing duct. The partition between male and female toilets can be moved, if necessary, on floors that have a predominance of male or female occupants. However, the location shown for the two square ducts is not convenient, as the access for any horizontal branches from them is restricted by the elevator shafts and the fire stair.

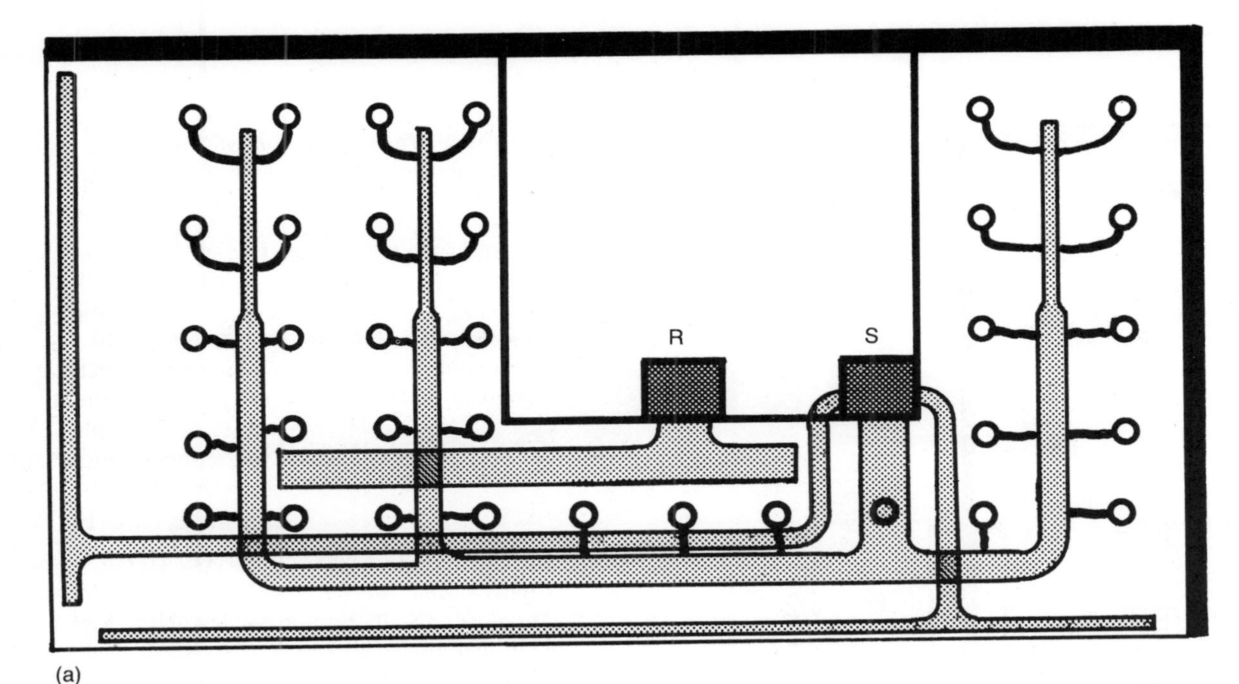

(a)

Fig. 13.12.3. A building similar to that shown in Fig. 13.12.1 requires air conditioning with two perimeter zones and one interior zone. Each outlet is connected to the main ductwork by a short length of flexible ducting.
(a) All the supply ducts and the return air duct are located in the core, in the two locations shown. The total length of ducting required is great, and major ducts cross each other in four places.
(b) By locating the perimeter supply ducts on the perimeter and splitting the interior zone supply into two vertical ducts, the layout is greatly simplified and much less ducting is needed. Two small, flexible ducts are shown crossing the return air duct, but it is possible to avoid even this if necessary. The depth of false ceiling required would be less than in (a), because ducts do not cross each other. The layout of ductwork in the plant room would be a little more complicated because of the greater number of vertical ducts, but this would be justified by the saving on each of the typical floors.

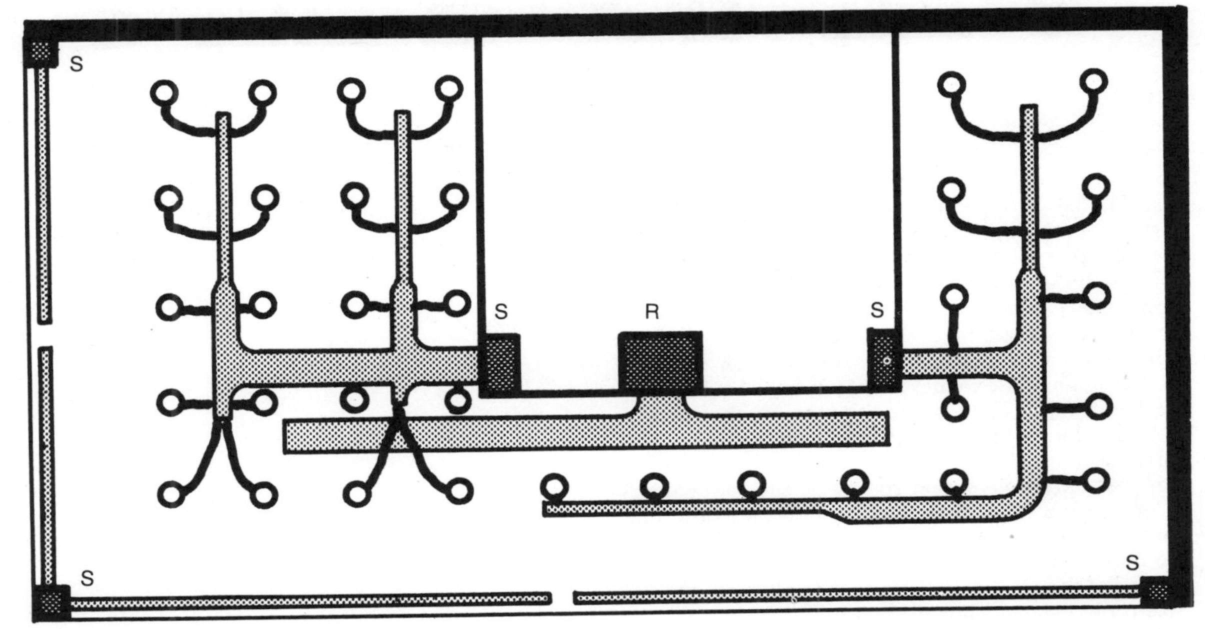

(b)

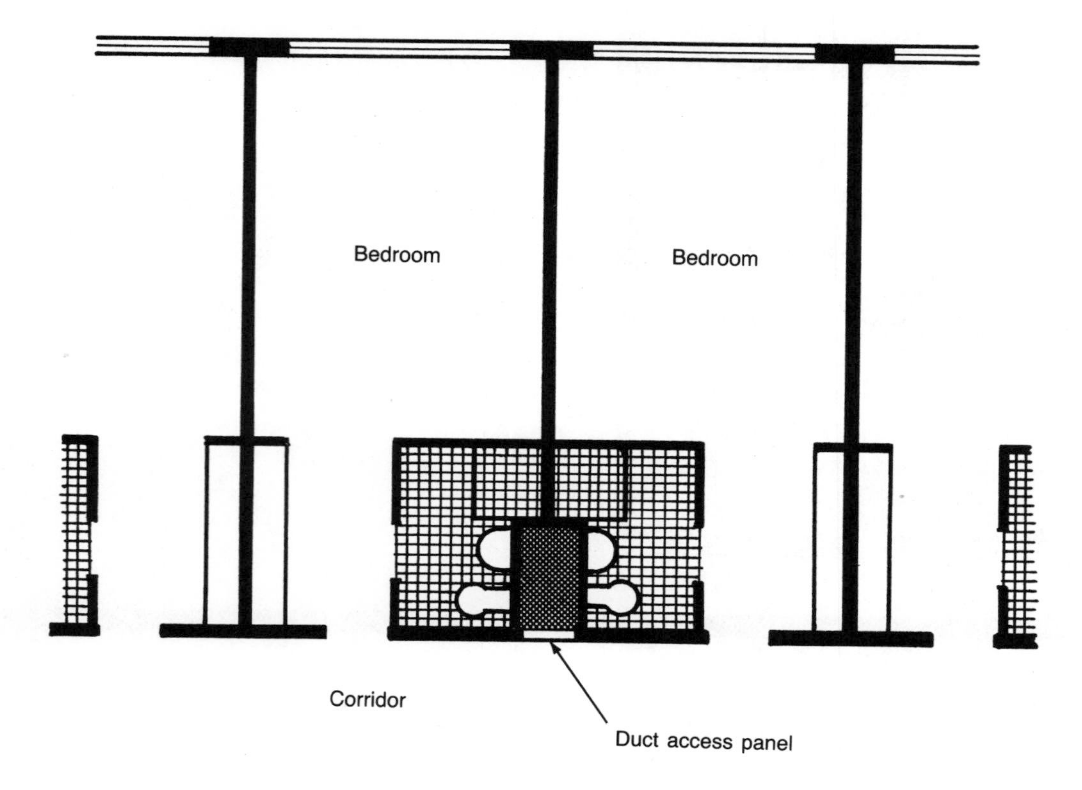

Fig. 13.12.4. A typical arrangement of hotel rooms, with the bathrooms back to back so that each pair uses the one duct. In this example access to the ducts is by a service door in the corridor, so that maintenance work can be carried out without disturbing the guests.

In addition to the plumbing stacks, each vertical duct contains an exhaust air stack for ventilating the bathrooms. The air conditioning for the rooms and corridors is usually carried in the ceiling of the corridors.

If there are convention rooms on the lower floors or on the top floor of a hotel, it is usual to put one floor of mechanical plant between the bedrooms and the floors that need large open spaces, to enable the pipework for all these ducts to be collected and taken into a few major ducts.

References

13.1 J. JAMES: *The Contractors of Chartres*. Mandorla Publications, Dooralong, Australia, 1979. 292 pp.

13.2 S. HUSER, E. GRANDJEAN, and M. SUCHANTKE: *Physiologische Grundlagen des Wohnungsbaues* (Physiological Basis for the Design of Residential Buildings). Eidgenössische Forschungskommission Wohnungsbau, Publication 14 d, Bern, 1971. 147 pp.

13.3 ETIENNE GRANDJEAN: *Ergonomics of the Home*. Taylor and Francis, London, 1973. 344 pp.

13.4 AMERICAN INSTITUTE OF ARCHITECTS: *Ramsey and Sleeper Architectural Graphic Standards,* Sixth Edition. Wiley, New York, 1970. 695 pp.

13.5 EDWARD D. MILLS (Ed.): *Planning,* Ninth Edition. Five Volumes. Newnes-Butterworths, London, 1977. About 800 pp.

13.6 RUDOLF HERZ: *Ernst Neufert's Architects' Data Sheets*. First English Edition. Crosby Lockwood, London, 1970. 354 pp.

13.7 ERNST J. McCORMICK: *Human Factors in Engineering and Design*. McGraw-Hill, New York, 1976. 491 pp.

13.8 G. UHLAND, H. DEIST, and E. STUBLER: *Untersuchungen über die Einrichtung von Küchen für den sozialen Wohnungsbau* (Experiments on kitchen layouts for public housing). Forschungsgemeinschaft Bauen and Wohnen, Report FBW 49, Stuttgart, 1958. 37 pp.

13.9 *American Standard Specifications for Making Buildings and Facilities Accessible to, and Usable by, the Physically Disabled. A 117.1* American National Standards Associations, New York, 1980. 68 pp.

13.10 *Access for the Disabled to Buildings. BS 5810.* British Standards Institution, London, 1979. 12 pp.

13.11 *Design Rules for Access by the Disabled. AS 1428.* Australian Standards Association, Sydney, 1977. 34 pp.

13.12 SELWYN GOLDSMITH: *Designing for the Disabled.* Second Edition. Royal Institute for British Architects, London, 1967. 207 pp.

13.13 B.M. LEVIN (Ed.): *Fire and Life Safety for the Handicapped.* Report of a Conference Held at the National Bureau of Standards. NBS Special Publication No. 585. U.S. Department of Commerce, Washington, D.C., 1980. 144 pp.

13.14 PETER TREGENZA: *The Design of Interior Circulation—People and Buildings.* Crosby Lockwood Staples, London, 1976. 159 pp.

13.15 W.P. MOORE, H.D. EBERHART, and H.J. COWAN (Eds.): *Tall Building Systems and Concepts.* Monograph on the Planning and Design of Tall Buildings, Volume SC. American Society of Civil Engineers, New York, 1980. 651 pp.

13.16 RODNEY R. ADLER: *Vertical Transportation for Buildings.* Elsevier, New York, 1970. 228 pp.

13.17 GEORGE R. STRAKOSCH: *Vertical Transportation: Elevators and Escalators.* Wiley, New York, 1967. 365 pp.

13.18 RICHARD J. HEALY: *Design for Security.* Wiley, New York, 1968. 309 pp.

13.19 LE CORBUSIER: *The City of Tomorrow.* Architectural Press, London, 1971. 302 pp. (First French edition by Editions Crés, Paris, 1924.)

Chapter 14 Epilogue

The interrelation between the environmental system, the building services, the structural system, and the surface finishes of the building is briefly examined. The book concludes with a note on the significance of energy consumption.

14.1 The Environmental System: Only One Aspect of the Design of the Building

We noted in Chapter 1 that building services prior to the eighteenth century were simple, and generally inadequate, except for an interlude of a few centuries in Ancient Rome. Some buildings consisted entirely of structural materials, such as natural stone and timber, without decorative finishes, so that the cost of the building was virtually identical with the cost of the structure.

During the nineteenth and twentieth centuries expectations of thermal comfort increased [Section 4.7], and the mechanical and electrical building services became more and more elaborate [Chapter 6], while the structure became lighter and more economical, so that for many modern buildings the cost of the building services was more than twice the cost of the structure. During the 1970s there was a reversal of this trend because of the rapid increase in the cost of energy, but building services still cost more than the structure for large buildings.

We could, and should, do more to integrate the environmental design of the building with the design of the building services, the structure, and the selection of the finishing materials. In many buildings erected in the earlier years of this century, particularly in the 1950s and 1960s, the established principles of optimizing the thermal environment of a building were neglected, not necessarily from ignorance but as a deliberate choice. Any required thermal environment could now be created by the building services, without relying on the contribution of the fabric of the building. This simplified design and freed the architect from the esthetic limitations imposed by sunshades and other passive solar devices. This trend also is now being reversed to conserve energy.

As this book is intended for architects, it discusses in detail the design of the fabric of the building to optimize the environmental conditions but deals only briefly with the building services, whose design remains the responsibility of the mechanical/electrical engineer (building services engineer). Consultation between the architect and the building services engineer is direct and generally works well.

14.2 The Environmental System and the Structural System

There is often no direct consultation between the structural engineer and the building services engineer, who are both consultants to the architect. The potential of the structure for improving the thermal environment is therefore rarely utilized.

We noted in Chapter 5 that both heat and coolness can be stored in a massive material, such as concrete, during the time when it is available in excess, for use later when it is needed. Concrete structures contain large quantities of massive material, whose structural performance would not be significantly affected by its use as a thermal store. Some pipes would need to be cast into the concrete to affect the thermal transfer, with allowance for any consequent loss of strength. Some structural concrete floors already contain hollow spaces for structural reasons, and these can be utilized.

Conversely, the massive walls required by many passive solar designs [Sections 5.15 and 5.16] could be employed to a much greater extent for structural support.

Because of its high mass, a structural concrete wall, floor, or roof also provides excellent insulation against airborne sound [Section 11.4].

14.3 The Environmental System and the Surface Finishes of the Building

The satisfactory performance of the environmental system depends to an appreciable extent on the choice of the correct surface finishes.

Collection of passive solar energy for storage requires noninsulated floor or wall surfaces, and thus carpeted floors are generally unsuitable for this purpose [Sections 5.15 and 5.16]. Reflection of thermal energy requires surfaces with a high reflectivity, and thermal absorption requires different finishes with a high absorptivity [Section 5.8].

The colors of surfaces are of particular importance for lighting, as light colors are needed to provide the reflection on which the natural lighting of spaces remote from windows depends. Light colors are also needed around light sources, both natural and artificial, to minimize glare.

Noise attenuation is aided by surfaces that provide sound absorption [Section 11.3], while auditoria depend for their satisfactory performance both on absorptive and on reflective surfaces in the locations where they are needed [Sections 12.4 to 12.7].

The materials used for surfaces and surface finishes are discussed in more detail in Ref. 14.2.

14.4 The Environmental System and Energy Consumption

We have passed through a period when energy was used very lavishly for building services, and another period when energy conservation was elevated to the status of a new morality. However, all that happened in the 1970s was a sharp increase in the cost of energy, which made some established design procedures of the 1960s uneconomical and made passive solar design much more attractive than it had been before. The supply of oil and natural gas will eventually be exhausted, but some new technologies will have become more economical by that time. For example, we can produce electricity directly from solar energy now [Section 3.1], but its cost is still far too high except for use in regions remote from major centers of human habitation.

During the period of cheap energy some energy-conservative environmental design procedures perfected in earlier periods fell into disuse, and many others developed by building researchers during the earlier years of this century were not utilized. This book aims to correct that deficiency. It is poor professional practice to ignore these methods and rectify the design by providing oversized building services that consume more energy than is necessary.

In some respects environmental design is more difficult than structural design, because too much is often as unsatisfactory as too little. A structural member that is oversized wastes material, but it does not endanger the structure. However, in thermal design a room that is overheated is as uncomfortable as one that is too cold. In acoustics too much sound may be as unsatisfactory as too little, and even an excess of quietness may sometimes be disturbing. On the other hand, we have a second line of defense, as we are able to utilize the building services to correct an inadequate environment produced by the fabric of the building on its own.

This book emphasizes that for some buildings a satisfactory environment can be created by using the fabric of the building, without building services. This is, in a sense, a return to the primitive conditions of architectural design that existed before the eighteenth century. However, it is possible to design many buildings in climatically favored regions, such as the coastal areas of California, Australia, and southern Europe, without installing building services. This has certain advantages. It obviates dependence on power stations, which may cease to function because of technical defects, of overloading in inclement weather, or because of industrial unrest, civil disturbance, or war. It also saves money that can be used to provide more floor space or better finishes.

References

14.1 H.J. COWAN and F. WILSON: *Structural Systems*. Van Nostrand Reinhold, New York, 1981. 256 pp.

14.2 H.J. COWAN and P.R. SMITH: *Materials of Building Construction*. Van Nostrand Reinhold, New York (in preparation).

Appendix A: Notation

The units of measurement are listed in Appendix B.

- a Absorptivity of solar radiation
- A Area
- b Width
- c_p Specific heat per unit mass
- d Thickness
- D Distance
- e Exponential constant = 2.7183
- E Illuminance
- f Flow rate in persons per second
- F Luminous flux
- h Boundary layer heat transfer coefficient
- H Height; enthalpy (quantity of heat)
- I Intensity of solar radiation; intensity of luminance
- k Thermal conductivity
- K Constant; temperature in degrees Kelvin
- L Length; luminance
- N Number of things or people
- p Ratio
- Q Quantity of heat per unit area
- r Reflectance of surface
- R Thermal resistance; heat loss by radiation
- S Acoustic absorption
- t Time
- T Temperature
- U Thermal transmittance
- v Velocity
- V Volume; Munsell color value
- z Angle between position of sun and zenith
- α Angle; thermal diffusivity
- ε Thermal emittance of surface
- θ Angle
- η Horizontal shadow angle
- λ Latitude
- π Circular constant = 3.1416
- ρ Density
- σ Stefan-Boltzman radiation constant
- ϕ Vertical shadow angle

Appendix B: Units of Measurement

The abbreviations used for units of measurement, both in the text and in this appendix, are listed in Section B.1. The SI system is explained in Sections B.2 and B.3.

In Sections B.4 to B.23 conversion figures are given between:

- the SI units used in this book;
- the customary British/American units;
- the old metric units;
- some specialized units, such as tons of refrigeration used in air conditioning, and apostilbs and lamberts used for luminance;
- units that are now considered obsolete but survive in recent books (including those listed in the references), for example, angstroms for radiation wavelength, and langleys for solar radiation records.

This appendix is selective. It is intended to assist the reader who wishes to consult other books on environmental systems, which are likely to employ non-SI units, unless they were published recently.

B.1 Abbreviations

%	percent
°	degree of temperature, and degree for angle measurement
′	minute for angle measurement
Btu	British thermal unit
°C	degree Celsius, also called degree centigrade
cd	candela
cd/m^2	candela per square meter
cu ft	cubic feet
db	decibel
°F	degree Fahrenheit
ft	feet
ft^2	square feet
ft/min	feet per minute
g	gram
g/kg	gram per kilogram
G	giga (10^9 = 1 thousand million)
h	hour
Hz	hertz (a single term for cycles per second)
in.	inch
J	joule
J/kgK	joule per kilogram per degree Kelvin
k	kilo (one thousand)
kg	kilogram
kg/m^2	kilogram per square meter
kg/m^3	kilogram per cubic meter
kJ	kilojoule
kJ/kg	kilojoule per kilogram

km	kilometer
kW	kilowatt
kWh	kilowatt-hour
kW/m^2	kilowatt per square meter
K	degree Kelvin (the symbol ° is omitted for degree Kelvin)
lb	pound
lm	lumen
lx	lux
m	meter; minute of time; prefix milli (one thousandth)
m^2	square meter
m^3	cubic meter
m^3/kg	cubic meter per kilogram
m^2K/W	meter squared Kelvin per watt
mm	millimeter
mm^2/s	square millimeter per second
m.p.h.	miles per hour
m^2/p	square meter per person
m/s	meter per second
m^3/s	cubic meter per second
M	mega (1 million)
MJ	megajoule (1 million joules)
MJ/m^3K	megajoules per cubic meter per degree Kelvin
MW	megawatt (1 million watt = 1000 kilowatt)
nm	nanometer (1 millionth of a millimeter)
p/m^2	persons per square meter
p/s	persons per second
Pa	pascal (the SI unit of pressure)
s	second
sq ft	square feet
T	tera (10^{12} = 1 million million)
W	watt
W/m^2	watt per square meter
W/mK	watt per meter per degree Kelvin
W/m^2K	watt per square meter per degree Kelvin
μm	micrometer (one thousandth of a millimeter)

B.2 The SI Metric System

There are three distinct systems of measurement in use at the present time. In customary British/American units length is measured in feet and inches, mass and force in pounds, electrical energy in joules, heat energy in British thermal units, and temperature in degrees Fahrenheit.

In the old metric system, which is also still widely used, length is measured in meters, mass and force in grams, electric energy in joules, heat energy in calories, and temperature in degrees Celsius.

In the new SI system (*Système International d'Unités*) length is also measured in meters, mass in kilograms, electrical energy in joules, and temperature in degrees Celsius; however, force is measured in newtons, and heat energy in joules.

Fortunately, most people throughout the world have switched to the SI system for calculations in acoustics and illumination, and they are doing so to an increasing extent for thermal problems.

B.3 Large and Small Units

In the SI systems the large units are obtained by the prefixes kilo (one thousand, or 10^3), mega (one million, or 10^6), and giga (one thousand million, or 10^9), and tera (one million million, or 10^{12}); for example,

1000 J $= 1 \times 10^3$ J = 1 kJ (kilojoule)
1000 kJ $= 1 \times 10^6$ J = 1 MJ (megajoule)
1000 MJ $= 1 \times 10^9$ J = 1 GJ (gigajoule)
1000 GJ $= 1 \times 10^{12}$ J = 1 TJ (terajoule)

The increases are in increments of one thousand. There are no special names for the multiples 10 and 100, as in the old metric system.

Small units are obtained by the prefixes milli (one thousandth or 10^{-3}), micro (one millionth, or 10^{-6}), and nano (10^{-9}); for example,

0.001 m $= 1 \times 10^{-3}$ m = 1 mm (millimeter)
0.001 mm $= 1 \times 10^{-6}$ m = 1 μm (micrometer)
0.001 μm $= 1 \times 10^{-9}$ m = 1 nm (nanometer)

The term "billion" should be avoided, as it means in America a multiple of 10^9, while in Europe and Australia it means a multiple of 10^{12}.

B.4 Angle

Angles are normally measured in degrees, minutes, and seconds. This is an arbitrary measure, and for absolute calculations radians are used; see Figure 8.2.1.

B.5 Temperature

In metric units temperature is measured in degrees Celsius; in customary British/American units it is measured in degrees Fahrenheit.

0°C = 32°F and 100°C = 212°F
°C $= \frac{5}{9}$ (°F − 32)

In some thermal and all lighting calculations, temperature is expressed in degrees Kelvin, which refer to absolute zero. The degree sign is omitted for degrees Kelvin.

Referred to absolute zero,

K = °C + 273.15

For temperature differences only the relative size of the degree is important:

1 K = 1°C = 1.8°F

B.6 Length

In metric units

1 km = 1000 m; 1 m = 1000 mm;
1 mm = 1000 μm; 1 μm = 1000 nm

The wavelength of light and radiation is sometimes stated in a superseded measure: 1 angstrom = 0.1 nm.

In customary British/American units 1 ft = 12 in.

1 ft = 0.3048 m; 1 in. = 25.4 mm
1 m = 3.281 ft = 39.37 in.

B.7 Area

1 m^2 = 10.764 square feet; 1 sq ft = 0.092 9 m^2

One square meter = 10 square feet is a common approximation.

B.8 Volume and Capacity

1 m^3 = 35.315 cubic feet;
1 cu ft = 0.028 32 m^3
1 liter = 0.001 m^3 = 0.220 Imperial gallons = 0.264 U.S. gallons = 1.057 U.S. quarts

B.9 Mass

1 kg = 1000 g;
1 kg = 2.205 lb;
1 lb = 0.453 6 kg

B.10 Density, or Unit Weight of Materials

1 kg/m^3 = 0.062 43 lb/cu ft;
1 lb/cu ft = 16.018 kg/m^3

B.11 Density of Persons

1 person per square meter is equal to 0.092 9 persons per square foot, or 1 person per 10.76 square feet.

B.12 Time and Frequency

1 hour = 60 minutes = 3600 seconds
1 day = 24 hours = 86 400 seconds

Frequency is the inverse of time. It is measured in cycles per second. In SI units this is called a hertz (Hz).

1 hertz = 1 cycle per second = 60 cycles per minute

B.13 Velocity

In SI calculations this is given in meters per second, or in kilometers per hour. In customary British/American units velocity is stated in miles per hour or in feet per minute.

1 m/s = 196.85 ft/min = 2.237 m.p.h. = 0.278 km/h
1 ft/min = 0.005 08 m/s; 1 m.p.h. = 0.447 m/s

B.14 Force, Pressure, and Stress

Although force and stress are very important in structural design, they are not mentioned in this book. However, pressure and sound pressure are discussed in Chapter 11.

In customary British/American units force is measured in pounds, and stress or pressure in pounds per square inch (psi).

In the old metric units, force is measured in kilograms, and stress or pressure in kilograms per square centimeter.

In SI units, force is measured in newtons, and stress or pressure in pascals, which are newtons per square meter.

1 Pa = 0.000 145 psi; 1 psi = 6 895 Pa
Standard atmospheric pressure is 101.3 kPa = 14.7 psi = 760 mm mercury approximately.

B.15 Energy

All systems of measurement use joules for electrical energy. Heat energy is also measured in joules in the SI system, but the old metric system uses calories, and the customary British/American units employ British thermal units (Btu).

1 MJ = 1000 kJ = 1 000 000 J
1 kJ = 0.948 Btu = 0.239 kilocalories; 1 Btu = 1.055 kJ
Another frequently employed unit is the kilowatt-hour:
1 kWh = 3.6 MJ

Some older books record solar radiation measurements in langleys; 1 langley = 41 868 J/m^2 (= 1 calorie per square centimeter).

B.16 Power and Heat Flow

All systems of measurement use the watt for electrical energy.

1 MW = 1000 kW = 1 000 000 W
1 W = 1 joule/second
A variety of other units are in use:
1 kW = 3 421 Btu/h = 860 kcal/h = 239 cal/s
= 0.284 tons of refrigeration
= 1.341 British/American horsepower
= 1.360 metric horsepower
1 Btu/h = 0.293 W

B.17 Thermal Conductivity

This is measured in composite units:
1 W/mK = 6.933 Btu in./ft^2h °F
= 0.578 Btu/ft h °F
= 0.860 kcal/m h °C

B.18 Thermal Transmittance

This is measured in composite units:
1 W/m^2K = 0.176 1 Btu/ft^2h °F
= 0.860 kcal/m^2h °C

B.19 Thermal Resistance

This is the reciprocal of thermal conductance:
1 m^2K/W = 5.678 ft^2h °F/Btu
= 1.163 m^2h °C/kcal

B.20 Illuminance

The metric unit is the lux, and the foot unit is the footcandle.

1 lux = 0.092 9 footcandles
1 footcandle = 10.76 lux
1 footcandle = 10 lux is a common approximation.

B.21 Luminance

In metric units this is measured in candela per square meter, and in foot units in candela per square foot or in footlambert. There are three other metric units: lambert, stilb, and apostilb. The apostilb was previously called a blondel, and the lambert has 1000 millilamberts.

1 cd/m^2 = 0.291 9 footlambert = 0.092 9 cd/ft^2
1 cd/ft^2 = 10.76 cd/m^2 = 3.382 millilambert
= 0.001 076 stilb = 33.82 apostilb (or blondel)

B.22 Sound Pressure and Sound Power

These are stated in decibel.

The decibel is a logarithmic ratio, not an absolute unit; but it can be made an absolute unit by referring it to a base measured in pascal for sound pressure, and in watt for sound power [Section 11.2].

B.23 Sound Absorption

Sound absorption is measured in sabin [Section 11.3]. The dimensions of the original sabin were square feet.

1 metric sabin = 10.76 foot sabin

Glossary

Words in italics are further defined in alphabetical order.

ABSOLUTE HUMIDITY The mass of water vapor per unit volume. See also *relative humidity*.

ABSORPTIVITY The ability of a surface to absorb solar radiation.

AC *Alternating current*.

ACRYLIC PLASTICS Thermoplastic materials, produced by the polymerization of monomeric derivatives of acrylic acid. They are obtainable in perfectly transparent form, and they have the best resistance to sunlight and outdoor weathering of all the transparent plastics.

ACTIVE SOLAR ENERGY A method for the utilization of solar energy that employs solar collectors or requires the use of electricity, as opposed to *passive solar energy*.

ADOBE Construction from large sunbaked, unburnt bricks, used in the southwestern United States and in other semiarid areas.

AIRBORNE NOISE Sound vibration transmitted to a part of a building by airborne pressure waves. See also *impact noise*.

AIR CONDITIONING Artificial ventilation with air at a controlled temperature and humidity.

ALTERNATING CURRENT A current that changes its direction of flow at regular intervals, commonly 50 or 60 times per second.

ALTITUDE The vertical angle subtended by an object, such as the sun, with the horizon. The horizontal angle is the *azimuth*.

ALUMINUM FOIL A thin sheet of aluminum, about 0.15 mm thick. It is commonly used for *reflective insulation*.

ANEMOMETER An instrument for measuring the velocity of the wind.

ANTHROPOMETRY The measurement of the human body.

APPARENT BRIGHTNESS The subjective response to the relationship between all the *luminances* in the visual field. Also called luminosity.

APPARENT SOLAR TIME The time according to the position of the sun in the sky. It differs from mean solar time shown by a watch, because the sun's motion is not entirely uniform [Section 3.3].

AQUEDUCT An artificial channel for the conveyance of water, particularly one built by the Ancient Romans.

ARTIFICIAL SKY Laboratory equipment for daylight studies. The most useful artificial sky is a hemisphere lit internally to resemble the distribution of luminance of the sky under consideration. However, a box-type sky can be used for models with vertical windows only [Section 9.5].

A-SCALE ON A SOUND LEVEL METER A filtering system that has characteristics roughly matching the response of the human ear at low sound levels.

ASHRAE American Society of Heating, Refrigerating, and Air Conditioning Engineers.

ATTENUATION Diminution or weakening, particularly of sound.

AZIMUTH The horizontal angle subtended by an object, such as the sun, with the standard meridian. The vertical angle is the *altitude*.

BEAUFORT SCALE A scale for wind speed that ranges from 0 for a complete calm to 12 for a hurricane or cyclone. The wind speed in km/h is $3B^{1.5}$, where B is the Beaufort number of the wind.

BEL The ratio of two measures of sound power, expressed as a logarithm to the base 10. See also *decibel*.

BIMETALLIC STRIP A strip fused together from two metals with widely differing coefficients of thermal expansion. It consequently deflects with a change of temperature, and it can be used as a temperature control element.

BLACK BODY The designation of a theoretical surface that absorbs all the radiation falling on it and does not transmit any radiation.

BOUNDARY LAYER The layer of a fluid, such as

air, adjacent to its boundary with a solid, for example, a building. Inside this layer the velocity of the fluid falls to zero at the boundary.

BRIGHTNESS See *apparent brightness*.

BRISE SOLEIL A sun break or sunshading device, particularly of the type used by Le Corbusier.

CANDELA The unit of luminous intensity [Section 8.2].

CANDLE POWER An obsolete unit of luminous intensity.

CF Configuration factor, a component of the *daylight factor* [Section 9.3].

CHECK VALVE A valve used in pipework to allow the fluid to flow in one direction only.

CIE Commission Internationale de l'Éclairage, the International Commission on Illumination.

CIE STANDARD OVERCAST SKY An idealized sky whose luminance at any point of *altitude* θ is $L_\theta = \frac{1}{3}L_z(1 + 2 \sin \theta)$, where L_z is the luminance at the *zenith*. Also called a Moon–Spencer sky (Fig. 9.2.2).

CLEANOUT Inspection opening in drainage pipework.

COLOR TEMPERATURE The color temperature of a lamp is the temperature of the *black body* that most closely resembles its color distribution. The lamp may operate at a temperature much lower than its color temperature.

CONDENSATION The formation of water on a surface because the air temperature falls below its *dewpoint*.

CONDUCTION See *thermal conduction*.

CONVECTION The transmission of heat by natural or forced motion of a liquid or a gas, that is, by movement of the particles, as opposed to *thermal conduction* or *radiation*.

COOL WHITE FLUORESCENT LAMP A lamp that contains more light in the blue-green part of the spectrum than an ordinary white lamp.

COOLING TOWER A device for cooling water to approximately the wet-bulb temperature of the outside air, to take heat away from the condenser of a refrigeration machine.

COSINE LAW OF ILLUMINANCE A law enunciated by J.H. Lambert in the eighteenth century, stating that the illuminance on a surface tilted at an angle θ is equal to the illuminance on a surface normal to the light sources times cosine θ [Section 8.2].

CYCLONE A *hurricane*.

DAYLIGHT Direct, diffused, or reflected sunlight, as opposed to artificial light. Also called natural light.

DAYLIGHT FACTOR A factor describing the efficiency of a window or skylight, used in the design of rooms for natural lighting [Section 9.3].

DBT *Dry-bulb temperature*.

DC *Direct current*.

DECIBEL One tenth of a *bel*.

DECIMAL LOGARITHM A logarithm to the base 10. This is the logarithm normally employed. It is denoted by the abbreviation *log*. The other commonly used logarithm is the natural logarithm, or logarithm to the base *e*.

DEGREE DAY A unit employed for estimating the heating load for a particular climate. For any one day when the temperature is below 18°C, the number of degrees below 18°C is noted. The number of degree days for a location is the total of the daily entries for one year.

DEWPOINT The temperature at which condensation of water vapor in the air takes place, that is, the temperature at which the air is fully saturated.

DF *Daylight factor*.

DIRECT CURRENT A current that flows continuously from negative to positive, as opposed to an *alternating current*.

DIRECT SOLAR GAIN Solar energy obtained directly through a window [Section 5.15].

DISABILITY GLARE Glare that impairs vision, without necessarily causing discomfort.

DISCOMFORT GLARE Glare that causes discomfort, without necessarily impairing vision.

DOWNFEED SYSTEM A system of water supply fed from a tank at the top of a building.

DRY-BULB TEMPERATURE The temperature shown on an ordinary thermometer, as opposed to the *wet-bulb temperature*.

EARTH-INTEGRATED CONSTRUCTION Building partly or wholly underground, or with an earth-covered roof.

EFFECTIVE TEMPERATURE The most commonly used criterion for determining the thermal comfort zone, introduced by Yaglou in 1924, and adopted with slight modifications by *ASHRAE*. It takes account of temperature, humidity, and air movement but ignores radiation.

ELEVATOR Lift.

EMISSIVITY The ratio of the rate of loss of radiant heat per unit area of a surface at a given temperature to the rate of loss of radiant heat per unit area of a *black body* at the same temperature and with the same surroundings.

ENTHALPY The heat content per unit mass, both due to *latent heat* and due to *sensible heat*.

EQUATION OF TIME The difference between *apparent solar time* and mean solar time.

EQUINOX The two days in the year when the length of the night is equal to the length of the day, due to the fact that the path of the sun crosses the equator. They occur on March 21 and September 23.

ERC Externally reflected component (of the *daylight factor*).

ET *Effective temperature*.

FAUCET A water tap.

FIRE COMPARTMENT A part of a building bounded by *fire-resistant* walls and fire-resistant doors that close automatically in case of a fire.

FIRE DAMPER A damper held open by a fusible link that melts at a predetermined temperature.

FIRE LOAD The amount of heat generated if the contents and combustible parts of a building were to be completely burnt.

FIRE RESISTANCE OF TWO HOURS Ability of a component to withstand a "standard" fire in a fire-testing furnace for two hours.

FIRE RESISTANT Attribute of a material that does not burn. The term has replaced "fireproof," since no material is completely proof against a fire.

FIRST LAW OF THERMODYNAMICS "Heat and mechanical energy are mutually convertible. There is a constant relation between the amount of heat lost and the energy gained, and vice versa."

FLASHOVER During a fire hot gases are formed that rise to the ceiling. When these have accumulated in sufficient quantities, a flashover occurs, after which a fire cannot be extinguished with simple equipment.

FLOCCULATE To cause microscopic particles to aggregate together to form larger particles that can more easily be removed by filtration.

FLUE A chimney used to take combustion products from a burner.

FLUORESCENT LAMP An electric lamp without a filament, consisting of a tube coated inside with a fluorescent powder. An *alternating current* passing through the mixture of mercury vapor and argon produces ultraviolet radiation. The fluorescent glass walls change this invisible radiation into visible light. See also *incandescent lamp*.

FREON A group of chemicals based on fluorine that are stable and physiologically harmless, used in the refrigeration units of air conditioning plants.

FREQUENCY The number of cycles of a periodic phenomenon that occur in a given time interval. The SI unit for frequency is the *hertz*.

FUSIBLE LINK A link that melts at a predetermined temperature, used for fire dampers, fire-resistant doors, and sprinklers.

GEOTHERMAL ENERGY Energy derived from the heat of the earth's interior, generally by tapping reservoirs of steam in geothermal regions to drive steam turbines.

GLARE See *disability glare* and *discomfort glare*.

GREAT CIRCLE The shortest possible line that can be drawn on the earth's surface between two points. It corresponds to a straight line on a plane surface. A great circle is also the circle with the largest diameter that can be drawn on the earth's surface. The equator and the meridians are great circles. Parallels of latitude are small circles.

GREENHOUSE A building or room containing large areas of glass, which transmits solar radiation, but not the long-wave radiation produced by the surfaces of a building after absorbing solar radiation. In consequence a greenhouse exposed to sunshine becomes much hotter than a room with smaller windows.

HEAT EXCHANGER A device in which heat is exchanged between two fluids while the fluids themselves are kept separate. An automobile radiator is a water-to-air heat exchanger.

HEAT TRANSFER A generic term for *thermal conduction, convection,* and *radiation*.

HELIODON A device for studying with the aid of models the sunlight penetration and the shadows to be cast by and on buildings. There are two types. In one the model is placed on a platform and a lamp at the end of a long arm is moved to imitate the position of the sun at various times of the day and the year. The other type consists of a platform that can be rotated in *altitude* and *azimuth;* these must be calculated from the sun's position. The sun is represented by a horizontal light at the end of a long room.

HELMHOLTZ RESONATOR A resonant absorber, named after the nineteenth-century scientist. It consists of a narrow neck connected to a larger volume of air, which vibrates. This sound energy can be fed back into the room, or it can be partly absorbed within the resonator, which then becomes an absorber.

HERTZ The unit of frequency; 1 hertz = 1 cycle per second.

HORIZON The *great circle* that has an *altitude* of zero. If the observer's horizon is obstructed by mountains or buildings, it may be necessary to use an artificial horizon.

HUMIDITY Water vapor within a given space. See also *absolute humidity* and *relative humidity*.

HURRICANE A wind of force 12 on the *Beaufort scale*. Also called a cyclone.

HYGROMETER An instrument for measuring the *humidity* in the air.

HYPOCAUST A central heating system used in Ancient Roman baths and occasionally in villas. Hot air and gases from a fire were passed through masonry chambers and flues under the floors.

IES Illuminating Engineering Society. The abbreviation is used both by the Illuminating Engineering Society of North America and by the (British) Illuminating Engineering Society.

ILLUMINANCE The luminous flux per unit area. It is measured in *lux*. See also *luminance*.

ILLUMINATION Until 1981 the Illuminating Engineering Society of North America used the term illumination for what is now called *illuminance*.

IMPACT NOISE Noise transmitted through building elements by impact, such as footsteps and vibrating bodies. See also *airborne noise*.

INCANDESCENT LAMP A lamp with a filament of tungsten that is made white-hot by the passage of an electric current. See also *fluorescent lamp*.

INTEGRATED CEILING A ceiling in which the *luminaires* and the air conditioning ducts are integrated so that the air is exhausted through the light fittings and cools the lamps.

INVERSE SQUARE LAW "As a spherical wave of light or sound travels outwards from a point source, its intensity decreases in inverse proportion to the square of its distance."

IRC Internally reflected component to the *daylight factor* [Section 9.3].

IRRADIANCE Flow of solar energy.

IRRADIATION Total solar energy over a period of time, such as an hour or a day.

JOULE The SI unit for energy, including heat energy.

KATA THERMOMETER An alcohol thermometer with a very large bulb that is heated to the temperature of the human body and then allowed to cool. The time required for the thermometer to cool gives an indication of the air's cooling power, and thus of the comfort conditions.

KELVIN The temperature scale referred to absolute zero, which is −273.15°C.

LATENT HEAT Thermal energy expended in changing the state of a body without changing the temperature, for example, converting water to steam at 100°C. See also *sensible heat*.

LATITUDE The angle subtended by any point on the earth's surface with the equator, measured along its *meridian*. See also *small circle*.

LAVATORY In the United States, a wash basin. In Britain and Australia, a toilet.

LIFE CYCLE COSTING Assessment of the cost of a design feature that allows for prime cost, running cost, and maintenance.

LITER The metric unit of capacity. A liter is 5% more than a U.S. quart. (In Europe and Australia, the spelling is litre).

LOUDNESS An observer's impression of the strength of sound.

LUMEN The unit of luminous flux [Section 8.2].

LUMEN METHOD The principal method for determining the number of lamps required [Section 10.7]. Also a method for the design of daylit rooms, particularly favored in the United States [Section 9.4].

LUMINAIRE A light fitting, together with the appropriate lamps.

LUMINANCE Luminance is what the eye sees reflected from a surface, whereas *illuminance* is what the light source produces. Luminance is measured in *candela* per square meter.

LUMINOSITY *Apparent brightness.*

LUMINOUS EFFICACY The ratio of the luminous flux emitted by a lamp in *lumen* to the power input in *watt.*

LUMINOUS INTENSITY The radiating capacity of a light, measured in *candela.*

LUX The unit of *illuminance.*

MASKING NOISE A noise that masks the sound from distant conversations, thus ensuring their privacy. The air conditioning plant frequently provides a masking noise, intentionally or unintentionally.

MASS LAW "Sound insulation is directly proportional to the mass of the insulating material."

MEAN SOLAR TIME See *apparent solar time.*

MERIDIAN The observer's meridian is the *great circle* passing through his location, the north pole, and the south pole.

MICROCLIMATE The climate of a small area.

MONITOR ROOF A roof with a series of highlight windows on both sides of a single-story factory roof that admit daylight and sometimes also provide ventilation. The monitor is a linear version of the lantern used in classical domes.

MOON-SPENCER SKY The *CIE Standard Overcast Sky* proposed by Parry Moon and Domenica Spencer.

MRT Mean radiant temperature.

MUNSELL COLOR CLASSIFICATION A commonly used method for classifying color, described in Section 8.6.

NANOMETER One millionth of a millimeter, or meter $\times 10^{-9}$.

NATURAL LIGHT *Daylight.*

NATURAL LOGARITHM Logarithm to the base *e*. See also *decimal logarithm.*

OCTAVE An interval between two tones, one of which has twice the frequency of the other.

OZONE An unstable form of oxygen that contains three atoms to the molecule (O_3), instead of two (O_2) as in ordinary oxygen.

PAL Permanent artificial lighting during daylight hours [Section 10.1].

PASSIVE SOLAR ENERGY Collection of solar energy, using only the fabric of the building, as opposed to *active solar energy.*

PHOTOELECTRIC CELL *Photovoltaic cell.*

PHOTOMETER An instrument for measuring *illuminance.* Photometers can be adapted for the measurement of *luminance.*

PHOTOSYNTHESIS The production of carbohydrates from carbon dioxide and water in the green cells of plants, using light as the source of energy.

PHOTOVOLTAIC CELL Also called photoelectric cell. A device that converts light into electricity. Most *photometers* and the lightmeters of cameras employ photovoltaic cells. By using a large enough array of photovoltaic cells it is possible to generate sufficient electricity to operate electric equipment and electric lights.

PLANE ANGLE An angle measured in two dimensions, as distinct from a *solid angle.*

PLENUM This term is used with two different meanings. It may denote a duct maintained at a pressure slightly above atmospheric, so that it can be used for the supply of air but return air is kept out by the excess pressure. The term is also used for the air space in an *integrated ceiling,* which may be above atmospheric pressure if used for the air supply or below atmospheric pressure if used for the air exhaust.

POPULATION FACTOR The average floor space occupied by a person in a building, that is, the total floor space available divided by the number of people who use it.

PSALI Permanent supplementary artificial lighting of interiors, to supplement daylight [Section 10.1].

PSYCHROMETRIC CHART A graphical representation of certain thermodynamic relations, used for the design of the thermal environment, particularly when air conditioning is used. The horizontal axis gives the *dry-bulb temperature,* and the vertical axis gives the water content of the atmosphere. Lines are drawn for the *relative humidity,* for the *wet-bulb temperature,* and for the *enthalpy.*

PSYCHROMETRICS The study of the properties of a mixture of air and water vapor.

PUNKAH A large, swinging fan formerly used in the hot-humid regions of Asia to provide air movement.

RADIATION Energy transmitted by electromagnetic waves. Solar radiation includes thermal radiation and visible light.

REFLECTANCE Proportion of light reflected by a surface.

REFLECTIVE INSULATION A metal sheet that reflects thermal radiation. The most common type is *aluminum foil.*

RELATIVE HUMIDITY The ratio of the quantity of water vapor actually present in the air to that present at the same temperature in a water-saturated atmosphere. It is commonly expressed as a percentage.

RESISTIVE INSULATION Conventional thermal insulation, as distinct from *reflective insulation.*

RETROFIT Making an improvement, for example, to thermal insulation or sunshading, to an old building.

REVERBERATION TIME The time required for a sound at a certain frequency to decay by 60 decibels, after its source has been silenced [Section 12.3].

RF Reflectance factor.

RH *Relative humidity.*

R-VALUE The numerical value of the *thermal resistance.*

SABIN The unit of sound absorption [Section 11.3].

SANITARY DRAINAGE The system of drainage that removes dirty water from kitchens, bathrooms, and laundries. It is usually kept separate from rainwater drainage. It may be combined with a *sewerage* system.

SAWTOOTH ROOF A sloping factory roof having one gentle slope without glazing, and a glazed roof pointing north (south in the Southern Hemisphere). In the temperate zone this has a slope to admit more daylight, but in the subtropics it is usually vertical to exclude direct sunlight.

SC Sky component of the *daylight factor.*

SELECTIVE SURFACE A surface having different

values of short-wave absorptance and long-wave emittance.

SENSIBLE HEAT The heat absorbed or emitted by a fluid or solid when the temperature changes without a change of state, as distinct from *latent heat.*

SEWAGE Human wastes, and the water used to carry them away. Any *sanitary drainage* or rainwater added to the sewage is, from that point on, regarded as sewage.

SEWERAGE A system of drainage used for removing *sewage.*

SHADOW ANGLE PROTRACTOR A transparent overlay for the *sunpath chart,* used for the design of sunshades [Section 3.5].

SKY LOBBY An elevator lobby at an upper floor [Section 13.8].

SLURRY A finely ground solid suspended in a liquid for the purpose of being transported in a pipeline.

SMALL CIRCLE A circle on the earth's surface that has a diameter smaller than that of a *great circle.* Parallels of latitude are small circles.

SOLID ANGLE An angle measured in three dimensions, as distinct from a *plane angle.*

SOLSTICE The longest day and the shortest day of the year, when the sun attains its greatest distance from the equator. This occurs on June 21 and December 22.

SOUND ATTENUATION The reduction of the energy or the intensity of sound.

SPECIFIC HEAT The quantity of heat required to raise a unit mass of a substance through a temperature range of one degree Kelvin.

SPECIFIC VOLUME The volume of a unit mass. It is the reciprocal of density.

SPRINKLER SYSTEM A system of pipes sealed with sprinkler heads that open automatically at a predetermined temperature and sprinkle water on the fire to keep it under control until the fire department arrives.

STACK A vertical drainage or vent pipe within a building.

STACK EFFECT Natural ventilation caused by air pressure differences due to variations in air density with height.

STEFAN-BOLTZMAN RADIATION CONSTANT The constant used to determine the quantity of heat radiated by a body [Eq. (4.3), Section 4.2].

STERADIAN The unit of *solid angle* (see Figure 8.2.1).

SUNPATH CHART A chart that shows the *altitude* and the *azimuth* of the sun at a particular location throughout the year. It can be used for the design of sunshades in conjunction with a *shadow angle protractor.*

THERMAL CONDUCTION The process of heat transfer through a substance in which heat is transferred from particle to particle, not as in *convection* by movement of particles, nor as in *radiation.*

THERMAL CONDUCTIVITY The rate of heat transfer along a body by *thermal conduction.* It is denoted by the letter k and measured in W/mK [Section 5.5].

THERMAL DIFFUSIVITY The property of a material, thermal conductivity divided by heat capacity per unit volume. A high value means that a sudden heat input is quickly diffused through the material [see Section 5.6].

THERMAL INERTIA The heat capacity of a material, measured in J/m^{3}K [Section 5.6].

THERMAL RESISTANCE The reciprocal of *thermal transmittance.* It is called the *R-value* and is measured in m^{2}K/W.

THERMAL STORAGE WALL Wall with high *thermal inertia* used for the storage of heat or coolness in passive solar design. See *passive solar energy.*

THERMAL TRANSMITTANCE The quantity of heat transmitted through a roof, wall, or floor due to a temperature difference in the air on both sides. It is called the *U-value* and is measured in W/m^{2}K [Section 5.5].

THERMISTOR A temperature-sensitive resistance element used in heat detectors.

THERMOCOUPLE An electric thermometer consisting of a pair of suitable wires of different material joined together, so that an electric current is produced with a change of temperature.

THERMOPILE An assembly of thermoelectric elements connected in series or in parallel, which can be used for measuring temperature.

THERMOSTAT A device for maintaining a constant temperature. Many thermostats employ a *bimetallic strip.*

THERMOSYPHON The circulation of a fluid due to the application of heat to one part of a system. The hotter fluid has a lower density and therefore tends to rise (see Figure 2.4.6).

TROPICS The parallels of *latitude* 23°26′ north and south that represent the farthest movement of the sun from the equator [Section 3.3]. The tropic zone is the region between the tropic circles.

UPFEED SYSTEM A system of water supply fed by mains pressure, or a pump, at the bottom of the building (see Figure 2.3.4).

U-VALUE The numerical value of the *thermal transmittance.*

VAPOR BARRIER An airtight skin that prevents moisture from the warm air in a building passing into a colder space where it may cause *condensation.*

WARM WHITE FLUORESCENT LAMP A lamp that contains less light in the blue-green part of the spectrum than an ordinary white lamp.

WATER SEAL A U-shaped section of pipework that holds water and therefore prevents the passage of gas from the drainage system into the room (see Figure 2.5.1).

WATT The SI unit of power and heat flow.

WBT *Wet-bulb temperature.*

WET-BULB TEMPERATURE Temperature indicated by a mercury-in-glass thermometer wrapped in a damp wick whose far end is dipped in water. This is lower than the *dry-bulb temperature* of a thermometer without a wick, because the evaporation of the water on the wick cools the mercury bulb. The difference depends on the *relative humidity,* which can be determined by reading both the dry-bulb temperature and the wet-bulb temperature of two identical thermometers mounted side by side.

WORKPLANE A term used in lighting design for the plane at which work is normally done. The illuminance for the workplane is specified, and the lighting is designed to provide this illuminance.

ZENITH The highest point in the sky, immediately overhead at the time of an observation. By definition its *altitude* is 90°.

Index